D0277868

J. O. Bird
B.Sc.(Hons), A.F.I.M.A., T.Eng.(CEI), M.I.T.E.

A. J. C. May
B.A., C.Eng., M.I.Mech.E., F.I.T.E., A.M.B.I.M.

Technician mathematics

Level 3

Longman London and New York

Longman Group Limited,
Longman House,
Burnt Mill, Harlow, Essex, U.K.

Published in the United States of America
by Longman Inc., New York

First published 1978
Second impression 1980
Third impression 1981

British Library Cataloguing in Publication Data

Bird, J O
Technician mathematics, level 3. – (Longman technician series: mathematics and sciences).
1. Shop mathematics
I. Title II. May, A J C
510′.2′46 TJ1165

ISBN 0-582-41613-2

Set in IBM 10/11pt Journal
and printed in Singapore by
Ban Wah Press

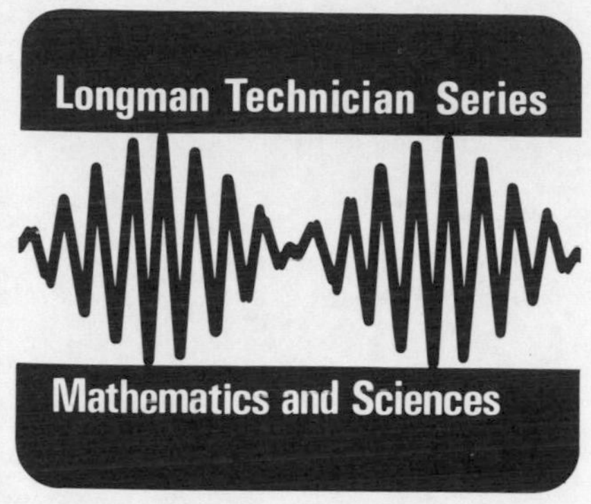

General Editor — Mathematics and Sciences

D. R. Browning, B.Sc., F.R.I.C., A.R.T.C.S.
Principal Lecturer and Head of Chemistry, Bristol Polytechnic

Books published in this sector of the series:

Technician mathematics, Level 1 **J. O. Bird** and **A. J. C. May**
Technician mathematics, Level 2 **J. O. Bird** and **A. J. C. May**
Technician mathematics, Levels 4 and 5 **J. O. Bird** and **A. J. C. May**
Mathematics for electrical and telecommunications technicians, Level 2 **J. O. Bird** and **A. J. C. May**
Mathematics for science technicians, Level 2 **J. O. Bird** and **A. J. C. May**
Mathematics for electrical technicians Level 3 **J. O. Bird** and **A. J. C. May**
Mathematics for electrical technicians Levels 4 and 5 **J. O. Bird** and **A. J. C. May**
Calculus for technicians **J. O. Bird** and **A. J. C. May**
Statistics for technicians **J. O. Bird** and **A. J. C. May**
Engineering science for technicians Level 1 **D. R. Browning** and **I. McKenzie Smith**
Physical sciences Level 1 **D. R. Browning** and **I. McKenzie Smith**
Fundamentals of chemistry **J. H. J. Peet**
Further studies in chemistry **J. H. J. Peet**
Mathematical formulae for TEC courses **J. O. Bird** and **A. J. C. May**
Science formulae for TEC courses **D.R. Browning**

Contents

Chapter 8 Application of integration to centroids and second moments of area 219

Chapter 9 Differential equations 270

Chapter 10 Probability 286

Chapter 11 The binomial and Poisson probability distributions 300

Chapter 12 The normal probability distribution 315

Preface

This textbook is the third in a series which deal simply and carefully with the fundamental mathematics essential in the development of technicians.

Treatment of this mathematics is developed soundly and logically by authors experienced in various fields of Engineering.

Technician mathematics, Level 3, provides a clear and comprehensive coverage of the new Technician Education Council Level III Mathematics Course (TEC U75/040).

Each topic considered in the text is presented in a way that assumes in the reader only the knowledge attained at TEC Level II Mathematics (i.e. TEC U75/012 and 039 syllabuses). This practical book contains over 130 illustrations, nearly 200 detailed worked problems, followed by some 800 further problems with answers.

The authors would like to thank Mr David Browning, Principal Lecturer, Bristol Polytechnic, for his valuable assistance in his capacity as General Editor of the Mathematics and Science Sector of the Longman Technician Series. They would also like to express their appreciation for the friendly cooperation and helpful advice given to them by the publishers.

Thanks are also due to Mrs Elaine Mayo, Mrs Janet Wright-Green and Mrs Margaret Hague for the excellent typing of the manuscript.

Finally, the authors would like to add a word of thanks to their wives, Elizabeth and Juliet, for their patience, help and encouragement during the preparation of this series of books.

J. O. Bird
A. J. C. May

Highbury College of Technology
Portsmouth, 1978

Acknowledgements

We are grateful to the following for permission to reprint copyright material: Macmillan, London and Basingstoke for Tables–Exponential Functions (extract) and Hyperbolic or Naperian Logarithms from *Four Figure Mathematical Tables* by Frank Castle; Luton College of Higher Education and H. A. Horner for question and answer from *ONC Science Basic Maths 1972;* Peoples College of Further Education and K.K.J. Hadley for question and answer from *ONC Science Basic Maths 1972;* Thurrock Technical College for question from *ONC Science Basic Maths 1972;* Waltham Forest College for question from *ONC Science Basic Maths 1972;* Warley College of Technology for questions and answers from *Supplementary Maths 1972.* Cover photograph by Paul Brierley.

Chapter 1

Trigonometry

1. Solution of scalene triangles

A scalene triangle is one in which all its sides and angles are unequal. Solving a triangle means finding all unknown sides and angles.

Use is made of the sine and cosine rules when solving scalene triangles. With reference to triangle ABC shown in Fig. 1:

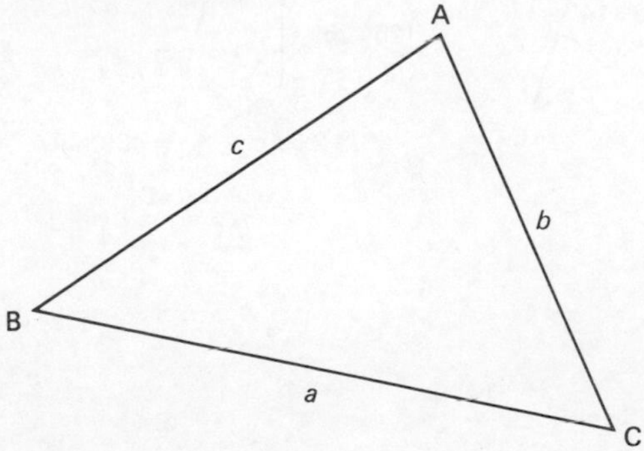

Figure 1

the **sine rule** states: $\frac{a}{\sin A} = \frac{b}{\sin B} = \frac{c}{\sin C}$

and the **cosine rule** states: $a^2 = b^2 + c^2 - 2bc \cos A$
or $b^2 = a^2 + c^2 - 2ac \cos B$
or $c^2 = a^2 + b^2 - 2ab \cos C$

Worked problems on the solution of scalene triangles

Problem 1. In a triangle ABC, C = 39° 14′, side c = 6.47 cm and side b = 8.23 cm. Solve the triangle.

Using the sine rule: $\frac{8.23}{\sin B} = \frac{6.47}{\sin 39° 14'}$

$$\sin B = \frac{8.23 \sin 39° 14'}{6.47}$$

Hence B = 53° 34′ or 126° 26′
If B = 53° 34′ then A = 180° − 53° 34′ − 39° 14′ = 87° 12′
If B = 126° 26′ then A = 180° − 126° 26′ − 39° 14′ = 14° 20′
Thus there are two possible solutions of triangle ABC. This occurrence is known as the ambiguous case.

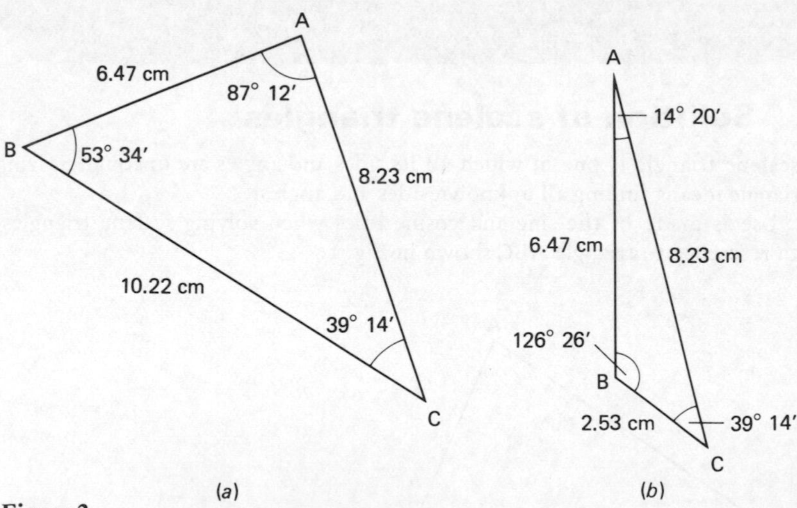

Figure 2

Case 1. When A = 87° 12′ then

$$\frac{a}{\sin 87° 12'} = \frac{6.47}{\sin 39° 14'}$$

$$a = \frac{6.47 \sin 87° 12'}{\sin 39° 14'} = \mathbf{10.22\ cm}$$

Case 2. When A = 14° 20′ then

$$\frac{a}{\sin 14^\circ\ 20'} = \frac{6.47}{\sin 39^\circ\ 14'}$$

$$a = \frac{6.47 \sin 14^\circ\ 20'}{\sin 39^\circ\ 14'} = \mathbf{2.532\ cm}$$

Figure 2 shows the two possible solutions of triangle ABC.

Both sets of answers are feasible since in each case the lengths of the sides are related directly to the sizes of the corresponding angles (i.e. the longest side is opposite the largest angle and so on).

Problem 2. Solve triangle DEF given F = 64° 55′, side d = 14.83 cm and side e = 10.72 cm.

Using the cosine rule:

$$\begin{aligned} f^2 &= d^2 + e^2 - 2de \cos F \\ &= (14.83)^2 + (10.72)^2 - [2(14.83)(10.72) \cos 64^\circ\ 55'] \\ &= 200.1 \\ f &= 14.14 \text{ cm} \end{aligned}$$

Using the sine rule:

$$\frac{14.14}{\sin 64^\circ\ 55'} = \frac{10.72}{\sin E}$$

$$\sin E = \frac{10.72 \sin 64^\circ\ 55'}{14.14}$$

Hence E = 43° 22′ or 136° 38′
when E = 43° 22′, D = 180° − 43° 22′ − 64° 55′ = 71° 43′
when E = 136° 38′, D = 180° − 136° 38′ − 64° 55′ = −21° 33′
This latter solution is impossible and is neglected.
Hence in triangle DEF, **f = 14.14 cm, D = 71° 43′ and E = 43° 22′**

Problem 3. Find the three angles of a triangular template having sides of 6.42 cm, 8.31 cm and 9.78 cm.

Let a triangle ABC represent the template with a = 6.42 cm, b = 8.31 cm and c = 9.78 cm.
The largest angle is C since 9.78 cm is the longest side.

Since $c^2 = a^2 + b^2 - 2ab \cos C$

$$\cos C = \frac{a^2 + b^2 - c^2}{2ab} = \frac{(6.42)^2 + (8.31)^2 - (9.78)^2}{2(6.42)\ (8.31)}$$

$$\therefore \quad \cos C = 0.137\ 1$$

$$C = \arccos 0.137\ 1 = 82^\circ\ 7'$$

Since cos C is positive the triangle is acute-angled. If the value of cos C had been negative C would have been an angle between 90° and 180° and the triangle would have been obtuse-angled.

4 By finding the largest angle first, when given the three sides of a triangle, an immediate indication of the type of triangle is given.

Using the sine rule: $\dfrac{9.78}{\sin 82^\circ\ 7'} = \dfrac{8.31}{\sin B}$

$\sin B = \dfrac{8.31 \sin 82^\circ\ 7'}{9.78}$

$B = 57^\circ\ 19'$

$A = 180^\circ - 82^\circ\ 7' - 57^\circ\ 19' = 40^\circ\ 34'$

Hence the three angles of the triangular template are $40^\circ\ 34'$, $57^\circ\ 19'$ and $82^\circ\ 7'$.

Problem 4. Two phasors are shown in Fig. 3. When $i_1 = 14.6$ amperes and $i_2 = 23.8$ amperes, calculate the value of their resultant (i.e. length AC) and the angle it makes with i_1.

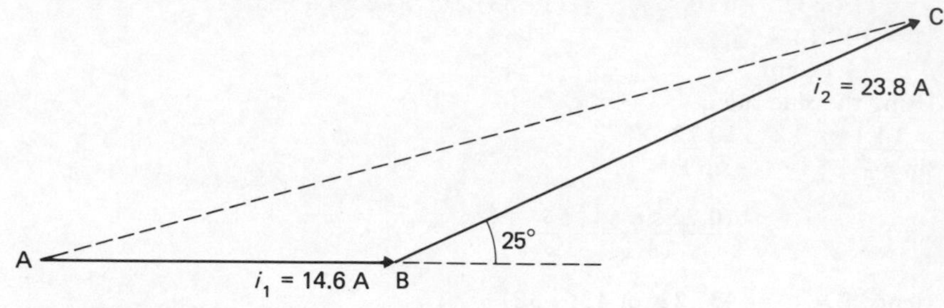

Figure 3

In triangle ABC shown in Fig. 3:

$B = 180^\circ - 25^\circ = 155^\circ$.

Using the cosine rule:

$$\begin{aligned} AC^2 &= AB^2 + BC^2 - [2\,(AB)\,(BC) \cos 155^\circ] \\ &= (14.6)^2 + (23.8)^2 - [2\,(14.6)\,(23.8)\,(-\cos 25^\circ)] \\ &= 1\,409 \end{aligned}$$

Here **the resultant AC** = $\sqrt{1\,409}$ = **37.54 amperes.**

Using the sine rule: $\dfrac{23.8}{\sin A} = \dfrac{37.54}{\sin 155^\circ}$

$\sin A = \dfrac{23.8 \sin 155^\circ}{37.54}$

$A = 15^\circ\ 32'$

Hence the resultant AC makes an angle of $15^\circ\ 32'$ with i_1.

Further problems on the solution of scalene triangles may be found in Section 13 (Problems 1–17).

2. The angle between a line and a plane

The angle between a line and a plane is defined as the angle between the line and its projection on the plane.

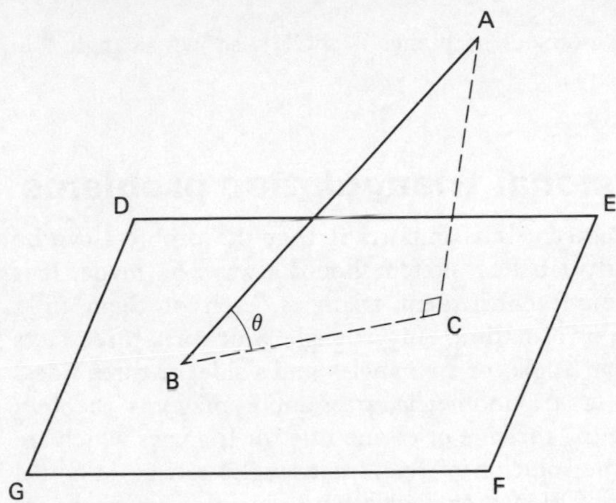

Figure 4

In Fig. 4 the line AB is shown making an angle with the plane DEFG. If the line AC is constructed perpendicular to plane DEFG then the projection of the line AB on the plane DEFG is given by the length BC (where BC lies in the plane DEFG and is thus perpendicular to AC). In Fig. 4 angle θ is the angle between the line AB and the plane DEFG.

3. The angle between two intersecting planes

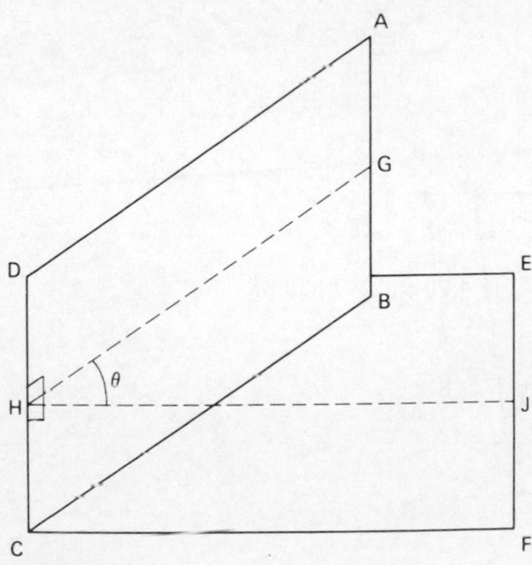

Figure 5

6 Figure 5 shows two planes ABCD and CDEF, having a common edge CD. To find the angle between these two intersecting planes it is necessary to establish a line that is perpendicular to the edge DC in each of the planes, i.e. GH is perpendicular to CD in plane ABCD and HJ is perpendicular to CD in plane CDEF.

The angle between the two intersecting planes is $\angle$ GHJ, shown as angle θ in Fig. 5.

4. Three-dimensional triangulation problems

In three-dimensional trigonometry it is important that the problem can be visualised and hence a clearly labelled sketch should always be made. It is often useful to redraw relevant constituent triangles, each of them fully dimensioned with the given information. Any triangle which has three facts given (either two sides and an angle, or two angles and a side, or three sides) may be solved using either: (a) trigonometric ratios and Pythagoras' theorem for right-angled triangles; or (b) the sine or cosine rule for triangles which are not right-angled. Provided the solution of relevant triangles can be obtained then angles between lines and planes or angles between intersecting planes may be found.

Worked problems on 'three-dimensional' trigonometry

Problem 1. The base of a right pyramid of vertex A is a rectangle BCDE.

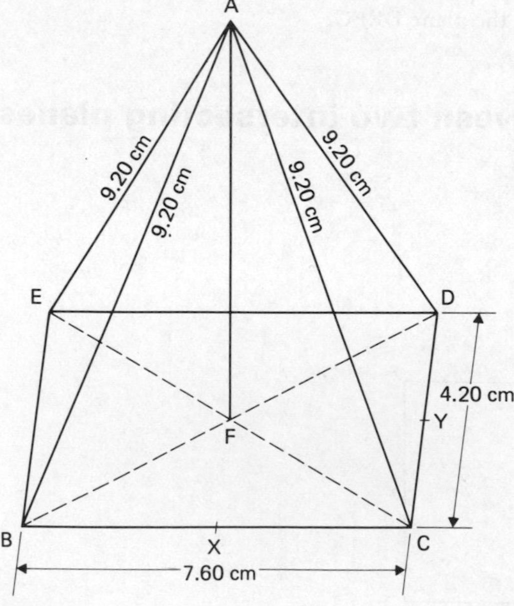

Figure 6

X and Y are the mid-points of sides BC and CD respectively. AB = 9.20 cm, BC = 7.60 cm and CD = 4.20 cm. Calculate: (a) the perpendicular height of the pyramid; (b) the angle edge AE makes with the base; (c) the angle faces ABC and ACD make with the base; and (d) the angle the plane AXY makes with the base.

The pyramid is shown in Fig. 6.

(a) Let the diagonals of the base, EC and BD, intersect at point F. Length of diagonal EC = $\sqrt{[(7.60)^2 + (4.20)^2]}$ = 8.683 cm

Length EF = $\frac{1}{2}$ EC = $\frac{8.683}{2}$ = 4.342 cm

Using the theorem of Pythagoras on the right-angled triangle AEF (see Fig. 7 (a)) gives:

AF = $\sqrt{[(9.20)^2 - (4.342)^2]}$ = 8.111 cm

Hence the height of the pyramid is 8.111 cm.

(b) The angle which a line makes with a plane is the angle which it makes with its projection on the plane. In the pyramid, EF is the projection of AE on to the base BCDE.

From triangle AEF, shown in Fig. 7 (a), cos E = $\frac{EF}{EA} = \frac{4.342}{9.20}$

Thus $\angle$ AEF = 61° 50′

Hence AE makes an angle of 61° 50′ with the base.

(c) Since point X is the mid-point of side BC, FX is perpendicular to BC. Also AX is the perpendicular height of the triangle ABC. Thus the angle between the two intersecting lines AX and XF is the angle between the face ABC and the base. In triangle AFX, shown in Fig. 7 (b), AF = 8.111 cm and

FX = $\frac{4.20}{2}$ = 2.10 cm

Hence **$\angle$ AXF** = arctan $\frac{8.111}{2.10}$ = **75° 29′**

Since Y is the mid-point of side CD, FY is perpendicular to CD. Also AY is the perpendicular height of triangle ACD. Thus the angle between the two intersecting lines AY and YF is the angle between the face ACD and the base.

In triangle AFY, shown in Fig. 7 (c), AF = 8.111 cm and FY = $\frac{7.60}{2}$ = 3.80 cm.

Hence **$\angle$ AYF** = arctan $\frac{8.111}{3.80}$ = **64° 54′**

(d) To find the angle between plane AXY and the base BCDE it is necessary to establish a perpendicular to the common edge XY in each of the planes.

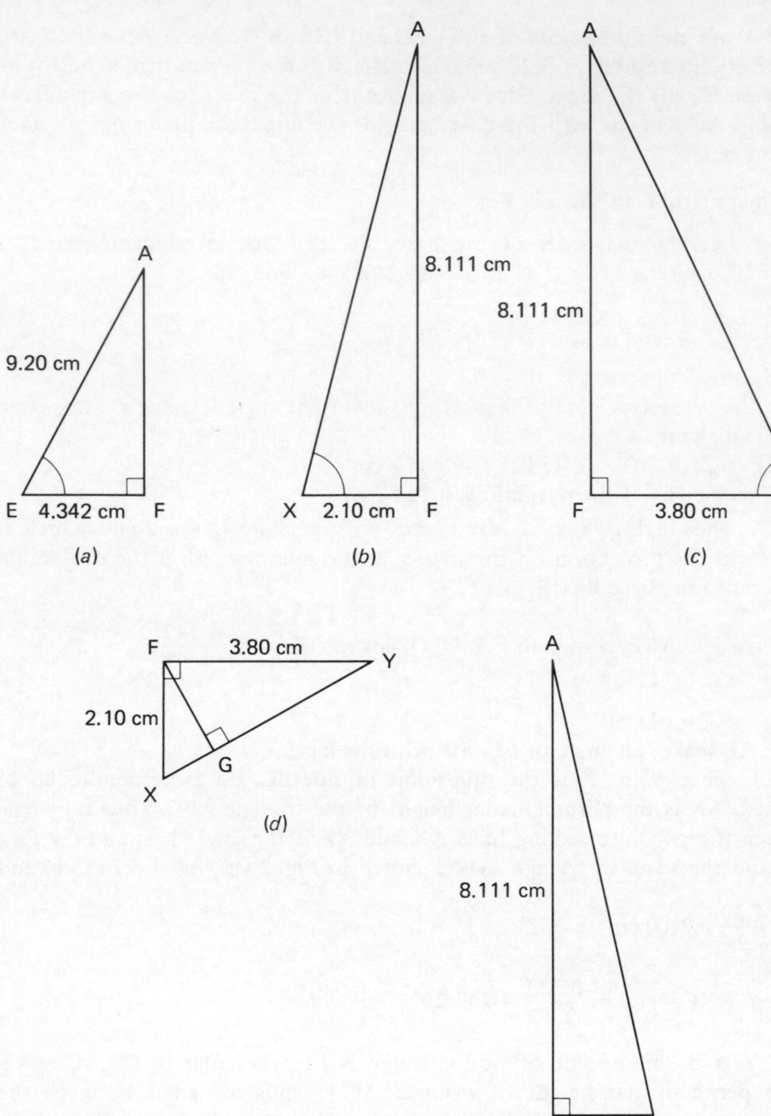

Figure 7

Triangle FXY is shown in Fig. 7 (d).

$\angle$ FXY = arctan $\dfrac{3.80}{2.10}$ = 61° 4′

Let FG be perpendicular to XY
Then FG = 2.10 sin 61° 4′ = 1.838 cm.
From the triangle AFG, shown in Fig. 7 (e),

$$\angle AGF = \arctan \frac{AF}{FG} = \arctan \frac{8.111}{1.838} = 77^\circ\ 14'$$

Hence **the angle between the plane AXY and the base is 77° 14′**

Problem 2. A 4.00-cm cube has top ABCD and base EFGH with vertical edges AE, BF, CG and DH. I and J are the mid-points of sides EH and HG respectively. Calculate:

(a) the angle between DF and CF;
(b) the angle between DI and GI;
(c) the angle between the line CE and the plane EFGH; and
(d) the angle between the plane BIJ and the plane EFGH.

The cube is shown in Fig. 8.

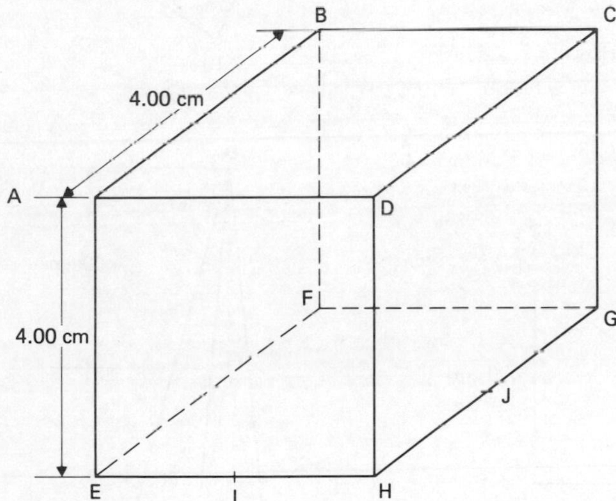

Figure 8

(a) To find the angle between DF and CF:

DF is one side of triangle DFH shown in Fig. 9 (a) where DH = 4.00 cm and FH = $\sqrt{[(4.00)^2 + (4.00)^2]}$ = 5.657 cm. Hence all the diagonals of the cube are 5.657 cm. Since ∠ DHF is 90°, DF = $\sqrt{[(4.00)^2 + (5.657)^2]}$, i.e. DF = 6.928 cm.
In triangle CFD, shown in Fig. 9 (b), FD = 6.928 cm, CF = 5.657 cm and CD = 4.00 cm.

$$\angle \mathbf{CFD} = \arctan \frac{4.00}{5.657} = \mathbf{35^\circ\ 16'}$$

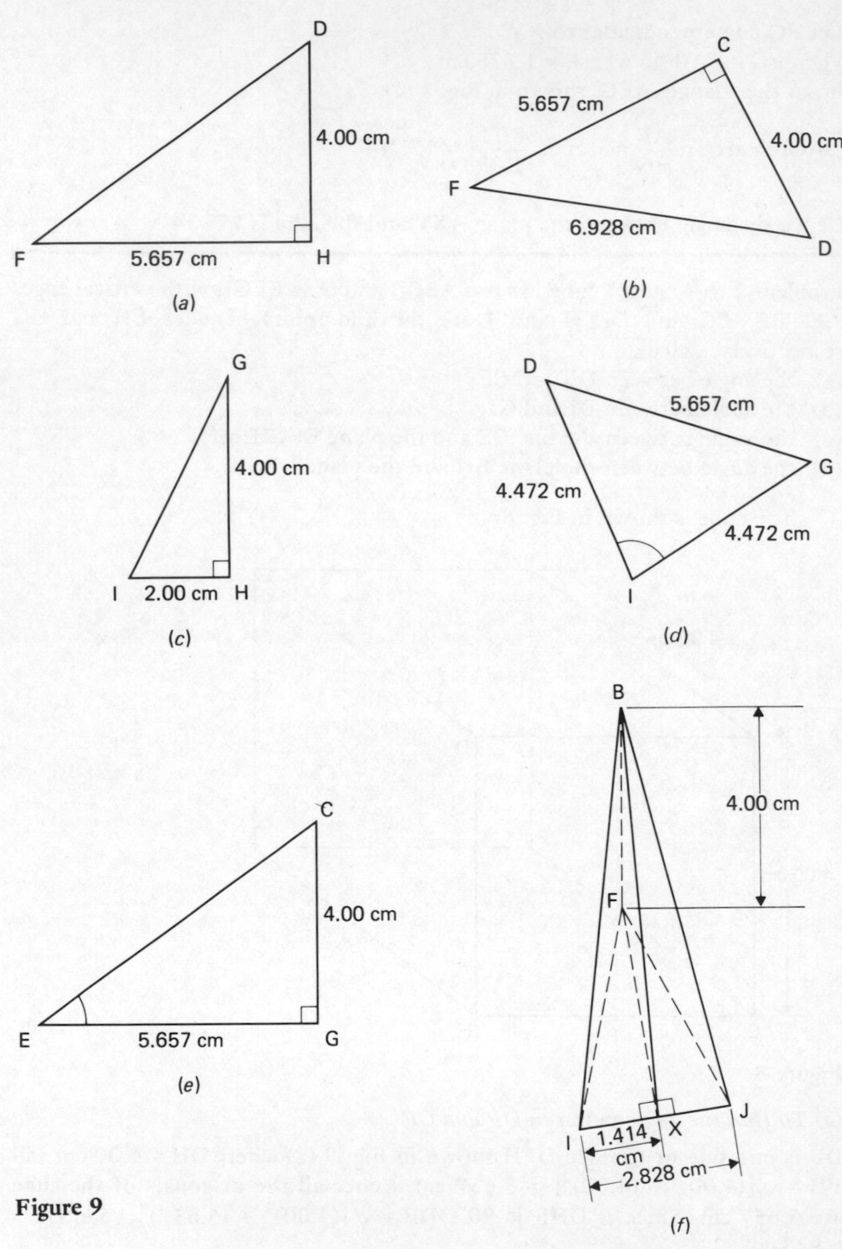

Figure 9

(b) To find the angle between DI and GI:

GI is one side of triangle GHI, shown in Fig. 9 (c), where GH = 4.00 cm, HI = 2.00 cm and ∠ GHI = 90°. Hence GI = $\sqrt{[(4.00)^2 + (2.00)^2]}$ = 4.472 cm.

Similarly, DI = 4.472 cm (i.e. the same length as GI).
Diagonal DG = 5.657 cm
In triangle GID, shown in Fig. 9 (d), ∠ DIG is given by the cosine rule.

$$\cos DIG = \frac{(DI)^2 + (IG)^2 - (DG)^2}{2\,(DI)\,(IG)}$$

$$= \frac{(4.472)^2 + (4.472)^2 - (5.657)^2}{2(4.472)\,(4.472)} = 0.199\,9$$

Hence ∠ **DIG = 78° 28′**

(c) To find the angle between the line CE and the plane EFGH:

The projection of the line CE on to the plane EFGH is given by the diagonal EG, where EG and CG are perpendicular to each other. In triangle CEG, shown in Fig. 9 (e), GE = 5.657 cm, ∠ CEG is the angle between line CE and plane EFGH. Hence ∠ **CEG** = arctan $\frac{CG}{EG}$ = arctan $\frac{4.00}{5.657}$ = **35° 16′**

(d) To find the angle between plane BIJ and plane EFGH:

From Fig. 9 (f), IJ is the common edge between planes BIJ and EFGH.
$IJ = \sqrt{[(2.00)^2 + (2.00)^2]} = 2.828$ cm
Let X be the mid-point of IJ, then IX = 1.414 cm, FI = 4.472 cm
(i.e. length FI is the same as length GI)
and $FX = \sqrt{[(4.472)^2 - (1.414)^2]} = 4.243$ cm
FX is perpendicular to IJ and BX is perpendicular to IJ. Thus the angle between planes BIJ and EFGH is given by angle BXF.

Hence ∠ **BXF** = arctan $\frac{4.00}{4.243}$ = **43° 19′**

Problem 3. A ship is observed from the top of a cliff 152 m high in a direction S 28° 19′ W at an angle of depression of 8° 46′. Six minutes later the same ship is seen in a direction W 17° 13′ N at an angle of depression of 9° 52′. Calculate the speed of the ship in km h^{-1}.

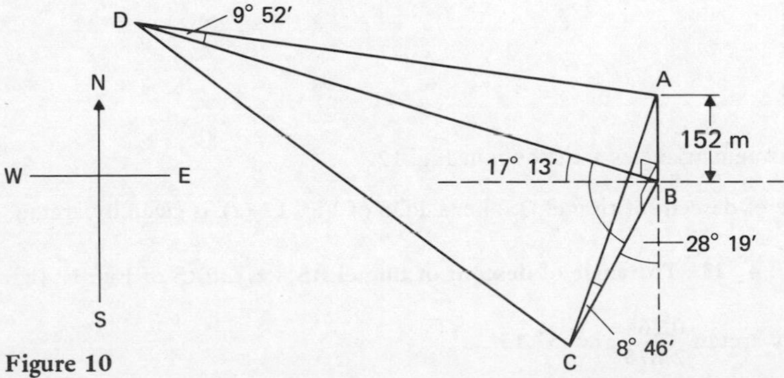

Figure 10

In Fig. 10, A represents the top of the cliff AB, C the initial position of the ship and D the position of the ship at the second observation. To find the speed of the ship the distance CD is required. To find CD, distances BC and BD need to be calculated.

$\angle$ CBD $= (90^\circ - 28^\circ\ 19') + 17^\circ\ 13' = 78^\circ\ 54'$.
From triangle ABC, BC = 152 cot $8^\circ\ 46'$ = 985.7 m
From triangle ABD, BD = 152 cot $9^\circ\ 52'$ = 873.9 m
Using the cosine rule on triangle BCD gives:

$$CD^2 = (985.7)^2 + (873.9)^2 - [2(985.7)(873.9)\cos 78^\circ\ 54']$$

Therefore CD = 1 185 m or 1.185 km

Speed of ship $= \dfrac{\text{distance travelled}}{\text{time}} = \dfrac{1.185 \text{ km}}{\frac{6}{60} \text{ h}}$

= 11.85 km h^{-1}

Problem 4. From a point P, at ground level, a mine shaft is constructed to a depth of 0.264 km. Tunnels are constructed from Q, 3.450 km due west of P, and R, 2.875 km S 40° E, to meet at the base S of the shaft. Assuming that P, Q and R are on horizontal ground and that the two tunnels descend uniformly find: (a) their angles of descent; and (b) the angle between the directions of the two tunnels.

The mine is shown in Fig. 11.

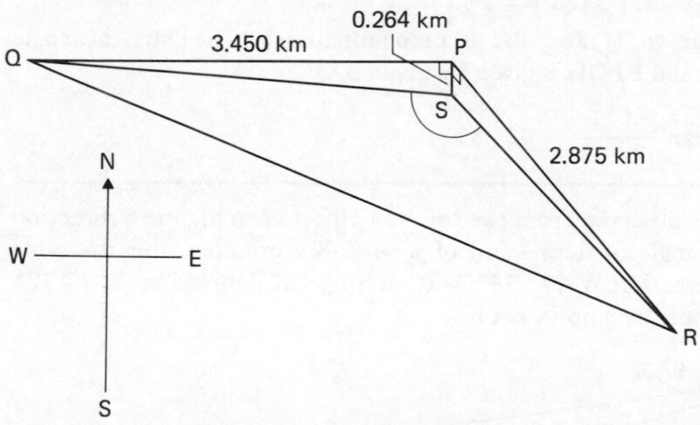

Figure 11

The constituent triangles are shown in Fig. 12.

The angle of descent of tunnel QS, i.e. $\angle$ PQS of Fig. 12 (a), is given by arctan $\dfrac{0.264}{3.450}$, i.e. **4° 23′**. **The angle of descent of tunnel RS**, i.e. $\angle$ PRS of Fig. 12 (b) is given by arctan $\dfrac{0.264}{2.875}$, i.e. **5° 15′**.

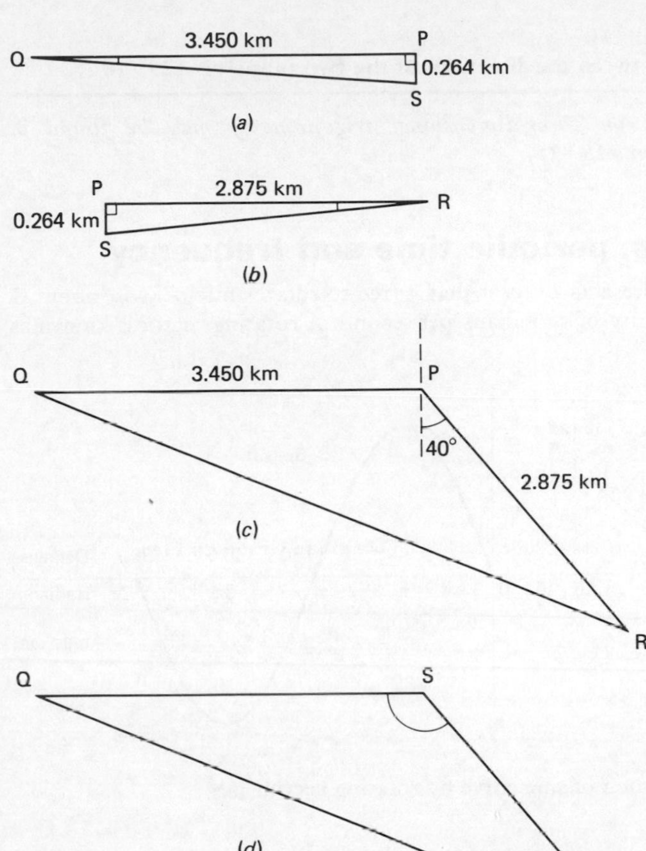

Figure 12

The angle between the directions of the two tunnels is given by ∠ RSQ of Fig. 12 (d).
From triangle PQS, $QS = \sqrt{[(3.450)^2 + (0.264)^2]} = 3.460$ km
From triangle PRS, $RS = \sqrt{[(2.875)^2 + (0.264)^2]} = 2.887$ km
From triangle PQR, shown in Fig. 12 (c), ∠ QPR = 130°. Using the cosine rule:
$QR^2 = (3.450)^2 + (2.875)^2 - [2(3.450)(2.875)\cos 130°]$
Hence QR = 5.738 m

$$\text{From triangle QRS, } \cos RSQ = \frac{(SQ)^2 + (SR)^2 - (QR)^2}{2(SQ)(SR)}$$

$$= \frac{(3.460)^2 + (2.887)^2 - (5.738)^2}{2(3.460)(2.887)}$$

$$= -0.631\,6$$

 Hence $\angle$ RSQ = 129° 10′
Hence the angle between the directions of the two tunnels is 129° 10′.

Further problems on three-dimensional trigonometry may be found in Section 13 (Problems 18–31).

5. Phasors, periodic time and frequency

In Fig. 13, OA represents a vector that is free to rotate anticlockwise about O at an angular velocity of ω radians per second. A rotating vector is known as a phasor.

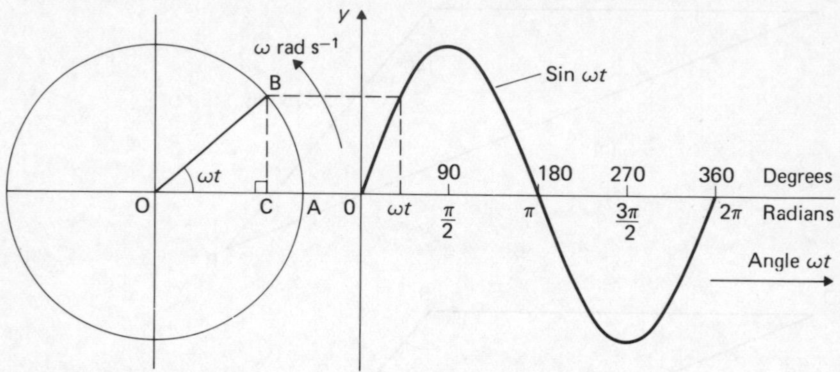

Figure 13 Production of sine curve by rotating vector

After a time t seconds the vector OA will have turned through an angle of ωt radians (shown as $\angle$ AOB in Fig. 13). When the line BC is constructed perpendicular to OA as shown then:

$$\sin \omega t = \frac{BC}{OB}$$

i.e. $\mathbf{BC = OB \sin \omega t}$

If all such vertical components are projected across on to the graph of y against angle ωt (in radians), a sine curve results of maximum value OA.

Periodic time

Let T seconds be the time for the rotating vector OA of Fig. 13 to make one revolution (i.e. 2π radians). Then

$$2\pi = \omega T$$

or $$T = \frac{2\pi}{\omega} \text{ seconds}$$

Time T is known as the **periodic time** and is defined as the time taken to complete one complete revolution of OA, i.e. one cycle of the oscillation.

Frequency

The number of complete oscillations occurring per second is called the frequency, f. The electrical unit of frequency is the hertz (Hz)

Frequency, $f = \dfrac{1}{T} = \dfrac{\omega}{2\pi}$ Hz

Angular velocity $\omega = 2\pi f$ **rad** s^{-1}

6. Graphs of trigonometric functions

Graphs of sin ωt and cos ωt

Figures 14 and 15 show graphs of $y = \sin \omega t$ and $y = \cos \omega t$ for values of ω of 1, 2, 3 and $\dfrac{1}{2}$.

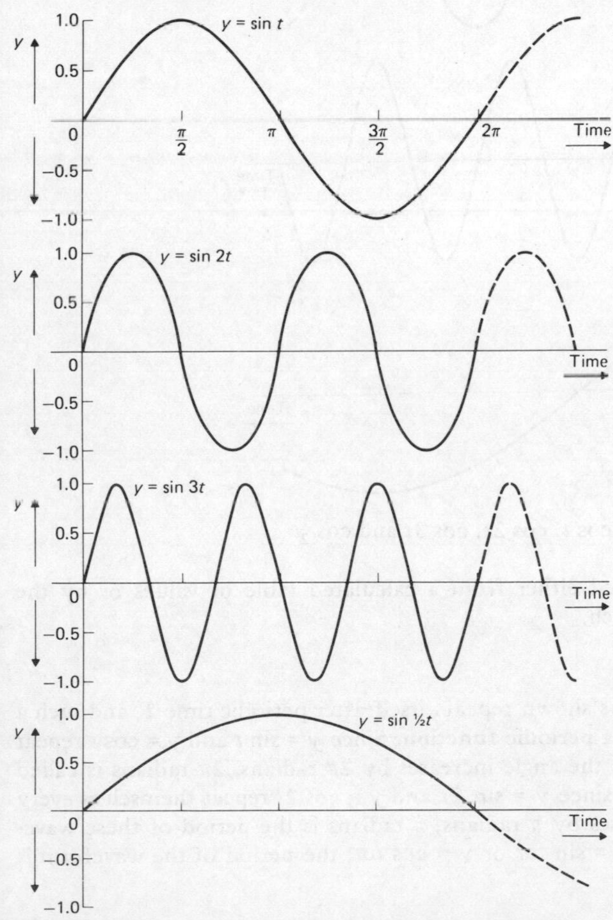

Figure 14 Graphs of sin t, sin $2t$, sin $3t$ and sin $\frac{1}{2}t$

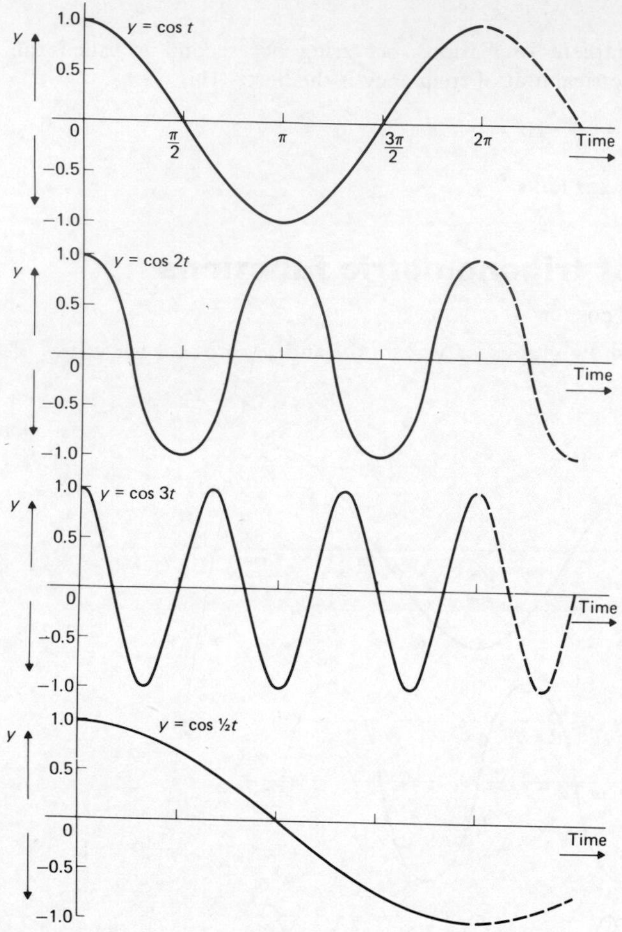

Figure 15 Graphs of cos t, cos $2t$, cos $3t$ and cos $\frac{1}{2}t$

The graphs are plotted either from a calculated table of values or by the rotating-vector approach.

Period

Each of the waveforms shown repeats itself after periodic time T, and such a function is known as a **periodic function.** Since $y = \sin t$ and $y = \cos t$ repeat themselves every time the angle increases by 2π radians, 2π radians is called the **period**. Similarly, since $y = \sin 2t$ and $y = \cos 2t$ repeat themselves every time the angle increases by π radians, π radians is the period of these waveforms. Generally, if $y = \sin \omega t$ or $y = \cos \omega t$, the period of the waveform is $\dfrac{2\pi}{\omega}$.

Graphs of $\sin^2\omega t$ and $\cos^2\omega t$

Figure 16 shows graphs of $y = \sin^2 t$ and $y = \cos^2 t$. Both are periodic functions of period π radians and display positive values only.

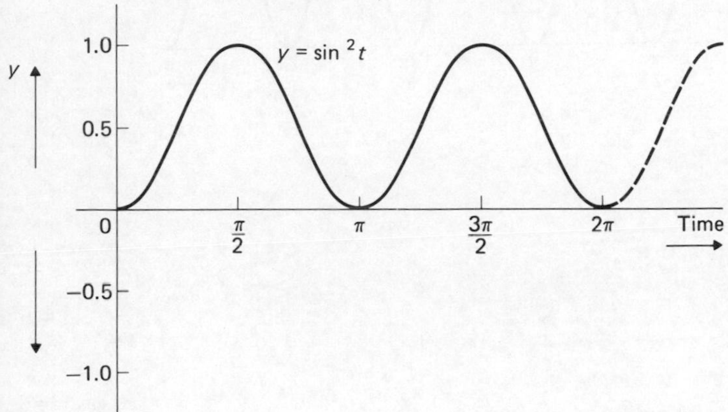

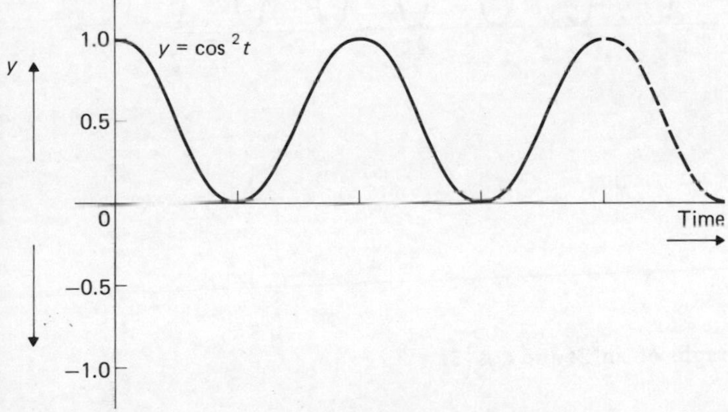

Figure 16 Graphs of $\sin^2 t$ and $\cos^2 t$

The graph of $y = \sin^2 2t$, shown in Fig. 17 (a), has a period of $\frac{\pi}{2}$ radians.

The graph of $y = \cos^2 3t$, shown in Fig. 17 (b) has a period of $\frac{\pi}{3}$ radians.

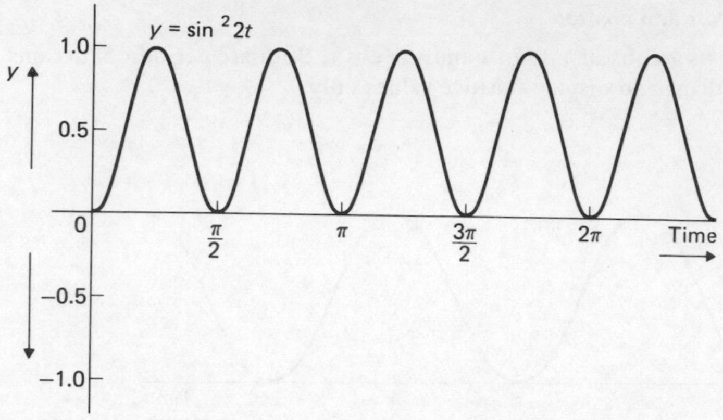

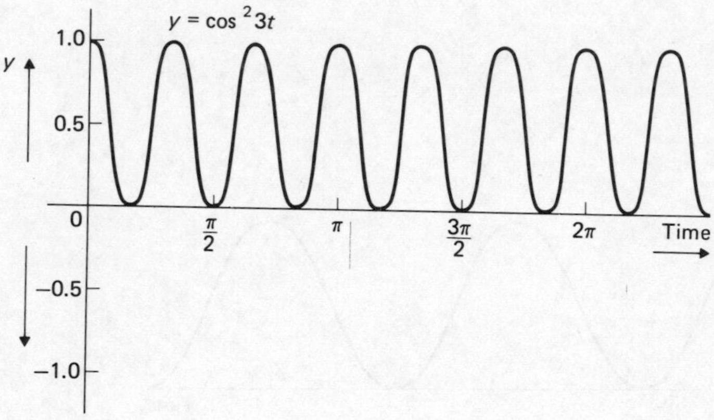

Figure 17 Graphs of $\sin^2 2t$ and $\cos^2 3t$

Generally, if $y = \sin^2\omega t$ or $y = \cos^2\omega t$ then the period of the waveform is $\dfrac{\pi}{\omega}$ radians. It should be noted that graphs of the form $y = \sin^2\omega t$ and $y = \cos^2\omega t$ are **not** sine waves and **cannot** be produced by the rotating-vector method. A summary of the information that can be obtained from the graphs shown in Figs. 14 to 17 is contained in Table 1.

Table 1 Summary of information obtained from graphs shown in Figs. 14 to 17

Function	Periodic time (seconds)	Frequency (Hz)	Number of cycles completed in 2π s
$\sin t$, $\cos t$	2π	$\frac{1}{2\pi}$	1
$\sin 2t$, $\cos 2t$	π	$\frac{1}{\pi}$	2
$\sin 3t$, $\cos 3t$	$\frac{2\pi}{3}$	$\frac{3}{2\pi}$	3
$\sin \frac{1}{2}t$, $\cos \frac{1}{2}t$	4π	$\frac{1}{4\pi}$	$\frac{1}{2}$
$\sin \omega t$, $\cos \omega t$	$\frac{2\pi}{\omega}$	$\frac{\omega}{2\pi}$	ω
$\sin^2 t$, $\cos^2 t$	π	$\frac{1}{\pi}$	2
$\sin^2 2t$, $\cos^2 2t$	$\frac{\pi}{2}$	$\frac{2}{\pi}$	4
$\sin^2 3t$, $\cos^2 3t$	$\frac{\pi}{3}$	$\frac{3}{\pi}$	6
$\sin^2 \omega t$, $\cos^2 \omega t$	$\frac{\pi}{\omega}$	$\frac{\omega}{\pi}$	2ω

7. The general form of a sine wave, $R \sin(\omega t \pm \alpha)$

Amplitude

Amplitude is the name given to the maximum value of a sine wave. For each of the graphs shown in Figs. 14 and 15 the maximum value is 1. However, if $y = 3 \sin t$, then each of the ordinate values is multiplied by 3 and the maximum value, and thus the amplitude, is 3. Similarly, if $y = 7 \cos 3t$ then the amplitude is 7. The period, the periodic time and the frequency are all unaffected by the amplitude.
Generally, if $y = R \sin \omega t$ then the amplitude is given by R.

Leading and lagging angles

A sine wave may not always be equal to zero at 0°. Similarly, a cosine wave may not always be equal to 1 at 0°. To show this, a periodic function is represented by $y = R \sin(\omega t \pm \alpha)$ or $y = R \cos(\omega t \pm \alpha)$ where α is the phase-angle difference compared with $y = R \sin \omega t$ or $y = R \cos \omega t$. Graphs of

$y = 2 \sin(\omega t + \frac{\pi}{3})$ and $y = 3 \cos(\omega t - \frac{\pi}{6})$ are shown in Fig. 18.

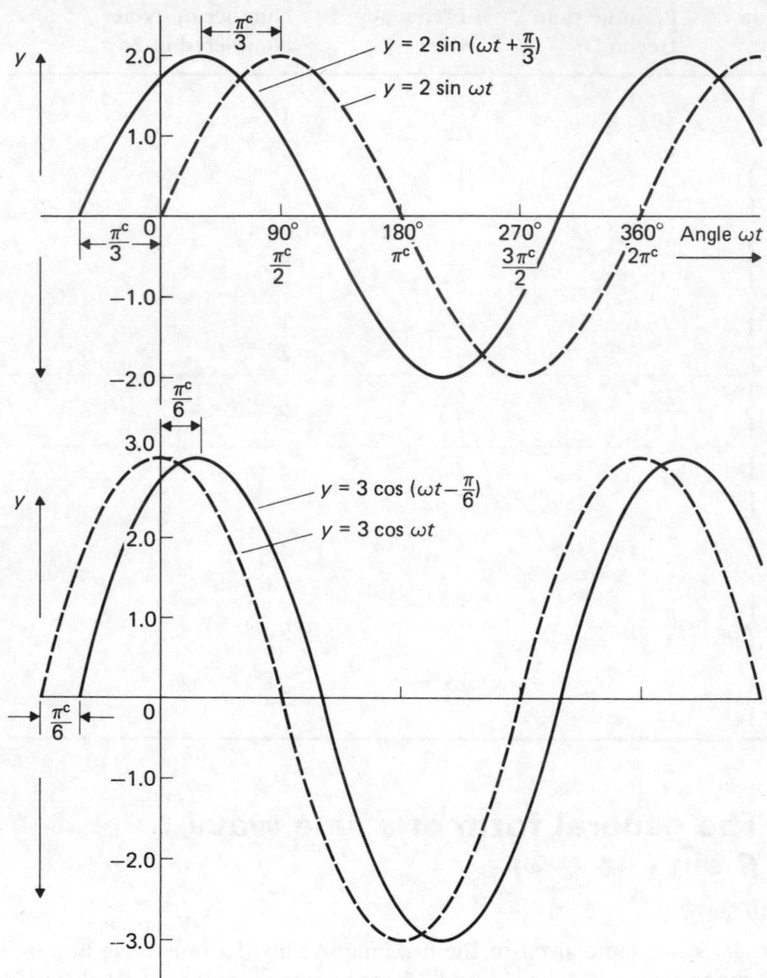

Figure 18 Graphs of $2 \sin(\omega t + \frac{\pi}{3})$ and $3 \cos(\omega t - \frac{\pi}{6})$

The graph of $y = 2 \sin \omega t$ is zero at 0°, and the graph of $y = 2 \sin(\omega t + \frac{\pi}{3})$ starts $\frac{\pi}{3}$ radians (i.e. 60°) earlier. The latter is said to **lead** the former by $\frac{\pi}{3}$ radians. Similarly, the graph of $y = 3 \cos \omega t$ commences at 0°, and the graph of $y = 3 \cos(\omega t - \frac{\pi}{6})$ starts $\frac{\pi}{6}$ radians (i.e. 30°) later. The latter graph is said to **lag** the former by $\frac{\pi}{6}$ radians.

In each of the above two examples the angles of lead or lag may also be seen on the graphs by comparing the positions of the maximum values, such that when one waveform leads another it reaches its maximum value earlier, and vice versa.

In general, a graph of $y = R \sin(\omega t + \alpha)$ leads $y = R \sin \omega t$ by angle α. A graph of $y = R \sin(\omega t - \alpha)$ lags $y = R \sin \omega t$ by angle α.

It may be seen from Figs. 14 and 15 that a cosine curve is the same shape as a sine curve except that it starts $\frac{\pi^c}{2}$ earlier, i.e. leads by $\frac{\pi^c}{2}$.

Hence $\mathbf{\cos \omega t = \sin(\omega t + \frac{\pi}{2})}$.

Since sine and cosine curves are of the same shape both are referred to generally as 'sine waves'.

Summary

If $y = R \sin(\omega t \pm \alpha)$ then:

(a) Amplitude $= R$;

(b) Angular velocity $= \omega$ radians per second;

(c) Periodic time, $T = \frac{2\pi}{\omega}$ seconds;

(d) Frequency, $f = \frac{\omega}{2\pi}$ hertz;

(e) Phase angle $= \alpha$ (positive when leading $R \sin \omega t$ and negative when lagging $R \sin \omega t$).

Worked problems on periodic functions of the form $R \sin(\omega t \pm \alpha)$

Problem 1. Sketch the following graphs from $t = 0$ to $t = 2\pi$ seconds showing relevant details: (a) $y = 2 \sin 2t$; (b) $y = 5 \cos 3t$; (c) $y = 4 \sin(t - \frac{\pi}{4})$; (d) $y = 3 \cos(t + \frac{\pi}{3})$; (e) $y = 4 \sin^2 t$; (f) $y = 3 \cos^2 2t$.

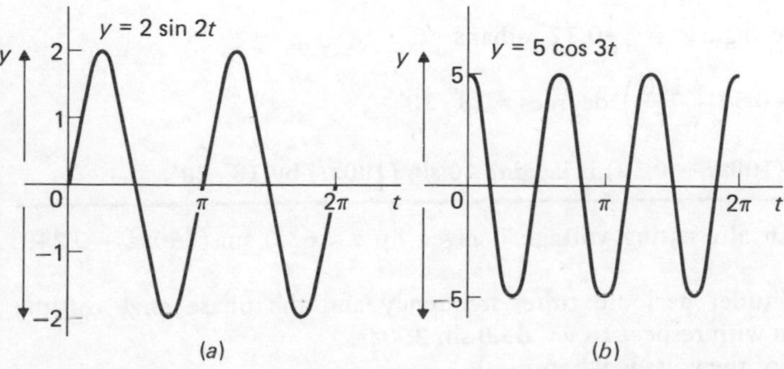

Figure 19

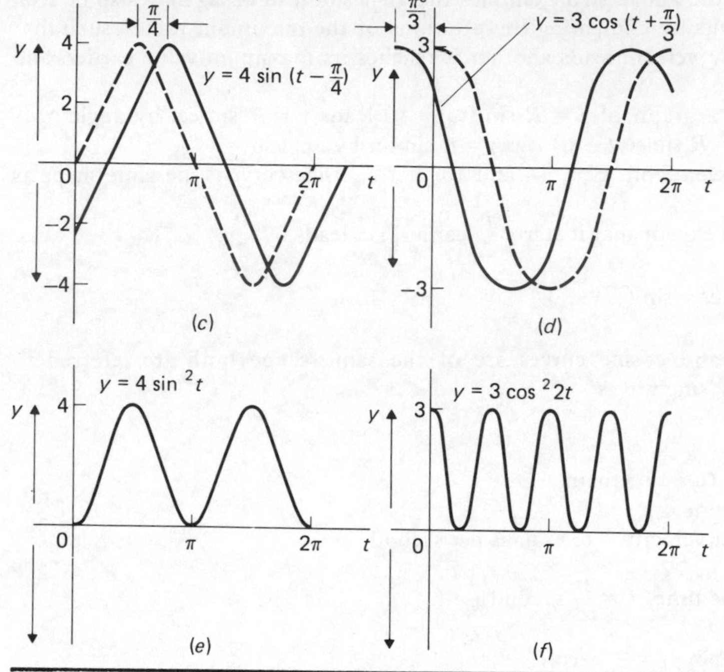

Problem 2. An alternating current is given by $i = 40 \sin (100\pi t - 0.32)$ amperes. Find: (a) the amplitude; (b) the periodic time; (c) the frequency; and (d) the phase angle of the oscillation compared with $i = 40 \sin 100\pi t$.

(a) **Amplitude = 40 amperes**
(b) Angular velocity $\omega = 100\pi$ radians per second

Periodic time $T = \dfrac{2\pi}{\omega} = \dfrac{2\pi}{100\pi} = \dfrac{1}{50}$ **= 0.02 seconds**

(c) **Frequency** $f = \dfrac{1}{T} = \dfrac{1}{0.02}$ **= 50 Hz.**

(d) Phase angle $\alpha = -0.32$ radians,

0.32 radians $= 0.32 \left(\dfrac{180}{\pi}\right)$ degrees $= 18° \ 20'$

Hence **40 sin (100πt − 0.32) is lagging 40 sin (100πt) by 18° 20′.**

Problem 3. An alternating voltage is given by $v = 63.0 \sin (250\pi t + 0.240)$ volts. Find:

(a) the amplitude, periodic time, frequency and the phase angle of the oscillation with respect to $v = 63.0 \sin 250\pi t$;
(b) the value of the voltage when $t = 0$;

(c) the value of the voltage when $t = 2$ ms;
(d) the time when the voltage first reaches 30.0 volts; and
(e) the time when the voltage is first a maximum.
Sketch one cycle of the oscillation.

(a) **Amplitude = 63.0 volts**

Periodic time, $T = \frac{2\pi}{\omega} = \frac{2\pi}{250\pi} = \frac{1}{125}$ **= 0.008 s or 8 ms**

Frequency, $f = \frac{1}{T}$ **= 125 Hz**

Phase angle α = 0.240 radians = 13° 45′
Hence 63.0 sin (250πt + 0.240) leads 63.0 sin (250πt) by 13° 45′.

(b) When $t = 0$, v = 63.0 sin (0.240) volts
= 63.0 sin 13° 45′
Hence the value of the voltage at t = 0 is 14.97 volts.

(c) When t = 2 ms, $v = 63.0 \sin\left[250\pi\left(\frac{2}{1\,000}\right) + 0.240\right]$

= 63.0 sin (1.810 8)
= 63.0 sin 103° 45′
Hence the value of the voltage at t = 2 ms is 61.19 volts.

(d) When v = 30 volts, 30.0 = 63.0 sin (250πt + 0.240)

$\frac{30.0}{63.0} = \sin(250\pi t + 0.240)$

0.476 2 = sin (250πt + 0.240)
(250πt + 0.240) radians = arcsin 0.476 2
= 28° 26′ = 0.496 3 radians

$\therefore 250\,\pi\, t + 0.240 = 0.496\,3$

$t = \frac{(0.496\,3 - 0.240)}{250\,\pi}$ seconds

= 0.326 3 ms
Hence the time when the voltage first reaches 30.0 volts is 0.326 3 ms.

(e) When the voltage is a maximum, v = 63.0 volts.
63.0 = 63.0 sin (250πt + 0.240)
1 = sin (250πt + 0.240)

(250πt + 0.240) radians = arcsin 1 = $\frac{\pi}{2}$ radians

$\therefore 250\pi t + 0.240 = 1.570\,8$

$t = \frac{1.570\,8 - 0.240}{250\,\pi}$ seconds

= 1.694 4 ms

 Hence the time when the voltage is first a maximum is 1.694 4 ms.
One cycle of the operation is shown in Fig. 20.

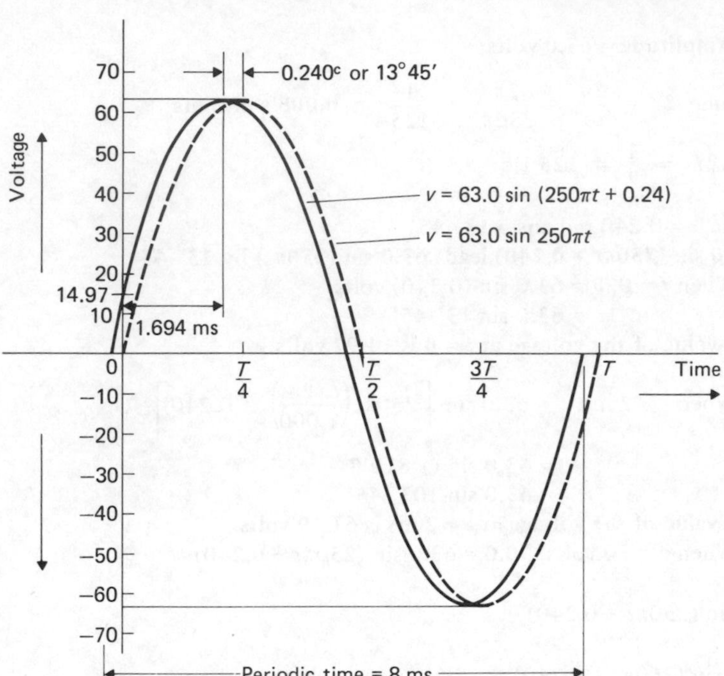

Figure 20

Further problems on periodic functions of the form R *sin* ($\omega t \pm \alpha$) *may be found in Section 13 (Problems 32–44).*

8. Compound angles

Angles such as $(A + B)$ or $(A - B)$ are called **compound angles** since they are the sum or difference of two angles A and B. Each expression of a compound angle has two components. It is useful to express the trigonometrical ratios of compound angles in terms of their two component angles. These are often called the **addition or subtraction formulae** and one method of deriving the formulae is shown below.
Consider the triangle XYZ shown in Fig. 21.

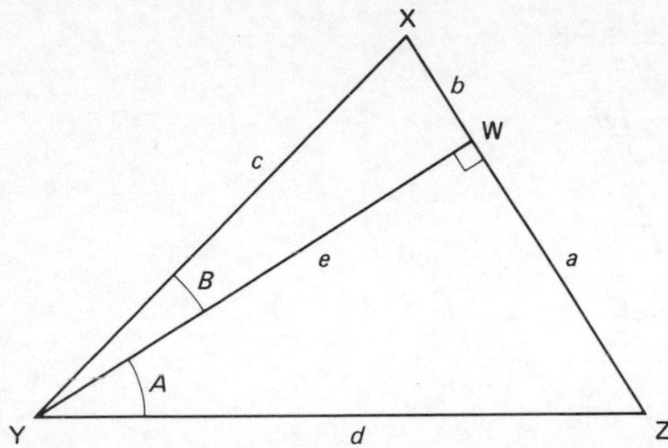

Figure 21

The line YW is constructed perpendicular to XZ. Thus two right-angled triangles, YZW and YWX, are produced. Let the sides of the two triangles be labelled a, b, c, d and e as shown, and let ∠WYZ be A and ∠XYW be B.

Area of triangle YZW $= \frac{1}{2}de \sin A$

Area of triangle YWX $= \frac{1}{2}ce \sin B$

Area of triangle XYZ $= \frac{1}{2}cd \sin (A + B)$

But area of triangle XYZ = area of triangle YZW + area of triangle YWX

i.e. $\frac{1}{2} cd \sin (A + B) = \frac{1}{2} de \sin A + \frac{1}{2} ce \sin B$

Dividing each term by $\frac{1}{2} cd$ gives:

$\sin (A + B) = \frac{e}{c} \sin A + \frac{e}{d} \sin B$

But $\frac{e}{c} = \cos B$ and $\frac{e}{d} = \cos A$

Hence $\sin (A + B) = \sin A \cos B + \cos A \sin B$ (1)

Applying the cosine rule to triangle XYZ gives:

$(a + b)^2 = c^2 + d^2 - 2cd \cos (A + B)$

$$\therefore \cos (A + B) = \frac{c^2 + d^2 - (a + b)^2}{2\,cd}$$

$$= \frac{c^2 + d^2 - [a^2 + 2ab + b^2]}{2cd}$$

$$= \frac{(c^2 - b^2) + (d^2 - a^2) - 2ab}{2cd}$$

But $c^2 - b^2 = e^2$ and $d^2 - a^2 = e^2$ by Pythagoras' theorem

$$\text{Hence } \cos(A+B) = \frac{e^2 + e^2 - 2ab}{2cd}$$

$$= \frac{2e^2 - 2ab}{2cd}$$

$$= \frac{e^2 - ab}{cd}$$

$$= \frac{e^2}{cd} - \frac{ab}{cd}$$

$$= \frac{e}{d}\frac{e}{c} - \frac{a}{d}\frac{b}{c}$$

Hence $\mathbf{\cos(A+B) = \cos A \cos B - \sin A \sin B}$ (2)
Replacing B by $-B$ in equation (1) and equation (2),
and remembering that $\sin(-B) = -\sin B$ and $\cos(-B) = +\cos B$,
$\mathbf{\sin(A-B) = \sin A \cos B - \cos A \sin B}$ (3)
and $\mathbf{\cos(A-B) = \cos A \cos B + \sin A \sin B}$ (4)
The addition and subtraction formulae (i.e. equations (1) to (4)), proved above for acute angles, are, in fact, true for all values of A and B. It is very important to realise that:
$\sin(A+B)$ is **not** equal to $(\sin A + \sin B)$
and $\cos(A-B)$ is **not** equal to $(\cos A - \cos B)$, and so on.

Summary

$\sin(A \pm B) = \sin A \cos B \pm \cos A \sin B$
$\cos(A \pm B) = \cos A \cos B \mp \sin A \sin B$
(In the cosine compound-angle formula care should be taken in particular with the signs).

Worked problems on compound angles

Problem 1. Verify from trigonometrical tables: (a) that the compound-angle addition formulae are true when $A = 30°$ and $B = 45°$; and (b) that the compound-angle subtraction formulae are true when $A = 203°$ and $B = 145°$. Use four-figure tables (i.e. three-significant-figure accuracy).

(a) $\sin(A+B) = \sin A \cos B + \cos A \sin B$
When $A = 30°$ and $B = 45°$,

$$\sin(30° + 45°) = \sin 30° \cos 45° + \cos 30° \sin 45°$$
$$= (0.500\,0)(0.707\,1) + (0.866\,0)(0.707\,1)$$
$$= 0.353\,6 + 0.612\,3$$

i.e. $\mathbf{\sin 75° = 0.965\,9}$

This may be verified from natural sine tables.
$\cos(A+B) = \cos A \cos B - \sin A \sin B$
When $A = 30°$ and $B = 45°$

$$\begin{aligned} \cos(30^\circ + 45^\circ) &= \cos 30^\circ \cos 45^\circ - \sin 30^\circ \sin 45^\circ \\ &= (0.866\,0)(0.707\,1) - (0.500\,0)(0.707\,1) \\ &= 0.612\,3 - 0.353\,6 \\ \text{i.e.} \quad \cos 75^\circ &= \mathbf{0.258\,7} \end{aligned}$$

This may be verified to three-significant-figure accuracy from natural cosine tables.

(b) $\sin(A - B) = \sin A \cos B - \cos A \sin B$

When $A = 203^\circ$ and $B = 145^\circ$,

$$\begin{aligned} \sin(203^\circ - 145^\circ) &= \sin 203^\circ \cos 145^\circ - \cos 203^\circ \sin 145^\circ \\ &= (-\sin 23^\circ)(-\cos 35^\circ) - (-\cos 23^\circ)(\sin 35^\circ) \\ &= (-0.390\,7)(-0.819\,2) - (-0.920\,5)(0.573\,6) \\ &= +0.320\,1 + 0.528\,0 \\ \text{i.e.} \quad \sin 58^\circ &= \mathbf{0.848\,1} \end{aligned}$$

This may be verified to three-significant-figure accuracy from natural sine tables.

$\cos(A - B) = \cos A \cos B + \sin A \sin B$

when $A = 203^\circ$ and $B = 145^\circ$,

$$\begin{aligned} \cos(203^\circ - 145^\circ) &= \cos 203^\circ \cos 145^\circ + \sin 203^\circ \sin 145^\circ \\ &= (-0.920\,5)(-0.819\,2) + (-0.390\,7)(0.573\,6) \\ &= +0.754\,1 - 0.224\,1 \\ \text{i.e.} \quad \cos 58^\circ &= \mathbf{0.530\,0} \end{aligned}$$

This may be verified to three-significant-figure accuracy from natural cosine tables.

For any chosen values of A and B, the compound-angle addition and subtraction formulae are valid.

Problem 2. Use the compound-angle addition and subtraction formulae to simplify the following expressions:

(a) $\sin 52^\circ \cos 29^\circ + \cos 52^\circ \sin 29^\circ$;

(b) $\cos 46^\circ \cos 11^\circ + \sin 46^\circ \sin 11^\circ$;

(c) $\sin(180^\circ + X)$;

(d) $\cos\left(\frac{3\pi}{2} + Y\right)$;

(e) $\cos(A - B) - \cos(A + B)$.

(a) Since $\sin A \cos B + \cos A \sin B = \sin(A + B)$

then $\sin 52^\circ \cos 29^\circ + \cos 52^\circ \sin 29^\circ = \sin(52^\circ + 29^\circ)$

Hence $\mathbf{\sin 52^\circ \cos 29^\circ + \cos 52^\circ \sin 29^\circ = \sin 81^\circ}$

(b) Since $\cos A \cos B + \sin A \sin B = \cos(A - B)$

then $\cos 46^\circ \cos 11^\circ + \sin 46^\circ \sin 11^\circ = \cos(46^\circ - 11^\circ)$

Hence $\mathbf{\cos 46^\circ \cos 11^\circ + \sin 46^\circ \sin 11^\circ = \cos 35^\circ}$

(c) $$\begin{aligned} \sin(180^\circ + X) &= \sin 180^\circ \cos X + \cos 180^\circ \sin X \\ &= (0)(\cos X) + (-1)(\sin X) \end{aligned}$$

Hence $\mathbf{\sin(180^\circ + X) = -\sin X}$.

(d) $\cos\left(\frac{3\pi}{2} + Y\right) = \cos\frac{3\pi}{2}\cos Y - \sin\frac{3\pi}{2}\sin Y$

$= (0)(\cos Y) - (-1)(\sin Y)$

Hence $\cos\left(\frac{3\pi}{2} + Y\right) = \sin Y$

(e) $\cos(A - B) - \cos(A + B) = [\cos A \cos B + \sin A \sin B] - [\cos A \cos B - \sin A \sin B]$
$= 2 \sin A \sin B$

Problem 3. (a) Given that $\cos A = \frac{4}{5}$ and $\sin B = \frac{15}{17}$ where A and B are acute angles, find without using trigonometrical tables the values of $\sin(A + B)$ and $\cos(A - B)$.

(b) Given that $\sin C = 0.400\,0$ and $\cos D = 0.600\,0$ where C and D are acute angles, find the values of $\sin(C - D)$ and $\cos(C + D)$.

(a) $\cos A = \frac{4}{5} = \frac{\text{adjacent side}}{\text{hypotenuse}}$. The opposite side is found by the theorem of Pythagoras, i.e. opposite side of triangle $= \sqrt{(5^2 - 4^2)} = 3$. Hence $\sin A = \frac{3}{5}$.

Similarly, since $\sin B = \frac{15}{17} = \frac{\text{opposite side}}{\text{hypotenuse}}$, the adjacent side of the triangle $= \sqrt{(17^2 - 15^2)} = 8$.

Hence $\cos B = \frac{8}{17}$

$\sin(A + B) = \sin A \cos B + \cos A \sin B$

$= \left(\frac{3}{5}\right)\left(\frac{8}{17}\right) + \left(\frac{4}{5}\right)\left(\frac{15}{17}\right)$

$= \frac{24}{85} + \frac{60}{85} = \frac{84}{85}$

$\cos(A - B) = \cos A \cos B + \sin A \sin B$

$= \left(\frac{4}{5}\right)\left(\frac{8}{17}\right) + \left(\frac{3}{5}\right)\left(\frac{15}{17}\right)$

$= \frac{32}{85} + \frac{45}{85} = \frac{77}{85}$

(b) If $\sin C = 0.400\,0$ then from natural sine tables, $C = 23^\circ\,35'$
Then $\cos C = \cos 23^\circ\,35' = 0.916\,5$
If $\cos D = 0.600\,0$ then from natural cosine tables, $D = 53^\circ\,8'$
Then $\sin D = \sin 53^\circ\,8' = 0.800\,0$

$$\sin(C-D) = \sin C \cos D - \cos C \sin D$$
$$= (0.400\ 0)(0.600\ 0) - (0.916\ 5)(0.800\ 0)$$
$$= 0.240\ 0 - 0.733\ 2$$
$$= \mathbf{-0.493\ 2}$$

$$\cos(C+D) = \cos C \cos D - \sin C \sin D$$
$$= (0.916\ 5)(0.600\ 0) - (0.400\ 0)(0.800\ 0)$$
$$= 0.549\ 9 - 0.320\ 0$$
$$= \mathbf{0.229\ 9}$$

Problem 4. Find the values of: (a) sin 15° and cos 15°; and (b) sin 75° and cos 75° without using trigonometric tables. Leave answers in surd form.

(a) $$\sin 15° = \sin(60° - 45°) = \sin 60° \cos 45° - \cos 60° \sin 45°$$
$$= \left(\frac{\sqrt{3}}{2}\right)\left(\frac{1}{\sqrt{2}}\right) - \left(\frac{1}{2}\right)\left(\frac{1}{\sqrt{2}}\right)$$
$$= \frac{\sqrt{3} - 1}{2\sqrt{2}}$$

$$\cos 15° = \cos(60° - 45°) = \cos 60° \cos 45° + \sin 60° \sin 45°$$
$$= \left(\frac{1}{2}\right)\left(\frac{1}{\sqrt{2}}\right) + \left(\frac{\sqrt{3}}{2}\right)\left(\frac{1}{\sqrt{2}}\right)$$
$$= \frac{1 + \sqrt{3}}{2\sqrt{2}}$$

(b) $$\sin 75° = \sin(45° + 30°) = \sin 45° \cos 30° + \cos 45° \sin 30°$$
$$= \left(\frac{1}{\sqrt{2}}\right)\left(\frac{\sqrt{3}}{2}\right) + \left(\frac{1}{\sqrt{2}}\right)\left(\frac{1}{2}\right)$$
$$= \frac{\sqrt{3} + 1}{2\sqrt{2}}$$

$$\cos 75° = \cos(45° + 30°) = \cos 45° \cos 30° - \sin 45° \sin 30°$$
$$= \left(\frac{1}{\sqrt{2}}\right)\left(\frac{\sqrt{3}}{2}\right) - \left(\frac{1}{\sqrt{2}}\right)\left(\frac{1}{2}\right)$$
$$= \frac{\sqrt{3} - 1}{2\sqrt{2}}$$

It may be seen by reference to tables of natural sine and cosine that $\cos\theta = \sin(90 - \theta)$ or $\sin\theta = \cos(90 - \theta)$. Thus $\cos 15° = \sin(90° - 15°) = \sin 75°$ and $\sin 15° = \cos(90° - 15°) = \cos 75°$, which is shown in this problem.

Further problems on compound angles may be found in Section 13 (Problems 45–62).

9. Conversion of $a \sin \omega t + b \cos \omega t$ into $R \sin(\omega t + \alpha)$

If $R \sin(\omega t + \alpha)$ is expanded using the compound-angle addition formula then:

$$R \sin(\omega t + \alpha) = R[\sin \omega t \cos \alpha + \cos \omega t \sin \alpha]$$
$$= R \sin \omega t \cos \alpha + R \cos \omega t \sin \alpha$$

Let $a \sin \omega t + b \cos \omega t = (R \cos \alpha) \sin \omega t + (R \sin \alpha) \cos \omega t$.
Equating the coefficients of $\sin \omega t$ gives:

$$a = R \cos \alpha, \text{ i.e. } \cos \alpha = \frac{a}{R}$$

Equating the coefficients of $\cos \omega t$ gives:

$$b = R \sin \alpha, \text{ i.e. } \sin \alpha = \frac{b}{R}$$

Therefore if the values of a and b are known the values of R and α can be calculated.
In Fig. 22 the relationships between the constants a, b, R and α are shown.

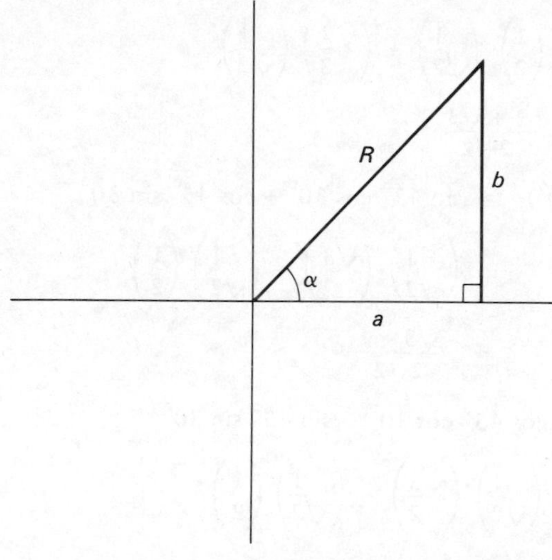

Figure 22

By the theorem of Pythagoras $R = \sqrt{(a^2 + b^2)}$

and from trigonometrical ratios $\alpha = \arctan \frac{b}{a}$

When the resultant of an expression such as $y = a \sin \omega t + b \cos \omega t$ is required, it may be expressed readily in the form $R \sin(\omega t + \alpha)$ by using the compound-angle addition formula.

In *Technician mathematics, Level 2*, Chapter 7, it was shown that the resultant of an expression such as $y = a \sin \omega t + b \cos \omega t$ could be obtained either: (i) by plotting graphs of $y_1 = a \sin \omega t$ and $y_2 = b \cos \omega t$ on the same axes and then adding ordinates at intervals; or (ii) by drawing or calculating the resultant of two phasors from their relative positions when the time is zero. The conversion of $a \sin \omega t + b \cos \omega t$ to $R \sin (\omega t + \alpha)$ gives a third method of obtaining the resultant of two sine waves. It is quite possible to convert $a \sin \omega t \pm b \cos \omega t$ into one of four forms, i.e. $R \sin (\omega t + \alpha)$, $R \sin (\omega t - \alpha)$, $R \cos (\omega t + \alpha)$ or $R \cos (\omega t - \alpha)$. This is achieved by expanding any one of the latter expressions by using addition or subtraction formulae, and then equating the coefficients of $\sin \omega t$ and $\cos \omega t$ (see Problem 3). Trigonometrical equations of the type $a \sin \omega t + b \cos \omega t = C$ may be solved by converting the left-hand side into the form $R \sin (\omega t + \alpha)$ (see Problems 5 and 6).

Worked problems on converting $a \sin \omega t \pm b \cos \omega t$ into the form $R \sin (\omega t \pm \alpha)$

Problem 1. Find an expression for $y = 4 \sin \omega t + 3 \cos \omega t$ in the general form $y = R \sin (\omega t \pm \alpha)$:

(a) by plotting graphs of $y_1 = 4 \sin \omega t$ and $y_2 = 3 \cos \omega t$ on the same axes and then adding ordinates;

(b) by phasor addition; and

(c) by using the compound-angle addition formula.

(a) Graphs of $y_1 = 4 \sin \omega t$ and $y_2 = 3 \cos \omega t$ are shown in Fig. 23 with their resultant $y = 4 \sin \omega t + 3 \cos \omega t$ obtained by adding ordinates at 15° intervals.

Hence, by drawing **$y = 5 \sin (\omega t + 37°)$**

(b) The relative positions of $y_1 = 4 \sin \omega t$ and $y_2 = 3 \cos \omega t$ at $t = 0$ are shown in Fig. 24 (a). From the phasor diagram shown in Fig. 24 (b) the resultant is obtained by drawing or by calculation.

By calculation $y = \sqrt{(4^2 + 3^2)} = 5$

$$\alpha = \arctan \frac{3}{4} = 36° \, 52'$$

Hence, by phasor addition, **$y = 5 \sin (\omega t + 36° \, 52')$**

(c) If $4 \sin \omega t + 3 \cos \omega t = R \sin (\omega t + \alpha)$

then
$$4 \sin \omega t + 3 \cos \omega t = R \, [\sin \omega t \cos \alpha + \cos \omega t \sin \alpha]$$
$$= (R \cos \alpha) \sin \omega t + (R \sin \alpha) \cos \omega t$$

Equating coefficients:

$$4 = R \cos \alpha, \text{ i.e. } \mathbf{\cos \alpha = \frac{4}{R}}$$

$$3 = R \sin \alpha, \text{ i.e. } \mathbf{\sin \alpha = \frac{3}{R}}$$

Hence R and α are calculated from Fig. 22 when $a = 4$ and $b = 3$.
$R = \sqrt{(4^2 + 3^2)} = 5$

and $\alpha = \arctan \frac{3}{4} = 36° \ 52'$ or $216° \ 52'$

From Fig. 22, R is in the first quadrant. Hence $\alpha = 216° \ 52'$ is neglected. If a diagram is always drawn, the quadrant in which R lies is established. There will thus be only one possible value of α.
Hence, by using the compound-angle addition formula,
$4 \sin \omega t + 3 \cos \omega t = 5 \sin (\omega t + 36° \ 52')$

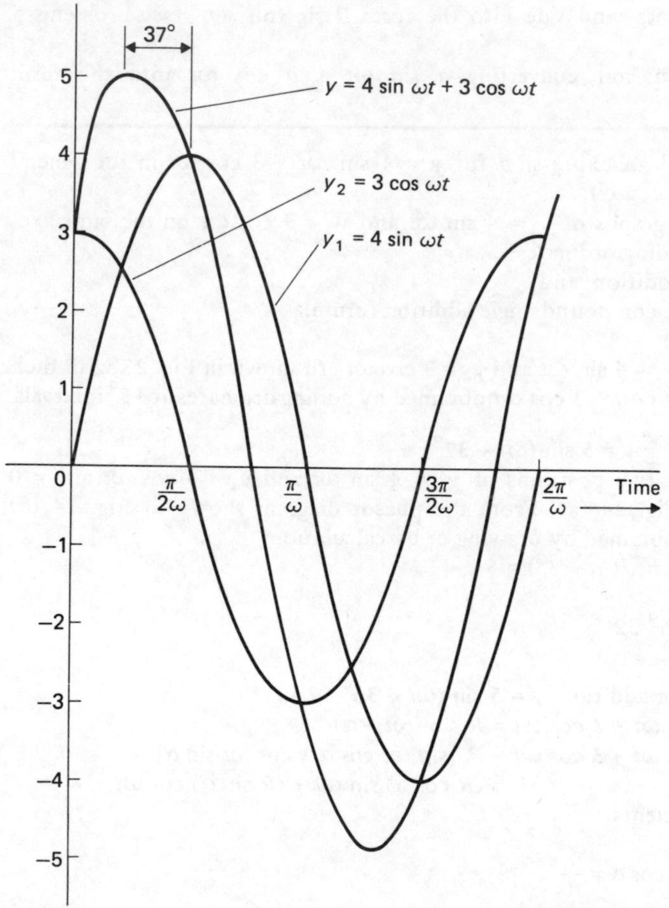

Figure 23 Graphs of $y_1 = 4 \sin \omega t$, $y_2 = 3 \cos \omega t$ and $y = 4 \sin \omega t + \cos \omega t$

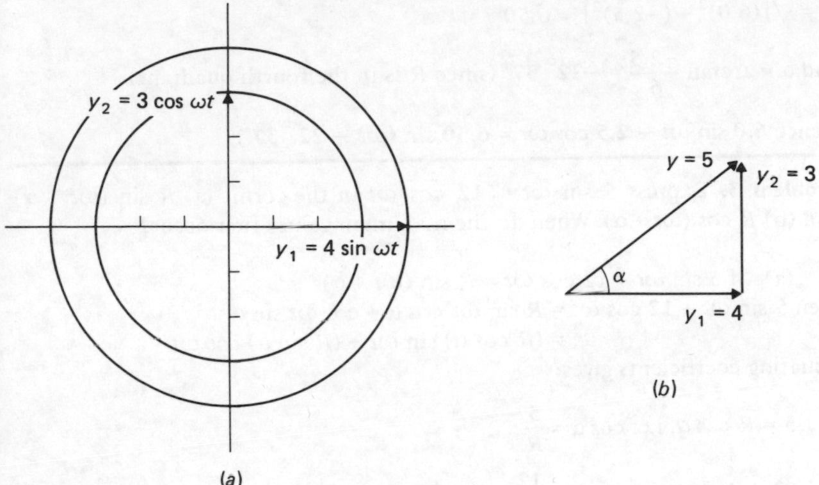

Figure 24

Problem 2. Express 6.0 sin ωt − 2.5 cos ωt in the form R sin ($\omega t + \alpha$).

If $6.0 \sin \omega t - 2.5 \cos \omega t = R \sin (\omega t + \alpha)$

then $6.0 \sin \omega t - 2.5 \cos \omega t = R\,[\sin \omega t \cos \alpha + \cos \omega t \sin \alpha]$

$$= (R \cos \alpha) \sin \omega t + (R \sin \alpha) \cos \omega t$$

Equating coefficients gives:

$$6.0 = R \cos \alpha, \text{ i.e. } \cos \alpha = \frac{6.0}{R}$$

$$\text{and } -2.5 = R \sin \alpha, \text{ i.e. } \sin \alpha = \frac{-2.5}{R}$$

Hence R and α are calculated from Fig. 25.

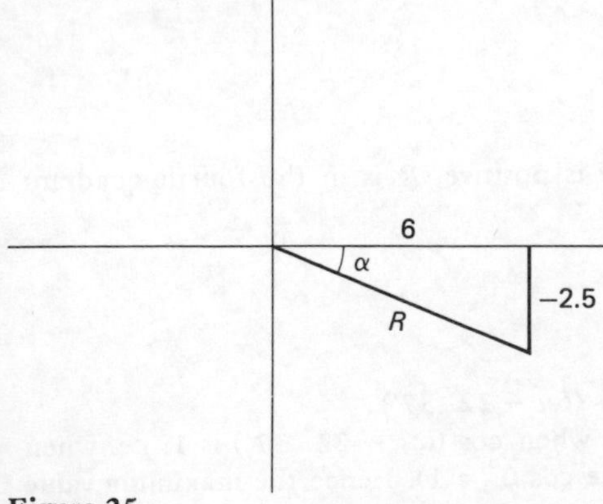

Figure 25

$R = \sqrt{[(6.0)^2 + (-2.5)^2]} = 6.50$

and $\alpha = \arctan \dfrac{-2.5}{6} = -22^\circ\ 37'$ (since R is in the fourth quadrant)

Hence **$6.0 \sin \omega t - 2.5 \cos \omega t = 6.50 \sin (\omega t - 22^\circ\ 37')$**

Problem 3. Express $5 \sin \omega t + 12 \cos \omega t$ in the form: (a) $R \sin (\omega t + \alpha)$; and (b) $R \cos (\omega t + \alpha)$. When do the maximum values first occur?

(a) If $5 \sin \omega t + 12 \cos \omega t = R \sin (\omega t + \alpha)$
then $5 \sin \omega t + 12 \cos \omega t = R \sin \omega t \cos \alpha + \cos \omega t \sin \alpha$
$= (R \cos \alpha) \sin \omega t + (R \sin \alpha) \cos \omega t$

Equating coefficients gives:

$5 = R \cos \alpha$, i.e. $\cos \alpha = \dfrac{5}{R}$

and $12 = R \sin \alpha$, i.e. $\sin \alpha = \dfrac{12}{R}$

Since $\cos \alpha$ and $\sin \alpha$ are both positive, R is in the first quadrant.
$R = \sqrt{(5^2 + 12^2)} = 13$

and $\alpha = \arctan \dfrac{12}{5} = 67^\circ\ 23'$

Hence **$5 \sin \omega t + 12 \cos \omega t = 13 \sin (\omega t + 67^\circ\ 23')$**
The maximum value of $5 \sin \omega t + 12 \cos \omega t$ is thus 13 and this occurs first when $(\omega t + 67^\circ\ 23')$ is equal to 90° (since $\sin 90^\circ = 1$). Hence **the maximum value first occurs when $\omega t = 22^\circ\ 37'$.**

(b) If $5 \sin \omega t + 12 \cos \omega t = R \cos (\omega t + \alpha)$
then $5 \sin \omega t + 12 \cos \omega t = R\ [\cos \omega t \cos \alpha - \sin \omega t \sin \alpha]$
$= (R \cos \alpha) \cos \omega t - (R \sin \alpha) \sin \omega t$

Equating coefficients gives:

$5 = -R \sin \alpha$, i.e. $\sin \alpha = -\dfrac{5}{R}$

and $12 = R \cos \alpha$, i.e. $\cos \alpha = \dfrac{12}{R}$

Since $\sin \alpha$ is negative and $\cos \alpha$ is positive, R is in the fourth quadrant (similar to Fig. 25).
$R = \sqrt{(12^2 + 5^2)} = 13$

and $\alpha = \arctan \dfrac{-5}{12} = -22^\circ\ 37'$

Hence **$5 \sin \omega t + 12 \cos \omega t = 13 \cos (\omega t - 22^\circ\ 37')$**
The maximum value of 13 occurs when $\cos (\omega t - 22^\circ\ 37')$ is 1, i.e. when $(\omega t - 22^\circ\ 37')$ is equal to 0° (since $\cos 0^\circ = 1$). Hence **the maximum value first occurs when $\omega t = 22^\circ\ 37'$.**

Problem 4. Express $-4.50 \sin \omega t - 2.90 \cos \omega t$ in the form $R \sin (\omega t + \alpha)$.

If $-4.50 \sin \omega t - 2.90 \cos \omega t = R \sin (\omega t + \alpha)$
then $-4.50 \sin \omega t - 2.90 \cos \omega t = R [\sin \omega t \cos \alpha + \cos \omega t \sin \alpha]$
$= (R \cos \alpha) \sin \omega t + (R \sin \alpha) \cos \omega t.$

Equating coefficients gives:

$$-4.50 = R \cos \alpha, \text{ i.e. } \cos \alpha = \frac{-4.50}{R}$$

and $-2.90 = R \sin \alpha$, i.e. $\sin \alpha = \dfrac{-2.90}{R}$

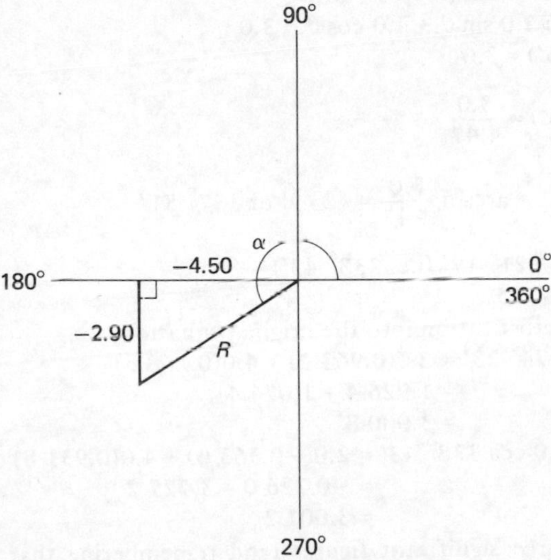

Figure 26

From Fig. 26, $R = \sqrt{[(-4.50)^2 + (-2.90)^2]} = 5.354$
and

$$\alpha = \arctan \frac{-2.90}{-4.50} = 212^\circ\ 48' \text{ (since } R \text{ is the third quadrant)}$$

Hence $-4.50 \sin \omega t -2.90 \cos \omega t = 5.354 \sin (\omega t + 212^\circ\ 48')$.
(This answer may also be stated as $5.354 \sin (\omega t - 147^\circ\ 12')$. Note that when $a \sin \omega t + b \cos \omega t$ is converted into the form $R \sin (\omega t \pm \alpha)$, the angle α is measured from 0°. Thus when R lies in the second or third quadrants α is **not** an acute angle.)

Problem 5. Solve the equation $2.0 \sin \theta + 4.0 \cos \theta = 3.0$ for $0^\circ \leqslant \theta \leqslant 360^\circ$.

If $2.0 \sin \theta + 4.0 \cos \theta = R \sin (\theta + \alpha)$
then $2.0 \sin \theta + 4.0 \cos \theta = R [\sin \theta \cos \alpha + \cos \theta \sin \alpha]$
$= (R \cos \alpha) \sin \theta + (R \sin \alpha) \cos \theta$

Equating coefficients gives:

$$2.0 = R \cos \alpha, \text{ i.e. } \cos \alpha = \frac{2.0}{R}$$

and $4.0 = R \sin \alpha$, i.e. $\sin \alpha = \dfrac{4.0}{R}$

Since $\cos \alpha$ and $\sin \alpha$ are both positive, R is in the first quadrant.
$R = \sqrt{[(2.0)^2 + (4.0)^2]} = 4.47$

and $\alpha = \arctan \dfrac{4.0}{2.0} = 63^\circ\ 26'$

Hence $2.0 \sin \theta + 4.0 \cos \theta = 4.47 \sin (\theta + 63^\circ\ 26')$
But from the original equation $2.0 \sin \theta + 4.0 \cos \theta = 3.0$
Therefore $4.47 \sin (\theta + 63^\circ\ 26') = 3.0$

$$\sin (\theta + 63^\circ\ 26') = \frac{3.0}{4.47}$$

$$\theta + 63^\circ\ 26' = \arcsin \frac{3.0}{4.47} = 42^\circ\ 9' \text{ or } 137^\circ\ 51'$$

Hence, $\theta = 42^\circ\ 9' - 63^\circ\ 26' = -21^\circ\ 17'$ (i.e. $338^\circ\ 43'$)
or $\theta = 137^\circ\ 51' - 63^\circ\ 26' = 74^\circ\ 25'$
The solutions are checked by substitution into the original equation.
Thus $2.0 \sin 74^\circ\ 25' + 4.0 \cos 74^\circ\ 25' = 2.0(0.963\ 2) + 4.0(0.268\ 6)$
$= 1.926\ 4 + 1.074\ 4$
$= 3.000\ 8$
Similarly, $2.0 \sin 338^\circ\ 43' + 4.0 \cos 338^\circ\ 43' = 2.0(-0.363\ 0) + 4.0(0.931\ 8)$
$= -0.726\ 0 + 3.727\ 2$
$= 3.001\ 2$
Taking answers correct to three significant figures (and remembering that values of θ are calculated correct to the nearest minute) then $2.0 \sin \theta + 4.0 \cos \theta = 3.0$ is satisfied by $\theta = 74^\circ\ 25'$ and $\theta = 338^\circ\ 43'$.
Hence **the solutions of the equation $2.0 \sin \theta + 4.0 \cos \theta = 3.0$ between 0° and 360° are $74^\circ\ 25'$ and $338^\circ\ 43'$.**

Problem 6. Solve the equation $7.0 \cos x - 9.0 \sin x - 7.6 = 0$ for values of x between 0° and 360°.

If $7.0 \cos x - 9.0 \sin x - 7.6 = 0$
then $7.0 \cos x - 9.0 \sin x = 7.6$
If $7.0 \cos x - 9.0 \sin x = R \sin (x + \alpha)$
then $7.0 \cos x - 9.0 \sin x = R\ [\sin x \cos \alpha + \cos x \sin \alpha]$
$= (R \cos \alpha) \sin x + (R \sin \alpha) \cos x$
Equating coefficients gives:

$$7.0 = R \sin \alpha, \text{ i.e. } \sin \alpha = \frac{7.0}{R}$$

and $-9.0 = R\cos\alpha$, i.e. $\cos\alpha = \dfrac{-9.0}{R}$

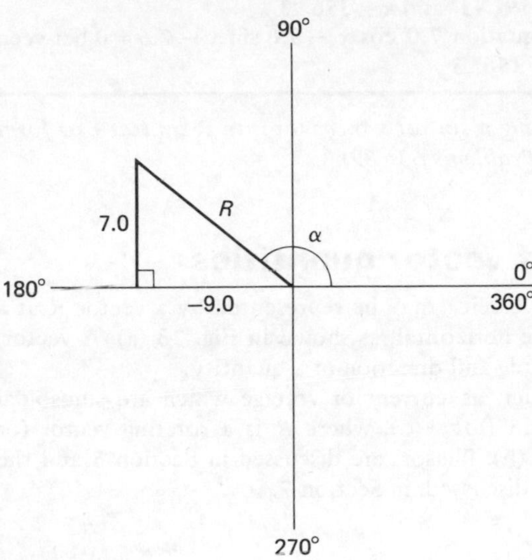

Figure 27

From Fig. 27, $R = \sqrt{[(-9.0)^2 + (7.0)^2]} = 11.4$

and $\alpha = \arctan \dfrac{7.0}{-9.0} = 142^\circ\ 8'$ (since R is in the second quadrant)

Hence $7.0\cos x - 9.0\sin x = 11.4\sin(x + 142^\circ\ 8') = 7.6$

$\sin(x + 142^\circ\ 8') = \dfrac{7.6}{11.4}$

$x + 142^\circ\ 8' = \arcsin\dfrac{7.6}{11.4} = 41^\circ\ 49'$ or $138^\circ\ 11'$

Hence $x = 41^\circ\ 49' - 142^\circ\ 8' = -100^\circ\ 19' = 259^\circ\ 41'$
or $\quad x = 138^\circ\ 11' - 142^\circ\ 8' = -3^\circ\ 57' = 356^\circ\ 3'$

The solutions are checked by substitution into the original equation.

$$
\begin{aligned}
\text{Thus } 7.0\cos x - 9.0\sin x - 7.6 &= 7.0\cos 259^\circ\ 41' - 9.0\sin 259^\circ\ 41' - 7.6\\
&= (7.0)(-0.179\ 1) - (9.0)(-0.983\ 8) - 7.6\\
&= (-1.253\ 7) - (-8.854\ 2) - 7.6\\
&= -1.253\ 7 + 8.854\ 2 - 7.6\\
&= 0.000\ 5
\end{aligned}
$$

Similarly, $7.0\cos 356^\circ\ 3' - 9.0\sin 356^\circ\ 3' - 7.6$

$$
\begin{aligned}
&= (7.0)(0.997\ 6) - (9.0)(-0.068\ 9) - 7.6\\
&= (6.983\ 2) - (-0.620\ 1) - 7.6\\
&= 6.983\ 2 + 0.620\ 1 - 7.6\\
&= 0.003\ 3
\end{aligned}
$$

Taking answers correct to three significant figures (and remembering that the values of x are calculated to the nearest minute), then $7.0 \cos x - 9.0 \sin x - 7.6 = 0$ is satisfied by $x = 259° 41'$ and $x = 356° 3'$.
Hence the solutions of the equation $7.0 \cos x - 9.0 \sin x - 7.6 = 0$ between $0°$ and $360°$ are $259° 41'$ and $356° 3'$.

Further problems on converting a *sin* ωt + b *cos* ωt *into* R *sin (*ωt + α*) form may be found in Section 13 (Problems 63–89).*

10. Resolution of vector quantities

Quantities such as force or velocity may be represented by a vector R at a particular angle α, say to the horizontal, as shown in Fig. 28 (a). A vector, in fact, represents the magnitude and direction of a quantity.

Alternating quantities such as current or voltage which are sinusoidal may be represented by $R \sin(\omega t + \alpha)$, where R is a rotating vector (or phasor), as shown in Fig. 28 (b). Phasors are discussed in Section 5 and the general form of a sine wave is discussed in Section 7.

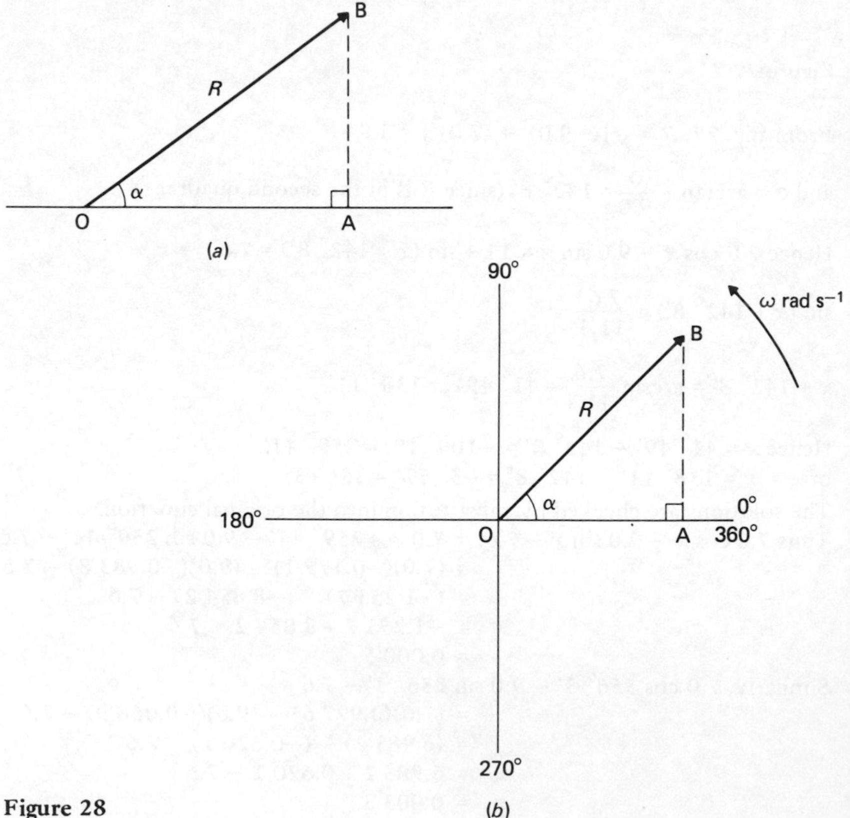

Figure 28

Each of the vectors in Fig. 28 may be replaced by two components, one in the horizontal direction and the other in the vertical direction, whose vector sum is equal to the original vector, R, in magnitude and direction. Such components are often referred to as **rectangular components** of R. With reference to Fig. 28:
The horizontal component of R is OA = $R \cos \alpha$
The vertical component of R is AB = $R \sin \alpha$

The resultant of a number of vectors, each having different magnitudes and directions, may be calculated by algebraically summing the horizontal components of the separate vectors, and the vertical components of the separate vectors and then using the theorem of Pythagoras.

The convention used for angular measurement is: anticlockwise is positive and clockwise is negative.

Worked problems on the resolution of vector quantities

Problem 1. A trolley on a horizontal track is being pulled by a cable at an angle of 25° with the direction of the track in the same plane as the track. The tension in the cable is 100 N. Calculate: (a) the force tending to move the trolley forward; and (b) the sideways thrust on the track.

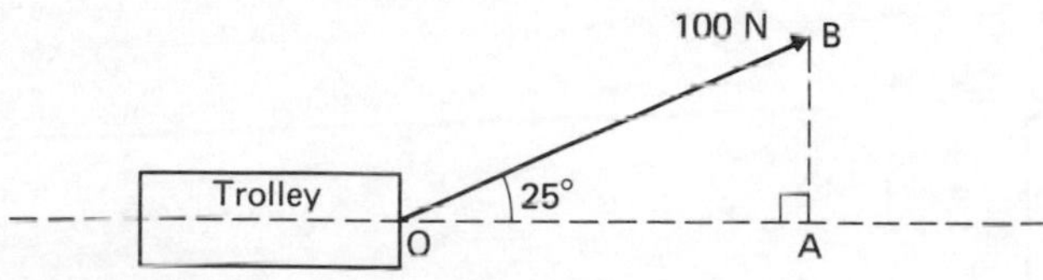

Figure 29

(a) The force tending to move the trolley forward is given by OA in the plan view shown in Fig. 29 (i.e. the horizontal component of 100 N).
OA = 100 cos 25° = **90.63 N**

(b) The sideways thrust on the track is given by AB in Fig. 29 (i.e. the vertical component of 100 N).
AB = 100 sin 25° = **42.26 N**

Problem 2. Further coplanar forces act at a point O as shown in Fig. 30. Calculate the value and the direction of the resultant force.

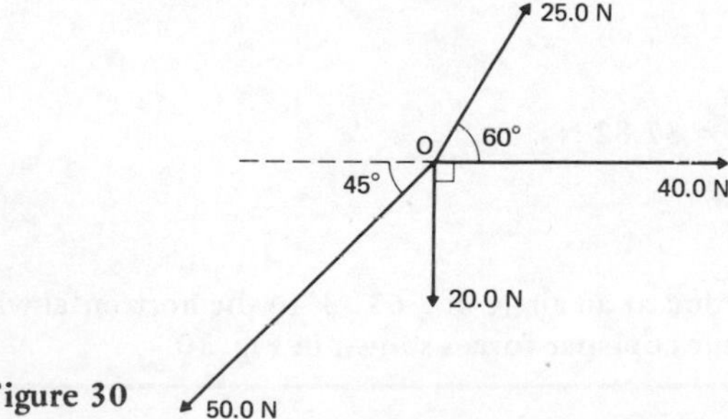

Figure 30

It will be assumed that, similar to the convention used in drawing graphs, the horizontal components of the forces are positive when they are acting towards the right of point O, and that vertical components are positive when they are acting upwards from O, and vice versa.

For the 40.0 N force:
Horizontal component = $40.0 \cos 0°$ = 40.0 N
Vertical component = $40.0 \sin 0°$ = 0
For the 25.0 N force:
Horizontal component = $25.0 \cos 60°$ = 12.50 N
Vertical component = $25.0 \sin 60°$ = 21.65 N
For the 50.0 N force:
Horizontal component = $-50.0 \cos 45°$ = −35.36 N
Vertical component = $-50.0 \sin 45°$ = −35.36 N
For the 20.0 N force:
Horizontal component = $20.0 \cos 90°$ = 0
Vertical component = $-20.0 \sin 90°$ = −20.00 N
Resultant horizontal component = $40.0 + 12.50 - 35.36 + 0$ = +17.14 N
Resultant vertical component = $0 + 21.65 - 35.36 - 20.00$ = −33.71 N
The resultant horizontal and vertical components are shown as OA and AB respectively in Fig. 31 and the resultant force *R* is represented by OB.

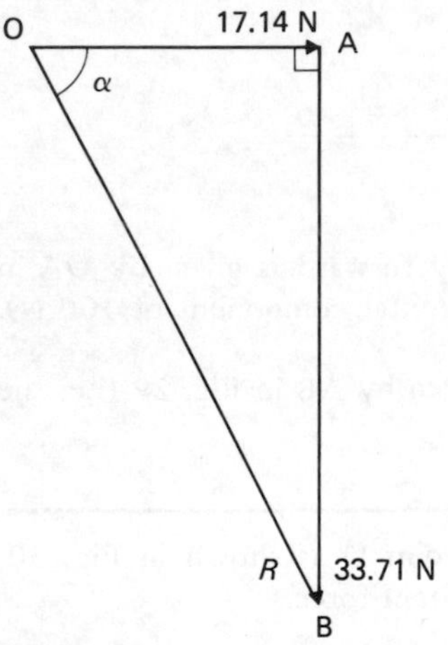

Figure 31

$$R = \sqrt{[(17.14)^2 + (-33.71)^2]} = 37.82 \text{ N}$$

$$\alpha = \arctan \frac{-33.71}{17.14} = -63° \, 3'$$

Hence a force of 37.82 N acting at an angle of −63° 3′ to the horizontal will have the same effect as the four coplanar forces shown in Fig. 30.

Problem 3. The instantaneous values of two alternating voltages are given by

$v_1 = 30.0 \sin(\omega t - \frac{\pi}{4})$ volts and $v_2 = 40.0 \cos(\omega t + \frac{\pi}{3})$ volts.

Calculate an expression for $v_1 + v_2$ by resolution of phasors.

The space diagram representing the alternating voltages is shown in Fig. 32. If the axis OX is assumed to represent $\sin \omega t$ then axis OY represents $\cos \omega t$

[since $\cos \omega t = \sin(\omega t + \frac{\pi}{2})$].

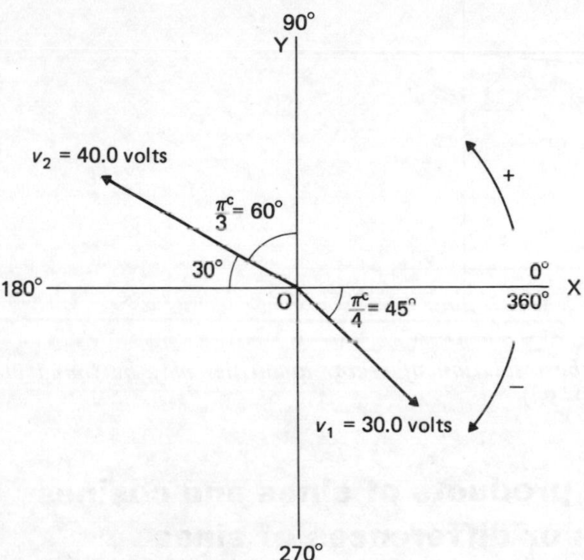

Figure 32

For v_1 = 30.0 volts:

Horizontal component	=	30.0 cos 45°	=	21.21 volts
Vertical component	=	−30.0 sin 45°	=	−21.21 volts

For v_2 = 40.0 volts:

Horizontal component	=	−40.0 cos 30°	=	−34.64 volts
Vertical component	=	40.0 sin 30°	=	20.0 volts
Resultant horizontal component	=	21.21 − 34.64	=	−13.43 volts
Resultant vertical component	=	−21.21 + 20.00	=	−1.21 volts

The resultant horizontal and vertical components are shown in Fig. 33.

$R = \sqrt{[(-1.21)^2 + (-13.43)^2]} = 13.48$ volts

$\alpha = \arctan \dfrac{-1.21}{-13.43} = 5° \, 9'$

Hence the obtuse angle α' (shown in Fig. 33) = $180^\circ - 5^\circ\ 9' = 174^\circ\ 51'$ = 3.052 radians

Hence $v_1 + v_2 = 13.48 \sin(\omega t - 174^\circ\ 51')$ volts or $13.48 \sin(\omega t - 3.052)$ volts

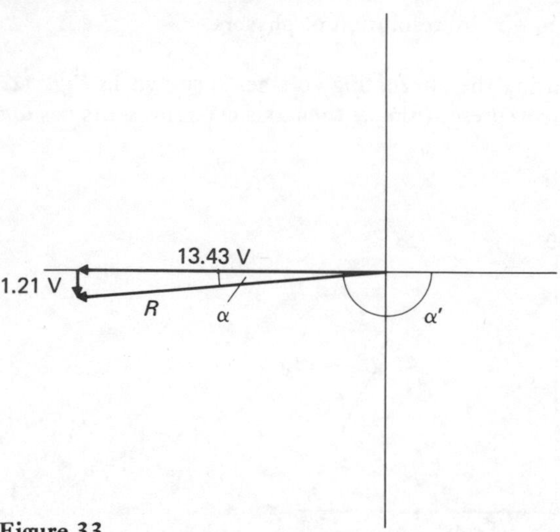

Figure 33

Further problems on the resolution of vector quantities may be found in Section 13 (Problems 90–98).

11. Changing products of sines and cosines into sums or differences of sines and cosines

From Section 8, $\sin(A + B) = \sin A \cos B + \cos A \sin B$

and $\sin(A - B) = \sin A \cos B - \cos A \sin B$

Therefore $\sin(A + B) + \sin(A - B) = 2 \sin A \cos B$

i.e. $\sin A \cos B = \frac{1}{2}[\sin(A + B) + \sin(A - B)]$ (1)

$\sin(A + B) - \sin(A - B) = 2 \cos A \sin B$

i.e. $\cos A \sin B = \frac{1}{2}[\sin(A + B) - \sin(A - B)]$ (2)

Similarly, $\cos(A + B) = \cos A \cos B - \sin A \sin B$

and $\cos(A - B) = \cos A \cos B + \sin A \sin B$

Therefore $\cos(A + B) + \cos(A - B) = 2 \cos A \cos B$

i.e. $\cos A \cos B = \frac{1}{2}[\cos(A + B) + \cos(A - B)]$ (3)

$$\cos(A+B) - \cos(A-B) = -2\sin A \sin B$$

$$\text{i.e. } \sin A \sin B = -\frac{1}{2}[\cos(A+B) - \cos(A-B)] \quad (4)$$

Worked problems on changing products of sines and cosines into sums or differences of sines and cosines

---

Problem 1. Express $\sin 3\theta \cos 2\theta$ as a sum or difference of sines or cosines.

From equation (1), $\sin 3\theta \cos 2\theta = \frac{1}{2}[\sin(3\theta + 2\theta) + \sin(3\theta - 2\theta)]$

$$= \mathbf{\frac{1}{2}[\sin 5\theta + \sin \theta]}$$

Problem 2. Express $\cos 4x \sin x$ as a sum or difference of sines or cosines.

From equation (2), $\cos 4x \sin x = \frac{1}{2}[\sin(4x + x) - \sin(4x - x)]$

$$= \mathbf{\frac{1}{2}[\sin 5x - \sin 3x]}$$

Problem 3. Express $\cos 5t \cos 3t$ as a sum or difference of sines or cosines.

From equation (3), $\cos 5t \cos 3t = \frac{1}{2}[\cos(5t + 3t) + \cos(5t - 3t)]$

$$= \mathbf{\frac{1}{2}[\cos 8t + \cos 2t]}$$

Problem 4. Express $\sin 45^\circ \sin 30^\circ$ as a sum or difference of sines or cosines.

From equation (4), $\sin 45^\circ \sin 30^\circ = -\frac{1}{2}[\cos(45^\circ + 30^\circ) - \cos(45^\circ - 30^\circ)]$

$$= -\frac{1}{2}[\cos 75^\circ - \cos 15^\circ]$$

$$= \mathbf{\frac{1}{2}[\cos 15^\circ - \cos 75^\circ]}$$

Further problems on changing products of sines and cosines into sums or differences of sines and cosines may be found in Section 13 (Problems 99–107).

12. Changing sums or differences of sines and cosines into products of sines and cosines

In the compound-angle formulae let $A + B = X$ and $A - B = Y$

Then $A = \dfrac{X+Y}{2}$ and $B = \dfrac{X-Y}{2}$

Thus instead of $\sin(A+B) + \sin(A-B) = 2\sin A\cos B$, we have

$$\sin X + \sin Y = 2\sin\left[\frac{X+Y}{2}\right]\cos\left[\frac{X-Y}{2}\right] \quad (1)$$

Similarly $$\sin X - \sin Y = 2\cos\left[\frac{X+Y}{2}\right]\sin\left[\frac{X-Y}{2}\right] \quad (2)$$

$$\cos X + \cos Y = 2\cos\left[\frac{X+Y}{2}\right]\cos\left[\frac{X-Y}{2}\right] \quad (3)$$

and $$\cos X - \cos Y = -2\sin\left[\frac{X+Y}{2}\right]\sin\left[\frac{X-Y}{2}\right] \quad (4)$$

Worked problems on changing sums or differences of sines and cosines into products of sines and cosines

---

Problem 1. Express $\sin 7\theta + \sin 3\theta$ as a product.

From equation (1), $\sin 7\theta + \sin 3\theta = 2\sin\left[\dfrac{7\theta + 3\theta}{2}\right]\cos\left[\dfrac{7\theta - 3\theta}{2}\right]$

$$= \mathbf{2\sin 5\theta\cos 2\theta}$$

Problem 2. Express $\sin 50^\circ - \sin 20^\circ$ as a product.

From equation (2), $\sin 50^\circ - \sin 20^\circ = 2\cos\left[\dfrac{50^\circ + 20^\circ}{2}\right]\sin\left[\dfrac{50^\circ - 20^\circ}{2}\right]$

$$= \mathbf{2\cos 35^\circ\sin 15^\circ}$$

Problem 3. Express $\cos 6x + \cos 2x$ as a product.

From equation (3), $\cos 6x + \cos 2x = 2\cos\left[\dfrac{6x + 2x}{2}\right]\cos\left[\dfrac{6x - 2x}{2}\right]$

$$= \mathbf{2\cos 4x\cos 2x}$$

Problem 4. Express cos 20° − cos 46° as a product.

From equation (4), $\cos 20^\circ - \cos 46^\circ = -2 \sin\left[\frac{20^\circ + 46^\circ}{2}\right] \sin\left[\frac{20^\circ - 46^\circ}{2}\right]$

$= -2 \sin 33^\circ \sin(-13^\circ)$

$= 2 \sin 33^\circ \sin 13^\circ$

Further problems on changing sums or differences of sines and cosines into products of sines and cosines may be found in the following Section (13) (Problems 108–115).

13. Further problems

Solution of scalene triangles

1. Solve the following triangles ABC:
   (a) A = 53°, B = 61°, a = 12.6 cm
   (b) B = 83° 16′, b = 16.48 mm, c = 12.92 mm
   (c) A = 37°, B = 73°, b = 4.3 m
   [(a) [C = 66°, b = 13.80 cm, c = 14.41 cm]
   (b) [A = 45° 36′, c = 51° 8′, a = 11.85(6) mm]
   (c) [a = 2.71 m, c = 4.23 m, C = 70°]]
2. Solve the following triangles DEF:
   (a) F = 41°, e = 9.30 cm, f = 7.21 cm
   (b) D = 58°, E = 69°, e = 7.4 cm
   (c) D = 114° 8′ F = 21° 5′, d = 14.6 cm
   [(a) [D = 81° 12′, E = 57° 48′, d = 10.86 cm
   or D = 16° 48′, E = 122° 12′, d = 3.177 cm]
   (b) [d = 6.72 cm, f = 6.33 cm, F = 53°]
   (c) [e = 11.27 cm, f = 5.755 cm, E = 44° 47′]]
3. Solve the following triangles GHI:
   (a) g = 8.52 m, i = 12.7 m, I = 24° 9′
   (b) h = 17.86 m, i = 12.67 m, H = 83° 46′
   (c) h = 45.3 mm, i = 35.7 mm, I = 36° 47′
   [(a) [G = 15° 56′, H = 139° 55′, h = 20.03 m]
   (b) [G = 51° 23′, I = 44° 51′, g = 14.03(8) m]
   (c) [H = 49° 21′, G = 93° 52′, g = 59.57 mm
   or H = 130° 39′, G = 12° 34′, g = 12.97 mm]]
4. Solve the following triangles JKL:
   (a) J = 71°, K = 36°, j = 23.7 mm
   (b) j = 37.2 cm, k = 31.6 cm, K = 37°
   (c) j = 19.53 cm, k = 15.96 cm, J = 102° 57′
   [(a) [k = 14.73 mm, l = 23.97 mm, L = 73°]
   (b) [J = 45° 7′, L = 97° 53′, l = 52.01 cm
   or J = 134° 53′, L = 8° 7′, l = 7.414 cm]
   (c) [K = 52° 47′, L = 24° 16′, l = 8.236(7) cm]]

5. Solve the following triangles MNP:
   (a) $m = 1.46$ cm, $p = 1.62$ cm, M = 26° 0′
   (b) N = 111° 49′, P = 15° 19′, $p = 43.81$ cm
   (c) $m = 45.0$ mm, $p = 35.0$ mm, M = 51° 19′

   (a) [P = 29° 6′, N = 124° 54′, $n = 2.732$ cm
   or P = 150° 54′, N = 3° 6′, $n = 0.180\,1$ cm]
   (b) [$m = 132.2(2)$ cm, $n = 153.9(7)$ cm, M = 52° 52′]
   (c) [P = 37° 23′, N = 91° 18′, $n = 57.63$ mm]

6. Solve the following triangles QRS:
   (a) R = 68° 15′, S = 57° 47′, $r = 1.47$ m
   (b) $r = 43.36$ cm, $s = 30.20$ cm, S = 29° 32′
   (c) $r = 1.00$ m, $q = 24.0$ cm, R = 53°

   (a) [$q = 1.280$ m, $s = 1.339$ m, Q = 53° 58′]
   (b) [R = 45° 3′, Q = 105° 25′, $q = 59.06(2)$ cm
   or R = 134° 57′, Q = 15° 31′, $q = 16.39(0)$ cm]
   (c) [Q = 11° 3′, S = 115° 57′, $s = 112.7$ cm]

7. Solve the following triangles TUV:
   (a) $t = 62.0$ mm, $u = 41.2$ mm, V = 62° 11′
   (b) T = 102° 8′, $u = 32.8$ cm, $v = 43.7$ cm
   (c) $t = 9$ cm, $u = 7$ cm, $v = 6$ cm

   (a) [$v = 56.81$ mm, T = 77° 55′, U = 39° 54′]
   (b) [$t = 59.90$ cm, U = 32° 22′, V = 45° 30′]
   (c) [T = 87° 16′, U = 50° 59′, V = 41° 45′]

8. Solve the following triangles WXY:
   (a) $w = 129$ mm, $x = 158$ mm, $y = 59$ mm
   (b) Y = 67°, $w = 16.4$ cm, $x = 11.8$ cm
   (c) W = 123° 17′, $y = 72.0$ mm, $x = 43.0$ mm

   (a) [W = 50° 43′, X = 108° 33′, Y = 20° 44′]
   (b) [W = 70° 21′, X = 42° 39′, $y = 16.03$ cm]
   (c) [X = 20° 37′, Y = 36° 6′, $w = 102.1$ mm]

9. Solve the following triangles ABC:
   (a) $b = 11$ cm, $c = 15$ cm, A = 55°
   (b) $a = 8.983$ m, $b = 12.460$ m, $c = 15.910$ m
   (c) A = 73°, $b = 7.20$ cm, $c = 9.00$ cm

   (a) [$a = 12.5$ cm, B = 46° 8′, C = 78° 52′]
   (b) [A = 34° 13′, B = 51° 20′, C = 94° 27′]
   (c) [$a = 9.746$ cm, B = 44° 57′, C = 62° 3′]

10. Solve the following triangles DEF:
   (a) $d = 19.47$ cm, $f = 17.63$ cm, E = 38° 29′
   (b) $d = 7.0$ m, $e = 5.0$ m, $f = 8.0$ m
   (c) $d = 25$ mm, $e = 32$ mm, $f = 39$ mm

   (a) [$e = 12.35(0)$ cm, D = 78° 50′, F = 62° 41′]
   (b) [D = 60°, E = 38° 13′, F = 81° 47′]
   (c) [D = 39° 43′, E = 54° 52′, F = 85° 25′]

11. Two phasors are shown in Fig. 34. If $v_1 = 60.0$ volts and $v_2 = 90.0$ volts calculate the value of their resultant (i.e. the length AB) and the angle it makes with v_1.

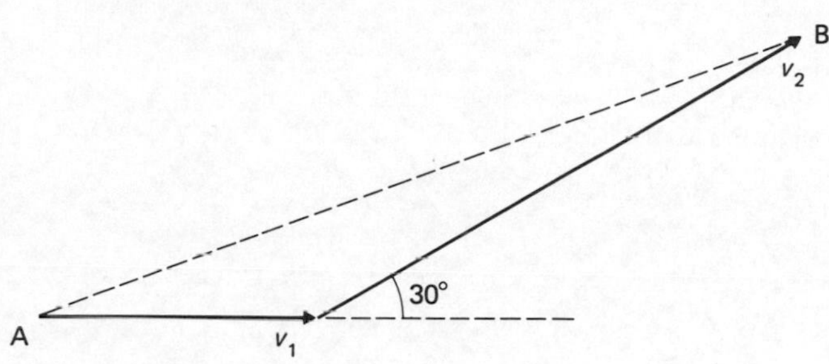

Figure 34

[145.1 volts, 18° 4′]

12. Two ships P and Q leave port at the same time. P sails at a steady speed of 52.0 km h^{-1} S 32° W and B at 38.0 km h^{-1} S 24° E. Find their distance apart after 2 hours 30 minutes. [110.1 km]
13. A man leaves a town walking at a steady speed of 6.00 km h^{-1} in a direction of S 30° W. Another man leaves the same town, cycling at a steady speed in a direction S 23° E. After 4.0 hours the two men are 100.00 km apart. Find the speed of the cyclist. [28.15 km h ^{1}]
14. A ship B sails from a port in direction N 41° W at an average speed of 30.0 km h^{-1}. Another ship D sails at the same time as B but in a direction N 32° E at an average speed of 24.0 km h^{-1}. Calculate their distance apart after 3.0 hours. [97.44 km]
15. A room 9.00 m wide has a span roof which slopes at 32° on one side and 41° on the other. Find the length of the roof slopes. [6.174 m, 4.987 m]
16. A jib crane consists of a vertical post PQ 5.2 m in length, the inclined jib QR, 12.8 m in length, and a tie PR. Angle QPR is 122°. Calculate: (a) the length of the tie; and (b) the inclination of the jib to the vertical. (a) [9.26 m] (b) [37° 51′]
17. A reciprocating engine mechanism is shown in Fig. 35, where XY represents the rotating crank, YZ the connecting rod and Z the piston which moves vertically along the broken line XZ. If the rotating crank is 1.240 cm in length and the connecting rod is 6.480 cm, calculate for the position shown: (a) the inclination of the connecting rod to the vertical; and (b) the distance XZ.

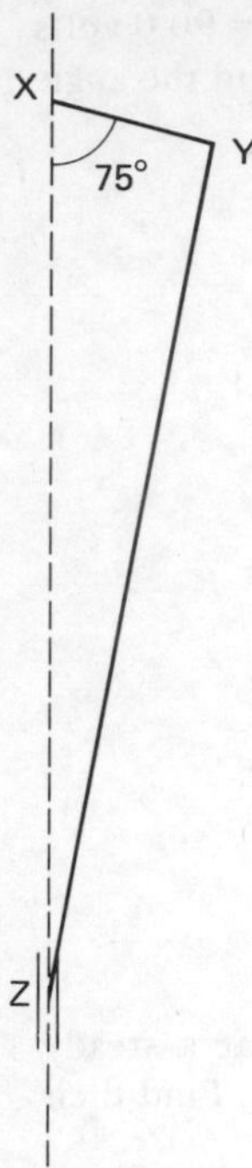

Figure 35

(a) [10° 39′] (b) [6.689 cm]

Three-dimensional trigonometry

18. A vertical aerial stands on horizontal ground. A surveyor standing due west of the aerial finds the angle of elevation of the top to be 51° 5′. He moves due south 20.0 m and finds the elevation of the top of the aerial to be 47° 19′. Calculate the height of the aerial.
[44.88 m]

19. The base of a right pyramid of vertex V is a rectangle ABCD. E and F are the mid-points of AB and BC respectively. VA = 15.00 cm, AB = 11.00 cm and BC = 8.00 cm. Calculate: (a) the perpendicular height of the pyramid; (b) the angle the edge VA makes with the base; (c) the angles which the faces VAB and VBC make with the base; and (d) the angle the plane VEF makes with the base ABCD.
(a) [13.37 cm] (b) [63° 3′] (c) [73° 21′, 67° 38′]
(d) [76° 24′]

20. An aeroplane is sighted due west from a radar station at an elevation of 38° and a height of 7 500 m and later at a elevation of 34° at height 5 000 m in direction W 48° S. If it is descending uniformly, find the angle of descent. Find the speed of the aeroplane (in m s^{-1} and in km h^{-1})

if the time between the two observations is 45.0 seconds.
[19° 9′, 169.4 m s^{-1}, 609.9 km h^{-1}]

21. From a point X at ground level a mine shaft is constructed to a depth of 300 m. Tunnels are constructed from Y, 5.0 km due south of X, and Z, 3.0 km N 30° W from X, to meet at the base A of the shaft. Find the angle between the directions of the two tunnels (i.e. ∠ YAZ) assuming that X, Y and Z are on horizontal ground. [148° 41′]

22. A tent is in the form of a regular octagonal pyramid with each side of the base 4 m in length. If the perpendicular height of the tent is 10 m find: (a) the angle between a sloping face and the base; and (b) the angle of inclination of an edge of the sloping face to the base.
(a) [64° 14′] (b) [62° 24′]

23. The base of a roof shown in Fig. 36 is a rectangle PQRS. The equal faces PSX and QRY each make an angle of 45° with the base; the equal faces PQYX and RSXY each make an angle of 50° with the base. Calculate: (a) the perpendicular height of the roof; (b) the length XY; (c) the length of the sloping edge RY; and (d) the angle the edge SX makes with the base.

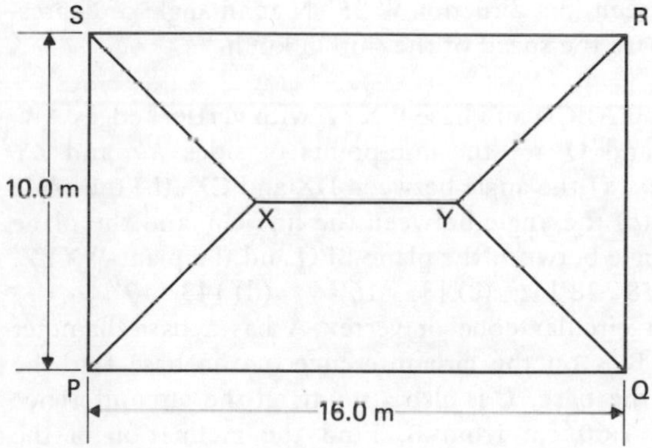

Figure 36

(a) [5.96 m] (b) [4.08 m] (c) [9.80 m] (d) [37° 27′]

24. A crane lifts a load from position X and deposits it at postiion Y on the same ground level, and in so doing the crane pivots through an angle of 65°, and the jib moves from an angle of 68° to the vertical to an angle of 48° to the vertical. Find the distance XY if the jib is of length 8.0 m. [7.29 m]

25. An aircraft is sighted due north from a radar station at an elevation of 49° 4′ at a height of 10 000 m, and later at an elevation of 37° 19′ and height 6 250 m in direction E 24° S. If it is descending uniformly, find the angle of descent and the speed of the aircraft if the time between the two observations is 72 seconds.
[14° 51′, 203.2 m s^{-1}, or 731.6 km h^{-1}]

26. From a ship situated at a point S 17° W of a vertical lighthouse, the angle of elevation of the top of the lighthouse is 13°. Later, when the ship is at a point S 2° E from the lighthouse, the angle of elevation is 9°. Find the height of the lighthouse if the distance travelled by the ship between the two observations is 146 m. [55.55 m]
27. A man standing W 30° S of a tower measures the angle of elevation of the top of the tower as 44° 19′. From a position E 28° S from the tower the elevation of the top is 36° 43′. Calculate the height of the tower if the distance between the two observations is 120 m. [57.86 m]
28. The access to an underground cave X may be made from three entrances – either in a vertical lift 0.42 km in length from point A at ground level, or by a tunnel 6.00 km long from a point B at ground level due west of A, or by another tunnel 4.20 km from point C at ground level E 41° S of A. Find the angle between the directions of the tunnels (i.e. ∠ BXC) assuming that A, B and C are on horizontal ground. [137° 55′]
29. A coastguard situated at the top of a cliff 200 m high observes a ship in a direction S 32° W at an angle of depression of 9° 14′. Five minutes later the same ship is seen in a direction W 25° N at an angle of depression of 10° 51′. Calculate the speed of the ship in km h^{-1}. [18.16 km h^{-1}]
30. A 3.00-cm cube has top ABCD and base WXYZ with vertical edges AW, BX, CY and DZ. P and Q are the mid-points of sides WZ and ZY respectively. Calculate: (a) the angle between DX and CX; (b) the angle between DP and YP; (c) the angle between the line CW and the plane WXYZ; and (d) the angle between the plane BPQ and the plane WXYZ. (a) [35° 16′] (b) [78° 28′] (c) [35° 16′] (d) [43° 19′]
31. A 15.30-cm-high right circular cone of vertex A has a base diameter of 9.80 cm. If point B is on the circumference of the base find the inclination of AB to the base. C is also a point on the circumference of the base, distance 7.60 cm from B. Find the inclination of the plane ABC to the base. [72° 15′, 78° 34′]

Periodic functions of the form $R \sin(\omega t \pm \alpha)$

In problems 32–36 find the amplitude, periodic time, frequency and phase angle (stating whether it is leading or lagging $R \sin \omega t$) for the alternating quantities shown.

32. $v = 30 \sin(50\pi t + 0.36)$ [30, 0.04 s, 25 Hz leading $v = 30 \sin 50\pi t$ by 20° 38′]
33. $i = 18 \sin(400\pi t - 0.231)$ [18, 0.005 s, 200 Hz lagging $i = 18 \sin 400\pi t$ by 13° 14′]
34. $y = 22.4 \sin(40t - 0.60)$ [22.4, 0.157 s, 6.37 Hz, lagging $y = 22.4 \sin 40t$ by 34° 23′]
35. $x = 15 \sin(314.2t + 0.468)$ [15, 0.02 s, 50 Hz leading $x = 15 \sin 314.2\,t$ by 26° 49′]

36. $v = 63.5 \sin (150\pi t - 0.132)$ [63.5, 0.013 3 s, 75 Hz lagging $v = 63.5 \sin 150\pi t$ by 7° 34′]

37. Sketch the following graphs from $t = 0$ to $t = 2\pi$ seconds showing relevant details.

(a) $y = 5 \sin 3t$
(b) $y = 7 \cos 2t$
(c) $y = 2 \sin \left[t + \frac{4\pi}{9}\right]$
(d) $y = 4 \cos \left[t - \frac{\pi}{5}\right]$
(e) $y = 6 \sin^2 2t$
(f) $y = 1.5 \cos^2 t$

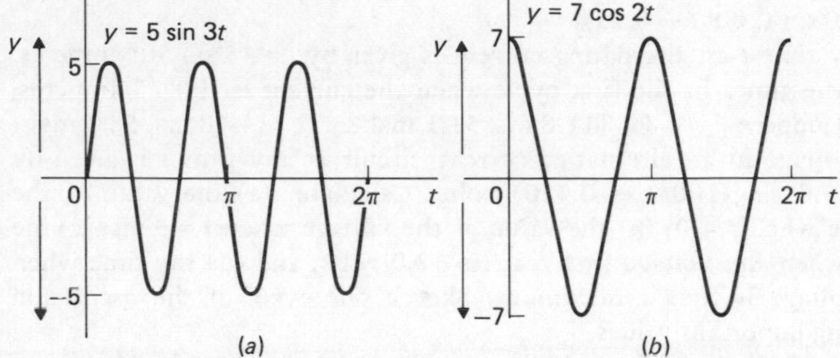

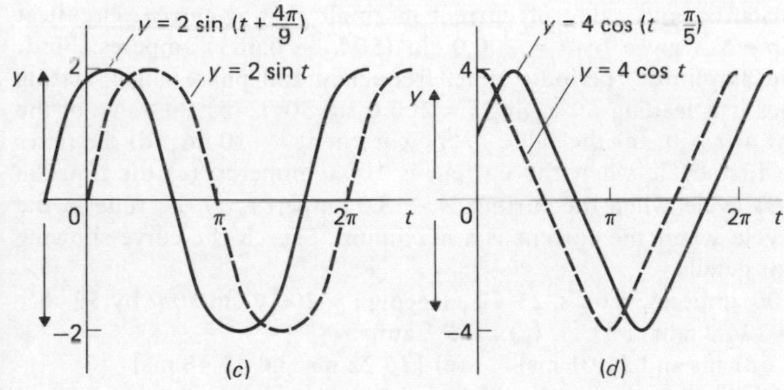

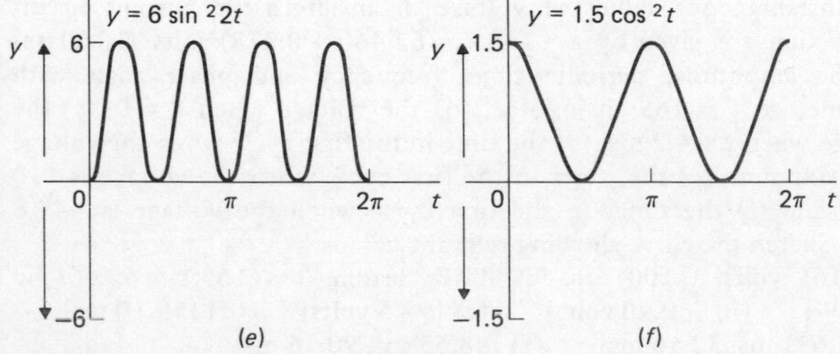

Figure 37 Solution to problem 37.

38. A sinusoidal voltage has a maximum value of 42 volts and a frequency of 60 Hz. At $t = 0$, the current is zero. Express the instantaneous voltage v in the form $R \sin \omega t$. $[v = 42 \sin 120\pi t]$
39. An oscillating mechanism has a maximum displacement of 3.60 m and a frequency of 50 Hz. At $t = 0$ the displacement is zero. Express the instantaneous displacement in the form $x = R \sin (\omega t \pm \alpha)$.
    $[x = 3.60 \sin 100\pi t]$
40. An alternating voltage has a periodic time of 0.01 s and a maximum value of 30 volts. When $t = 0$, the voltage is -20 volts. Express the instantaneous voltage v in the form $v = R \sin (\omega t \pm \alpha)$.
    $[v = 30 \sin (200\pi t - 0.73)]$
41. At any time t an alternating current is given by $i = 45 \sin 50t$ amperes. Find the times in the first cycle when the current is: (a) 25 amperes; (b) 10 amperes. (a) [11.8 ms, 51.1 ms] (b) [4.48 ms, 58.3 ms]
42. The voltage in an alternating-current circuit at any time t is given by $v = 55.0 \sin (100\pi t + 0.410)$ volts. Calculate: (a) the value of the voltage when $t = 0$; (b) the value of the voltage when $t = 5$ ms; (c) the time when the voltage first reaches 32.0 volts; and (d) the time when the voltage is first a maximum. Sketch one cycle of the oscillation showing important values.
    (a) [21.92 V] (b) [50.44 V] (c) [0.672 ms] (d) [3.69 ms]
43. The instantaneous value of current in an alternating-current circuit at any time t is given by $i = 200.0 \sin (50\pi t - 0.683)$ amperes. Find: (a) the amplitude, periodic time, frequency and phase angle, stating whether it is leading or lagging $i = 200.0 \sin 50\pi t$; (b) the value of the current at $t = 0$; (c) the value of the current at $t = 10$ ms; (d) the times in the first cycle when the current is 100.0 amperes; (e) the times in the first cycle when the current is -58.0 amperes; (f) the time in the first cycle when the current is a maximum. Sketch the curve showing relevant details.
    (a) [200 amperes, 0.04 s, 25 Hz, lagging $i = 200.0 \sin 50\pi t$ by $39^\circ 8'$]
    (b) [$-$ 126.2 amperes] (c) [155.1 amperes]
    (d) [7.681 ms and 21.01 ms] (e) [26.22 ms and 42.48 ms]
    (f) [14.35 ms]
44. The instantaneous values of voltage in an alternating-current circuit at any time t is given by $v = 165 \sin (62.46\, t + 0.378)$ volts. Calculate: (a) the amplitude, periodic time, frequency and phase angle with reference to $v = 165 \sin 62.46t$; (b) the voltage when $t = 0$; (c) the voltage when $t = 4.2$ ms; (d) the time in the first cycle when the voltage is a maximum; (e) the times in the first cycle when the voltage is 110 volts; and (f) the times in the first cycle when the voltage is -44.6 volts. Sketch the curve showing relevant details.
    (a) [165 volts, 0.100 6 s, 9.941 Hz, leading $v = 165 \sin 62.46t$ by $21^\circ 39'$] (b) [60.90 volts] (c) [98.6 volts] (d) [19.10 ms]
    (e) [5.631 ms, 32.56 ms] (f) [48.63 ms, 90.16 ms]

45. Verify from trigonometrical tables that $\sin(A + B) = \sin A \cos B + \cos A \sin B$ and $\cos(A + B) = \cos A \cos B - \sin A \sin B$, when: (a) $A = 15^\circ$ and $B = 35^\circ$; (b) $A = 26^\circ\ 37'$ and $B = 41^\circ\ 17'$; (c) $A = 148^\circ$ and $B = 74^\circ$.

46. Verify from trigonometrical tables that $\sin(A - B) = \sin A \cos B - \cos A \sin B$ and $\cos(A - B) = \cos A \cos B + \sin A \sin B$, when: (a) $A = 11^\circ$ and $B = 52^\circ$; (b) $A = 36^\circ\ 13'$ and $B = 45^\circ\ 56'$; (c) $A = 97^\circ$ and $B = 114^\circ$.

47. Reduce the following to the sine of one angle:
(a) $\sin 41^\circ \cos 34^\circ + \cos 41^\circ \sin 34^\circ$
(b) $\sin 3x \cos 4x + \cos 3x \sin 4x$
(c) $\sin 49^\circ \cos 53^\circ + \cos 49^\circ \sin 53^\circ$
(a) [$\sin 75^\circ$] (b) [$\sin 7x$] (c) [$\sin 102^\circ$ ($= \sin 78^\circ$)]

48. Reduce the following to the sin of one angle:
(a) $\sin 73^\circ \cos 24^\circ - \cos 73^\circ \sin 24^\circ$
(b) $\sin 151^\circ \cos 82^\circ - \cos 151^\circ \sin 82^\circ$
(c) $\sin 7\theta \cos 2\theta - \cos 7\theta \sin 2\theta$
(a) [$\sin 49^\circ$] (b) [$\sin 69^\circ$] (c) [$\sin 5\theta$]

49. Reduce the following to the cosine of one angle:
(a) $\cos 35^\circ \cos 27^\circ - \sin 35^\circ \sin 27^\circ$
(b) $\cos 4t \cos 2t - \sin 4t \sin 2t$
(c) $\cos 64^\circ \cos 48^\circ - \sin 64^\circ \sin 48^\circ$
(a) [$\cos 62^\circ$] (b) [$\cos 6t$] (c) [$\cos 112^\circ$ ($= -\cos 68^\circ$)]

50. Reduce the following to the cosine of one angle:
(a) $\cos 64^\circ \cos 39^\circ + \sin 64^\circ \sin 39^\circ$

(b) $\cos \frac{2\pi}{5} \cos \frac{\pi}{4} + \sin \frac{2\pi}{5} \sin \frac{\pi}{4}$

(c) $\cos 164^\circ \cos 71^\circ + \sin 164^\circ \sin 71^\circ$

(a) [$\cos 25^\circ$] (b) [$\cos \frac{3\pi}{20}$ ($= \cos 27^\circ$)] (c) [$\cos 93^\circ$ ($= -\cos 87^\circ$)]

In Problems 51–54 use the addition and subtraction formulae to simplify the expressions.

51. (a) $\sin 66^\circ \cos 41^\circ - \cos 66^\circ \sin 41^\circ$
(b) $\cos 2\,\omega t \cos 3\,\omega t - \sin 2\omega t \sin 3\omega t$
(a) [$\sin 25^\circ$] (b) [$\cos 5\omega t$]

52. (a) $\sin 4x \cos x + \cos 4x \sin x$
(b) $\cos 78^\circ \cos 22^\circ + \sin 78^\circ \sin 22^\circ$
(a) [$\sin 5x$] (b) [$\cos 56^\circ$]

53. (a) $\sin \frac{\pi}{4} \cos \frac{\pi}{6} + \cos \frac{\pi}{4} \sin \frac{\pi}{6}$

(b) $\cos 8\alpha \cos 3\alpha + \sin 8\alpha \sin 3\alpha$

(a) [$\sin \frac{5\pi}{12}$ ($= \sin 75^\circ$)] (b) [$\cos 5\alpha$]

54. (a) $\sin \omega t \cos (\omega t - \alpha) - \cos \omega t \sin (\omega t - \alpha)$
(b) $\cos (A - B) \cos (A + B) - \sin (A - B) \sin (A + B)$
(a) [$\sin \alpha$] (b) [$\cos 2A$]

55. Prove that:

(a) $\sin \left[\frac{\pi}{2} + x\right] = \cos x$

(b) $\sin (y + 60^\circ) + \sin (y + 120^\circ) = \sqrt{3} \cos y$
(c) $\cos (90^\circ + \alpha) = -\sin \alpha$

56. Prove that:
(a) $\cos (\theta + 45^\circ) - \cos (\theta - 135^\circ) = \sqrt{2} (\cos\theta - \sin\theta)$
(b) $-\sin (\pi + x) = \sin x$

(c) $\dfrac{\cos (2\pi - \phi)}{\cos \left(\frac{3\pi}{2} + \phi\right)} = \cot \phi$

57. If $\sin A = \frac{40}{41}$ and $\cos B = \frac{5}{13}$ find $\sin (A + B)$ and $\sin (A - B)$ without using trigonometrical tables. $\left[\frac{308}{533} (= 0.577\,9), \frac{92}{533} (= 0.172\,6)\right]$

58. If $\cos E = 0.500\,0$ and $\cos F = 0.300\,0$ find $\sin (E + F)$ and $\cos (E + F)$. [0.736 8, −0.676 1]

59. If $\sin C = 0.843\,2$ and $\cos D = 0.732\,8$ find $\cos (C + D)$ and $\cos (C - D)$. [−0.179 8, 0.967 7]

60. Find the value of $\sin 70^\circ$ given $\sin 36^\circ = 0.587\,8$ and $\cos 34^\circ = 0.829\,0$. [0.939 7]

61. Find the value of $\cos 56^\circ$ given $\sin 24^\circ = 0.406\,7$ and $\sin 32^\circ = 0.529\,9$. [0.559 2]

62. Find the values of $\sin 105^\circ$ and $\cos 105^\circ$ without using trigonometrical tables. Assume that $\sin 45^\circ = \cos 45^\circ = \frac{1}{\sqrt{2}}$, $\sin 60^\circ = \frac{\sqrt{3}}{2}$ and $\cos 60^\circ = \frac{1}{2}$. $\left[\frac{1 + \sqrt{3}}{2\sqrt{2}} (= 0.965\,9), \frac{1 - \sqrt{3}}{2\sqrt{2}} = (-0.258\,8)\right]$

Conversion of $a \sin \omega t + b \cos \omega t$ into $R \sin (\omega t \pm \alpha)$

63. (a) Plot graphs of $y_1 = 5 \sin \omega t$ and $y_2 = 12 \cos \omega t$ over one cycle on the same axes and using the same scales. By adding ordinates plot $y = 5 \sin \omega t + 12 \cos \omega t$ and express the resultant y_r in the form $y_r = R \sin (\omega t \pm \alpha)$.
(b) With reference to the relative positions of y_1 and y_2 at $t = 0$, obtain the resultant y_r by phasor addition.
(c) Obtain the resultant y_r by converting $5 \sin \omega t + 12 \cos \omega t$ into the form $R \sin (\omega t \pm \alpha)$ using the trigonometrical addition formula.
[$13 \sin (\omega t + 67^\circ\,23')$]

In Problems 64–70 change the functions into the form $R \sin (\omega t \pm \alpha)$.

64. $8 \sin \omega t + 15 \cos \omega t$ [$17 \sin (\omega t + 61° 56')$]
65. $3 \sin \omega t + 4 \cos \omega t$ [$5 \sin (\omega t + 53° 8')$]
66. $3 \sin \omega t - 4 \cos \omega t$ [$5 \sin (\omega t - 53° 8')$]
67. $-2.00 \sin \omega t + 3.00 \cos \omega t$ [$3.606 \sin (\omega t + 123° 41')$]
68. $-5 \sin \omega t - 12 \cos \omega t$ [$13 \sin (\omega t + 247° 23')$ or $13 \sin (\omega t - 112° 37')$]
69. $6.60 \sin \omega t + 11.80 \cos \omega t$ [$13.52 \sin (\omega t + 60° 47')$]
70. $-12.62 \sin \omega t - 6.92 \cos \omega t$ [$14.39 \sin (\omega t + 208° 44')$ or $14.39 \sin (\omega t - 151° 16')$]

In Problems 71–75 change the functions into the form $R \cos (\omega t \pm \alpha)$.

71. $8.00 \cos \omega t - 5.00 \sin \omega t$ [$9.434 \cos (\omega t + 32° 1')$]
72. $4 \cos \omega t + 3 \sin \omega t$ [$5 \cos (\omega t - 36° 52')$]
73. $6.00 \sin \omega t - 4.00 \cos \omega t$ [$7.211 \cos (\omega t + 236° 19')$ or $7.211 \cos (\omega t - 123° 41')$]
74. $-19.6 \cos \omega t - 12.4 \sin \omega t$ [$23.19 \cos (\omega t + 147° 41')$]
75. $13.00 \sin \omega t - 5.00 \cos \omega t$ [$13.93 \cos (\omega t + 248° 58')$ or $13.93 \cos (\omega t - 111° 2')$]
76. Solve the equations: (a) $3 \sin \theta - 5 \cos \theta = 4$; and (b) $15 \sin \theta + 11 \cos \theta = 7$ for $0° \leqslant \theta \leqslant 360°$.
(a) [102° 21′ or 195° 43′] (b) [121° 39′ or 345° 51′]
77. Solve the following equations for all values of A between 0° and 360°:
(a) $4 \cos A + 3 \sin A = 5$
(b) $42 \cos A - 19 \sin A = 24$
(c) $23 \sin A + 14 \cos A = 12$
(a) [36° 52′] (b) [34° 18′ or 277° 2′] (c) [122° 12′ or 355° 8′]
78. Solve the equations: (a) $5 \sin \phi + 7 \cos \phi - 3 = 0$
(b) $17 \cos \phi = 4 + 8 \sin \phi$
for values of ϕ between 0° and 360°.
(a) [105° 7′ or 325° 57′] (b) [52° 31′ or 257° 5′]
79. Find the roots of the equation $7.32 \cos x - 5.62 = 3.81 \sin x$ in the range 0° to 360°. [19° 34′ or 285° 26′]
80. Solve the equation $100 \sin y + 250 \cos y = 190.4$ for values of y between 0° and 360°. [66° 48′ or 336° 48′]
81. Find the maximum value of $8 \sin \phi + 7 \cos \phi$ and find the smallest positive value of ϕ at which it occurs. [10.63, 48° 49′]
82. Alternating currents are given by $i_1 = 3.00 \sin (10\pi t - \frac{\pi}{6})$ and $i_2 = 8.00 \sin (10\pi t + \frac{\pi}{3})$. Express $i_1 + i_2$ in the form $y = R \sin (10\pi t \pm \alpha)$
Find also the frequency of the resultant function.
[$8.544 \sin (10\pi t + 39° 27')$, 5 Hz]
83. Express $2.00 \sin (\phi + \frac{\pi}{4}) + 3.00 \cos (\phi - \frac{\pi}{6})$ in the form $a \sin \phi + b \cos \phi$ and then convert this into the form $R \cos (\phi \pm \alpha)$.
[$2.914 \sin \phi + 4.012 \cos \phi$; $4.959 \cos (\phi - 36°)$]

84. The third harmonic of a wave motion is given by $5.5 \cos 3\theta - 7.2 \sin 3\theta$. Express this in the form $A \cos (3\theta + \alpha)$. [$9.06 \cos (3\theta + 52^\circ\ 37')$]

85. A voltage V is given by $V = I (R \sin at - aL \cos at)$. Express the voltage in the form $IZ \sin (at - \phi)$.

$$\left[I\sqrt{[R^2 + (aL)^2]} \sin \left(at - \arctan \frac{aL}{R}\right)\right]$$

86. The displacement x metres of a body from a fixed point about which it is oscillating is given by the expression $x = 3.6 \sin 2t + 4.2 \cos 2t$, where t is the time in seconds. Express x in the form $R \sin (2t + \alpha)$. [$5.53 \sin (2t + 49^\circ\ 24')$]

87. Find the sum of the voltages $v_1 = 50 \sin 100\pi t$ and $v_2 = 30 \cos 100\,\pi\, t$ in the form $R \sin (100\pi t + \alpha)$. [$58.3 \sin (100\pi t + 30^\circ\ 58')$]

88. Alternating currents are given by $i_1 = 5.0 \sin \omega t$ and $i_2 = 15.0 \cos \omega t$. Calculate the maximum value of $i_1 + i_2$ and its phase angle relative to i_1. [15.8; $71^\circ\ 34'$]

89. Two voltages $4 \cos \omega t$ and $-3 \sin \omega t$ are inputs to an analogue circuit. Find an expression for the output voltage if this is given by the addition of the two inputs. [$5 \sin (\omega t + 126^\circ\ 52')$]

Resolution of vector quantities

90. A barge is being towed along a canal by a rope inclined at 30° to the direction of the canal. The tension in the rope is 175 N. Calculate the values of the rectangular components of this tension, one component being in the direction of the canal.
[151.6 N in direction of canal, 87.50 N at right angles to direction of canal]

91. Calculate, using resolution of forces, the magnitude and direction of the resultant of the three coplanar forces given below, when they are acting at a point. Force A, 10.0 N acting horizontally to the right, force B, 6.0 N inclined at an angle of 70° to force A, and force C, 13.0 N inclined at an angle of 135° to force A.
[15.1 N, 79° 5′ to the horizontal]

92. Find the magnitude and direction of the resultant of the coplanar forces listed below, which are acting at a point:
Force A, 3.0 N acting vertically upwards;
Force B, 2.0 N acting at an angle of 110° to force A;
Force C, 5.0 N acting at an angle of 290° to force A;
Force D, 7.0 N acting horizontally to the right.
[10.6 N, 22° 18′ to the horizontal]

93. Forces of: 200 kN acting horizontally to the right;
300 kN inclined at 60° to the horizontal;
600 kN inclined at 120° to the horizontal;
400 kN inclined at 180° to the horizontal; and
500 kN inclined at 300° to the horizontal;
are acting at a point. Find the magnitude and direction of the resultant

of these forces by resolution of forces.
[360.6 kN, 106° 6′ to the horizontal]

94. Find the magnitude and direction of the resultant of the following coplanar forces which are acting at a point:
Force A, 3.0 kN inclined at 30° to the horizontal to the right;
Force B, 5.0 kN inclined at 105° to force A;
Force C, 6.0 kN inclined at 45° to force B; and
Force D, 4.0 kN inclined at 120° to force C.
[5.18 kN, 162° 20′ to the horizontal]

95. The voltage in a circuit is the resultant of the three voltages shown in Fig. 38. By resolution of the voltages calculate the magnitude of the resultant.

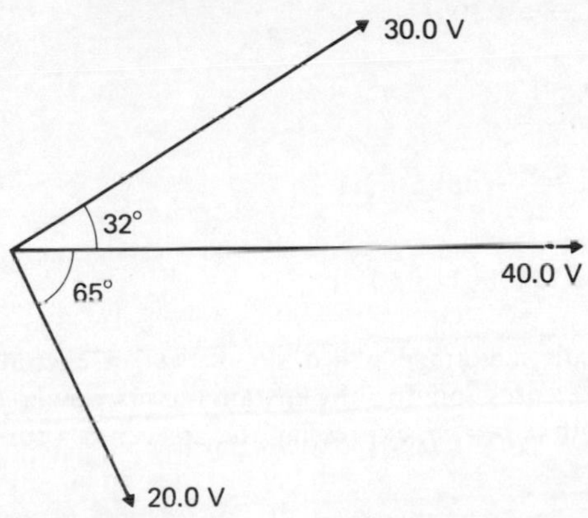

Figure 38

[73.93 volts]

96. The instantaneous values of two alternating currents are given by

$i_1 = 15.0 \sin(\omega t + \frac{\pi}{3})$ amperes and $i_2 = 9.0 \sin(\omega t - \frac{\pi}{6})$ amperes.

By resolving each current into its horizontal and vertical components, obtain a sinusoidal expression for $i_1 + i_2$ in the form $R \sin(\omega t + \alpha)$.
[$17.49 \sin(\omega t + 29° 2')$ amperes or $17.49 \sin(\omega t + 0.507)$ amperes]

97. The instantaneous values of two alternating voltages are given by

$v_1 = 42.5 \sin(\omega t - \frac{\pi}{3})$ volts and $v_2 = 22.5 \sin(\omega t - \frac{\pi}{4})$ volts. Calculate,

by resolving the voltages horizontally and vertically, an expression for $v_1 + v_2$. [$64.50 \sin(\omega t - 0.956\,8)$ volts]

98. Obtain, by resolution, a sinusoidal expression in the form $R \sin(\omega t + \alpha)$ for the resultant of the alternating-current expression

$i = 7.2 \sin(\omega t - \frac{\pi}{5}) + 5.5 \cos(\omega t + \frac{4\pi}{9}) + 6.8 \sin(\omega t + \frac{4\pi}{3})$ amperes.

[$9.64 \sin(\omega t - 1.89)$ amperes]

Changing products into sums or differences

Express as sums or differences the following:

99. $\sin 6t \cos t$ $\quad [\frac{1}{2}(\sin 7t + \sin 5t]$

100. $\cos 5x \sin 3x$ $\quad [\frac{1}{2}(\sin 8x - \sin 2x)]$

101. $4 \cos 4\theta \cos 2\theta$ $\quad [2(\cos 6\theta + \cos 2\theta)]$

102. $\sin 7\alpha \sin 3\alpha$ $\quad [-\frac{1}{2}(\cos 10\alpha - \cos 4\alpha)]$

103. $2 \cos 69^\circ \sin 30^\circ$ $\quad [\sin 90^\circ - \sin 30^\circ]$

104. $\sin \frac{\pi}{3} \cos \frac{\pi}{6}$ $\quad [\frac{1}{2}(\sin \frac{\pi}{2} + \sin \frac{\pi}{6})]$

105. $\sin 72^\circ \sin 22^\circ$ $\quad [-\frac{1}{2}(\cos 94^\circ - \cos 50^\circ)]$

106. $3 \cos 78^\circ \cos 48^\circ$ $\quad [\frac{3}{2}(\cos 126^\circ + \cos 30^\circ)]$

107. In an alternating-current circuit a voltage $v = 6 \sin \omega t$ and a current $i = 5 \sin (\omega t - \frac{\pi}{4})$. Find an expression for the instantaneous power p in the circuit in time t, given that $p = vi$, expressing the answer as a sum or difference of sines or cosines.

$$\left[p = 15 \left\{\cos \frac{\pi}{4} - \cos \left(2\omega t - \frac{\pi}{4}\right)\right]$$

Changing sums or differences into products

Express as products the following:

108. $\sin 4x + \sin x$ $\quad \left[2 \sin \frac{5x}{2} \cos \frac{3x}{2}\right]$

109. $\sin 11\theta - \sin 7\theta$ $\quad [2 \cos 9\theta \sin 2\theta]$

110. $\cos 6t + \cos 4t$ $\quad [2 \cos 5t \cos t]$

111. $\frac{1}{4}(\cos 5\alpha - \cos 2\alpha)$ $\quad \left[-\frac{1}{2} \sin \frac{7\alpha}{2} \sin \frac{3\alpha}{2}\right]$

112. $\sin 40^\circ - \sin 22^\circ$ $\quad [2 \cos 31^\circ \sin 9^\circ]$

113. $\cos \frac{\pi}{4} + \cos \frac{\pi}{6}$ $\quad \left[2 \cos \frac{5\pi}{24} \cos \frac{\pi}{24}\right]$

114. $\frac{1}{2}(\sin 75^\circ + \sin 15^\circ)$ $\quad [\sin 45^\circ \cos 30^\circ]$

115. $\cos 82^\circ - \cos 38^\circ$ $\quad [-2 \sin 60^\circ \sin 22^\circ]$

Chapter 2

The binomial expansion

1. The expansion of $(a+b)^n$, where n is a small positive integer, using Pascal's triangle

Table 1

Term	Expansion
$(a+b)^0 =$	1
$(a+b)^1 =$	$1a + 1b$
$(a+b)^2 =$	$1a^2 + 2ab + 1b^2$
$(a+b)^3 =$	$1a^3 + 3a^2b + 3ab^2 + 1b^3$
$(a+b)^4 =$	$1a^4 + 4a^3b + 6a^2b^2 + 4ab^3 + 1b^4$
$(a+b)^5 =$	$1a^5 + 5a^4b + 10a^3b^2 + 10a^2b^3 + 5ab^4 + 1b^5$
$(a+b)^6 =$	$1a^6 + 6a^5b + 15a^4b^2 + 20a^3b^3 + 15a^2b^4 + 6ab^5 + 1b^6$

The word binomial indicates a 'two-number' expression. When n is a small positive integer, up to, say, 10, the expansion of $(a + b)^n$ can be done by multiplying $(a + b)$ by itself n times. The result of doing this is shown in Table 1 for values of n from 0 to 6.

An examination of Table 1 shows that patterns are forming and the following observations can be made.

(i) The power of a, when looking at each term in any of the expansions and moving from left to right, follows the pattern:

$n, n-1, n-2, \dots, 2, 1, 0$ (since $a^0 = 1$).

Thus for $n = 5$, the 'a' part of each term is:

a^5, a^4, a^3, a^2, a and 1.

(ii) The power of b, when looking at each term in any of the expansions and moving from left to right, follows the pattern:

0, (since $b^0 = 1$), $1, 2, 3, \dots, n-2, n-1, n$.

For $n = 5$, the 'b' part of each term is:

$1, b, b^2, b^3, b^4$ and b^5.

(iii) The values of the coefficients of each of the terms in any of the expansions are symmetrical about the middle coefficient when n is even and symmetrical about the middle two coefficients when n is odd. This can be seen in Table 1, where for n being an even number, say 4, the coefficients of $(a + b)^4$ are 1, 4, 6, 4, 1, i.e. symmetrical about 6, the middle coefficient. When n is odd, say 5, the coefficients of $(a + b)^5$ are 1, 5, 10, 10, 5, 1, i.e. symmetrical about the two 10's, the middle two coefficients.

Table 2 shows the coefficients of the expansions of $(a + b)^n$, where n is a positive integer and varies from 0 to 6.

Table 2

Term	Values of coefficients in the expansion
$(a+b)^0$	1
$(a+b)^1$	1 1
$(a+b)^2$	1 2 1
$(a+b)^3$	1 3 3 1
$(a+b)^4$	1 4 6 4 1
$(a+b)^5$	1 5 10 10 5 1
$(a+b)^6$	1 6 15 20 15 6 1

The coefficient of, say, the fourth term in the expansion of $(a + b)^6$ is obtained by adding together the 10 and 10 immediately above it, giving a result of 20, this being shown by the triangle in the table. An examination of Table 2 shows that the first and last coefficients of any expansion of the form $(a+b)^n$ are 1's. To obtain the coefficients of, say, $(a + b)^4$, the first and last coefficients are 1's. The second coefficient, 4, is obtained from adding the 1 and 3 from the line above it. Similarly the third coefficient, 6, is obtained by adding the 3 and 3 from the line above it, and so on. The configuration shown in Table 2 is called **Pascal's triangle.** It is used to determine the coefficients of the expansion $(a + b)^n$ when n is a relatively small positive integer.

Worked problems on the expansion of the type $(a + b)^n$, where n is a small positive integer, using Pascal's triangle

---

Problem 1. Find the expansion of $(a + b)^7$.

The coefficients of the terms are determined by producing Pascal's triangle as far as the seventh power and selecting the last line.
From Table 2, the coefficients of $(a + b)^6$ are:
1, 6, 15, 20, 15, 6, 1
The coefficients of $(a + b)^7$ are therefore:
1, since the first and last coefficients are always 1's;
7, obtained by adding the first and second coefficients of $(a + b)^6$, i.e., 1 + 6;
21, obtained by adding the second and third coefficients of $(a + b)^6$, i.e., 6 + 15;
and so on, giving the coefficients of $(a + b)^7$ as:
1, 7, 21, 35, 35, 21, 7, 1
The 'a' terms are $a^7, a^6, a^5, \ldots, a, 1$
The 'b' terms are $1, b, b^2, \ldots, b^6, b^7$
Combining these results gives:
$(a+b)^7 = a^7 + 7a^6b + 21a^5b^2 + 35a^4b^3 + 35a^3b^4 + 21a^2b^5 + 7ab^6 + b^7$
A check for blunders can be made by adding the powers of a and b for each term. These should always be equal to n. In this problem, adding the powers of a and b together for each term gives:
7+0 = 7, 6+1 = 7, 5+2 = 7, 4+3 = 7, 3+4 = 7, 2+5 = 7, 1+6 = 7, and 0+7 = 7.
Thus no blunder has been made in determining the powers of a and b.

Problem 2. Find the expansion of $(1 - 3x)^{12}$ as far as the term in x^4.

Comparing $(1 - 3x)^{12}$ with $(a + b)^n$ shows $a = 1$, $b = (-3x)$, (note the minus sign) and $n = 12$. Hence the 'a' terms are
$1^{12}, 1^{11}, 1^{10}, \ldots, 1^1, 1^0$
The 'b' terms are $(-3x)^0$, $(-3x)^1$, $(-3x)^2$, $(-3x)^3$, $(-3x)^4$, hence only five terms of the expansion are required since the expansion is as far as the term in x^4. Also, only the first five coefficients of each line of Pascal's triangle are necessary. Taking the first five coefficients of the seventh-power expansion from Problem 1 and calculating the first five coefficients as far as the twelfth power gives:

				1		7		21		35		35	
			1		8		28		56		70		
		1		9		36		84		126			
	1		10		45		120		210				
1		11		55		165		330					
1	12		66		220		495						

Thus $(1-3x)^{12} = 1(1)^{12} + 12\,(1)^{11}(-3x) + 66\,(1)^{10}(-3x)^2 + 220\,(1)^9(-3x)^3 +$ $+\,495\,(1)^8\,(-3x)^4$ as far as the term in x^4, i.e.

$$\mathbf{(1-3x)^{12} = 1 - 36x + 594\,x^2 - 5\,940\,x^3 + 40\,095\,x^4} \text{ as far as the term in } x^4.$$

Problem 3. Expand $\left(-1 - \frac{3}{y}\right)^7$ to five terms.

Comparing $\left(-1 - \frac{3}{y}\right)^7$ with $(a + b)^n$ shows that $a = (-1)$, $b = \left(-\frac{3}{y}\right)$ and $n = 7$. Note that the minus signs must be included for both the 'a' and 'b' terms. The first five coefficients of the seventh power are obtained using Pascal's triangle as shown in Problem 1 and are

$$1, \quad 7, \quad 21, \quad 35 \text{ and } 35$$

The 'a' terms are $(-1)^7$, $(-1)^6$, $(-1)^5$, $(-1)^4$ and $(-1)^3$

The 'b' terms are $\left(\frac{-3}{y}\right)^0$, $\left(\frac{-3}{y}\right)^1$, $\left(\frac{-3}{y}\right)^2$, $\left(\frac{-3}{y}\right)^3$ and $\left(\frac{-3}{y}\right)^4$

Hence the first five terms of the expansion of $\left(-1 - \frac{3}{y}\right)^7$ are:

$$(1)(-1)^7\left(\frac{-3}{y}\right)^0 + (7)(-1)^6\left(\frac{-3}{y}\right)^1 + (21)(-1)^5\left(\frac{-3}{y}\right)^2 + (35)(-1)^4\left(\frac{-3}{y}\right)^3 +$$
$$+\,(35)(-1)^3\left(\frac{-3}{y}\right)^4$$

that is, $(1)(-1)(1) + (7)(1)\left(\frac{-3}{y}\right) + (21)(-1)\left(\frac{9}{y^2}\right) + (35)(1)\left(\frac{-27}{y^3}\right) +$ $+\,(35)(-1)\left(\frac{81}{y^4}\right)$

i.e. $\mathbf{-1 - \frac{21}{y} - \frac{189}{y^2} - \frac{945}{y^3} - \frac{283\,5}{y^4}}$

Problem 4. Determine the expansion of $\left(2x + \frac{y}{2}\right)^9$ as far as the term in y^5.

Taking the first six coefficients of the seventh-power expansion from Problem 1 and calculating the first six coefficients as far as the ninth power gives:

1 7 21 35 35 21

1 8 28 56 70 56

1 9 36 84 126 126

The 'a' terms in the $(a + b)^n$ expansion are replaced by $2x$ and are:
$(2x)^9, (2x)^8, (2x)^7, (2x)^6, (2x)^5, (2x)^4$

The 'b' terms in the $(a + b)^n$ expansion are replaced by $\frac{y}{2}$ and are:

$$\left(\frac{y}{2}\right)^0, \left(\frac{y}{2}\right)^1, \left(\frac{y}{2}\right)^2, \left(\frac{y}{2}\right)^3, \left(\frac{y}{2}\right)^4 \text{ and } \left(\frac{y}{2}\right)^5$$

Combining these three results gives:

$$\left(2x + \frac{y}{2}\right)^9 = 1(2x)^9 \left(\frac{y}{2}\right)^0 + 9\,(2x)^8\left(\frac{y}{2}\right)^1 + 36\,(2x)^7\left(\frac{y}{2}\right)^2 + 84\,(2x)^6\left(\frac{y}{2}\right)^3 + 126\,(2x)^5\left(\frac{y}{2}\right)^4 + 126\,(2x)^4\left(\frac{y}{2}\right)^5$$

as far as the term in y^5.

Thus, $$\left(2x + \frac{y}{2}\right)^9 = 2^9\,x^9 + 9\,(2)^7x^8y + 36(2)^5x^7y^2 + 84(2)^3x^6y^3 + 126(2)x^5y^4 + \frac{126}{2}\,x^4y^5$$

as far as the term in y^5. That is,

$$\left(2x + \frac{y}{2}\right)^9 = 512\,x^9 + 1\,152\,x^8y + 1\,152\,x^7y^2 + 672\,x^6y^3 + 252\,x^5y^4 + 63\,x^4y^5, \text{ as far as the term in } y^5.$$

Further problems on the expansion of the type $(a + b)^n$, *where* n *is a small positive integer, using Pascal's triangle, may be found in Section 4 (Problems 1–10).*

2. The general expansion of $(a + b)^n$, where n is any positive integer

The value of the coefficients of the expansion of $(a + b)^n$ for integer values of n from 0 to 6 are shown in Table 2. This table shows that the coefficients for $(a + b)^6$ are:

1, 6, 15, 20, 15, 6 and 1

Instead of using Pascal's triangle to derive these coefficients, they could have been obtained using a factor method from the relationships:

$$1, \frac{6}{1} = 6, \frac{(6)(5)}{(1)(2)} = 15, \frac{(6)(5)(4)}{(1)(2)(3)} = 20, \frac{(6)(5)(4)(3)}{(1)(2)(3)(4)} = 15,$$

$$\frac{(6)(5)(4)(3)(2)}{(1)(2)(3)(4)(5)} = 6 \text{ and } \frac{(6)(5)(4)(3)(2)(1)}{(1)(2)(3)(4)(5)(6)} = 1$$

Replacing $(a + b)^6$ by $(a + b)^n$ and building up the coefficients by a factor method using those for $(a + b)^6$ as a pattern, gives:

$$1, n, \frac{n(n-1)}{(1)(2)}, \frac{n(n-1)(n-2)}{(1)(2)(3)}, \frac{n(n-1)(n-2)(n-3)}{(1)(2)(3)(4)}, \text{ and so on.}$$

For example, the value of the third coefficient of $(a + b)^5$ is obtained from $\frac{n(n-1)}{(1)(2)}$ where n is 5, and is $\frac{(5)(4)}{(1)(2)}$, i.e. 10. Similarly, the value of the fourth coefficient of $(a + b)^4$ is determined using $\frac{n(n-1)(n-2)}{(1)(2)(3)}$ where n is equal to 4, and is $\frac{(4)(3)(2)}{(1)(2)(3)}$, i.e. 4.

Combining this factorial method of writing coefficients with the observations previously made for $(a + b)^n$ shows that the terms in 'a' are $a^n, a^{n-1}, a^{n-2}, \ldots$, and the terms in '$b$' are $b^0, b, b^2, b^3, \ldots$ Thus the general expansion of $(a + b)^n$ is:

$$(a+b)^n = a^n + n\,a^{n-1}b + \frac{n(n-1)}{(1)(2)}a^{n-2}b^2 + \frac{n(n-1)n-2)}{(1)(2)(3)}a^{n-3}b^3 \text{ and so on.}$$

The product (1)(2)(3) is usually denoted by 3!, called 'factorial 3'. In general, (1)(2)(3)(4) . . . (n) is denoted by $n!$, (factorial n). Hence,

$$(a+b)^n = a^n + n\,a^{n-1}\,b + \frac{n(n-1)}{2!}a^{n-2}\,b^2 + \frac{n(n-1)(n-2)}{3!}a^{n-3}\,b^3 + \ldots$$

This expansion is the **general binomial expansion of $(a + b)^n$**.

Practical problems can arise, for example, in the binomial distribution in statistics, where it is required to find the value of just one or two terms of a binomial expansion. The fifth term of the expansion of $(a + b)^n$ is

$$\frac{n(n-1)(n-2)(n-3)}{4!}a^{n-4}\;b^4$$

It can be seen that in the fifth term of any expansion the number 4 is very evident. There are four products of the type $n(n-1)(n-2)(n-3)$; 'a' is raised to the power $(n-4)$; 'b' is raised to the power of 4, and the denominator of the coefficient is 4!. For any term in a binomial expansion, say the rth term, $r-1$, is very evident. The value of the coefficient of the rth term is given by:

$$\frac{n(n-1)(n-2)\ldots \text{ to } (r-1) \text{ terms}}{(r-1)!}$$

The power of 'a' for the rth term is $n - (r-1)$ and the power of 'b' is $(r-1)$. Thus the rth term of the expansion of $(a + b)^n$ is:

$$\frac{n(n-1)(n-2)\ldots\text{ to }(r-1)\text{ terms}}{(r-1)!}\,a^{n-(r-1)}\,b^{(r-1)}$$

For example, to find the fifth term: in the expansion of $(a + b)^{15}$, n is 15 and r is 5 and $(r-1)$ is 4. Hence the fifth term is

$$\frac{(15)(14)(13)(12)}{4!}\,a^{15-4}\,b^4,\ \text{i.e. } 1\,365\,a^{11}\,b^4$$

Worked problems on the general expansion of $(a + b)^n$, where n is any positive integer

---

Problem 1. Expand $(x + y)^{20}$ as far as the fifth term.

The general binomial cxpansion for $(a + b)^n$ is

$$a^n + na^{n-1}\,b + \frac{n(n-1)}{2!}\,a^{n-2}b^2 + \frac{n(n-1)(n-2)}{3!}\,a^{n-3}b^3 + \ldots$$

Substituting in this general formula, $a = x$, $b = y$ and $n = 20$ gives:

$$(x+y)^{20} = x^{20} + 20x^{(20-1)}y + \frac{20(20-1)}{(2)(1)}\,x^{(20-2)}y^2 + \frac{20(20-1)(20-2)}{(3)(2)(1)}\,x^{(20-3)}y^3 + \frac{20(20-1)(20-2)(20-3)}{(4)(3)(2)(1)}\,x^{(20-4)}y^4 + \ldots$$

That is:

$$(x+y)^{20} = x^{20} + 20x^{19}y + \frac{20(19)}{2}\,x^{18}y^2 + \frac{20(19)(18)}{6}\,x^{17}y^3 + \frac{20(19)(18)(17)}{24}\,x^{16}y^4 + \ldots$$

Or $(x+y)^{20} = x^{20} + 20x^{19}y + 190x^{18}y^2 + 1\,140x^{17}y^3 + 4\,845x^{16}y^4$ when expanded as far as the fifth term.

Problem 2. Determine the expansion of $\left(p - \frac{4}{p^2}\right)^{15}$ as far as the term containing p^3.

Substituting $a = p$, $b = \left(\frac{-4}{p^2}\right)$ and $n = 15$ in the general expansion of $(a + b)^n$ gives:

$$\left(p - \frac{4}{p^2}\right)^{15} = (p)^{15} + 15(p)^{14}\left(\frac{-4}{p^2}\right) + \frac{15(14)}{(2)(1)}\,(p)^{13}\left(\frac{-4}{p^2}\right)^2 +$$

(P.T.O.)

$$+ \frac{15(14)(13)}{(3)(2)(1)}(p)^{12}\left(\frac{-4}{p^2}\right)^3 + \frac{15(14)(13)(12)}{(4)(3)(2)(1)}(p)^{11}\left(\frac{-4}{p^2}\right)^4 + \ldots$$

i.e., $\left(p - \frac{4}{p^2}\right)^{15} = p^{15} + 15p^{14}\left(\frac{-4}{p^2}\right) + 105p^{13}\left(\frac{16}{p^4}\right) + 455p^{12}\left(\frac{-64}{p^6}\right) +$

$+ 1\,365p^{11}\left(\frac{256}{p^8}\right) + \ldots$

i.e. $\left(p - \frac{4}{p^2}\right)^{15} = p^{15} - 60p^{12} + 1\,680p^9 - 29\,120p^6 + 349\,440p^3$ when expanded as far as the term in p^3.

Problem 3. Determine the sixth term of the expansion of $\left(\frac{1}{m} + \frac{m^2}{2}\right)^{14}$.

The rth term of the expansion of $(a + b)^n$ is given by

$$\frac{n(n-1)(n-2)\ldots \text{to } (r-1) \text{ terms}}{(r-1)!}\, a^{n-(r-1)}\, b^{(r-1)}$$

Substituting $a = \frac{1}{m}$, $b = \frac{m^2}{2}$, $n = 14$, and $(r-1) = 5$ (since $r = 6$), in this expression gives:

$$\frac{(14)(13)(12)(11)(10)}{(5)(4)(3)(2)(1)}\left(\frac{1}{m}\right)^{14-5}\left(\frac{m^2}{2}\right)^5$$

$$= 2\,002\left(\frac{1}{m}\right)^9\left(\frac{m^2}{2}\right)^5$$

$$= \frac{1\,001}{16}\,m$$

Thus the sixth term of the expression of $\left(\frac{1}{m} + \frac{m^2}{2}\right)^{14}$ is $\frac{1\,001}{16}\,m$.

Problem 4. Find the middle term of the expansion of $\left(3u - \frac{1}{3v}\right)^{18}$.

In any expansion of the form $(a + b)^n$ there are $(n + 1)$ terms. Hence, in the expansion of $\left(3u - \frac{1}{3v}\right)^{18}$ there are 19 terms. The middle term is the 10th term. Using the general expression for the rth term, where $a = 3u$, $b = \left(-\frac{1}{3v}\right)$,

$n = 18$ and $(r-1) = 9$, gives:

$$\frac{18\ (17)(16)(15)(14)(13)(12)(11)(10)}{9\ \ (8)\ (7)\ (6)\ (5)\ (4)\ (3)\ (2)\ (1)}(3u)^9\left(-\frac{1}{3v}\right)^9$$

$$= 48\,620\ (3)^9\ (u^9)\ \frac{(-1)^9}{3^9 v^9} = -\ 48\,620\left(\frac{u}{v}\right)^9$$

Thus the middle term of the expansion of $\left(3u - \frac{1}{3v}\right)^{18}$ is $-48\,620\left(\frac{u}{v}\right)^9$.

Problem 5. Derive the term containing y^{12} in the expansion of $\left(y^2 - \frac{x}{4}\right)^{10}$.

The y terms are $(y^2)^{10}$, $(y^2)^9$, $(y^2)^8$, $(y^2)^7$, $(y^2)^6$, and so on. Hence the term involving y^{12} is the fifth term. Using the expression for the rth term, where

$a = y^2$, $b = \left(-\frac{x}{4}\right)$, $n = 10$ and $(r-1) = 4$, we obtain

$$\frac{10(9)(8)(7)}{4(3)(2)(1)}\,(y^2)^{10-4}\left(-\frac{x}{4}\right)^4$$

i.e. $\frac{105}{128}y^{12}x^4$.

Thus the term containing y^{12} in the expansion of $\left(y^2 - \frac{x}{4}\right)^{10}$ is $\frac{105}{128}y^{12}x^4$.

Further problems on the expansion of $(a + b)^n$ *where n is any positive integer may be found in Section 4 (Problems 11–20).*

3. The application of the binomial expansion to determining approximate values of expressions

The general binomial expansion of $(a + b)^n$ is:

$$(a + b)^n = a^n + n\,a^{n-1}b + \frac{n(n-1)}{2!}\,a^{n-2}b^2 + \frac{n(n-1)(n-2)}{3!}\,a^{n-3}b^3 + \ldots$$

When $a = 1$ and $b = x$, then

$$(1 + x)^n = (1)^n + n(1)^{n-1}x + \frac{n(n-1)}{2!}(1)^{n-2}x^2 + \frac{n(n-1)(n-2)}{3!}(1)^{n-3}x^3 + \ldots$$

i.e. $(1 + x)^n = 1 + nx + \frac{n(n-1)}{2!}x^2 + \frac{n(n-1)(n-2)}{3!}x^3 + \ldots$

If an expression is written in the form $(1 + x)^n$ where x is small compared with 1, then terms such as x^2, x^3, x^4, . . . become very small and can be ignored if only an approximate result is required. Approximate values of expressions which could be written in this form used to be found in this way before electronic calculators came into widespread use. However solving problems of this sort using the binomial expansion, assists the understanding and provides practice in the expansion of two numbers into a series. A series of the form $1 + ax + bx^2 + cx^3 + \ldots$ where $a, b, c, \ldots$ are constants is called a **power series** since it is expressed in terms of powers of x. Thus the binomial expansion is used to produce a power series for a two-number expansion. Using this method to find the value of, say, $(1.002)^7$ correct to four decimal places, the expression is written as $(1 + 0.002)^7$ and since 0.002 is small compared with 1, only a few terms of the binomial expansion are required. Thus

$$(1 + 0.002)^7 \triangleq 1 + 7\,(0.002) + \frac{7\,(6)}{2}(0.002)^2 + \ldots$$

$$\triangleq 1 + 0.014 + 21\,(0.000\,004) + \ldots$$
$$\triangleq 1 + 0.014 + 0.000\,084 + \ldots$$
$$= 1.014\,1 \text{ correct to four decimal places}$$

The fourth term of the expansion is $\frac{(7)(6)(5)}{(1)(2)(3)}\,(0.002)^3$ and does not affect the result, to the accuracy required.

In experimental work, measurements are taken in the workshop or laboratory under the conditions prevailing at the time and corrections are subsequently made to enable results to be obtained more accurately. For example, the radius and height of a cylinder are measured and the volume is calculated. Later on, it changes due to temperature fluctuations or inherent inaccuracies within the measuring devices. The measured value of the radius has an error of 2½ per cent too large and the measured value of the height has an error of 1½ per cent too small. The binomial expansion can be used to find an approximate value of the error made in calculating the volume, when the other errors are known.

Let the correct values be volume V, radius r and height h. Then the correct value of volume is given by $V = \pi\, r^2 h$ for a cylinder. The uncorrected value of the radius is $\frac{102.5}{100}\,r$ or $(1 + 0.025)r$, since the radius is 2½ per cent too large. The uncorrected value of the height is $\frac{98.5}{100}\,h$ or $(1 - 0.015)h$ since the measured value of the height is 1½ per cent too small. Thus the uncorrected value of the volume, V_1, based on these measurements is given by:

$$V_1 = \pi\,[(1 + 0.025)r]^2\,(1 - 0.015)h$$
$$= (1 + 0.025)^2\,(1 - 0.015)\,\pi\, r^2 h$$

Using the binomial expansion to evaluate $(1 + 0.025)^2$ and ignoring the term containing $(0.025)^2$, since $(0.025)^2 = 0.000\,625$, which is small compared with 1, gives:

$$V_1 \doteqdot (1 + 2\,(0.025))\,(1 - 0.015)\,\pi r^2 h$$
$$\doteqdot (1 + 0.05)\,(1 - 0.015)\,\pi r^2 h$$
$$\doteqdot [1 + 0.05 - 0.015 + (0.05)(-0.015)]\,\pi r^2 h$$

When approximate values are required, it is also usual to ignore the products of small terms. In general, in any binomial expansion, both products of small terms and powers of small terms can be ignored. This is because numbers less than unity get progressively smaller both when multiplied together and when they are raised to larger powers.

Hence $V_1 \doteqdot (1 + 0.05 - 0.015)\,\pi r^2 h$

$$\doteqdot 1.035\,\pi r^2 h \text{ or } 1.035\,V \text{ or } \frac{103.5}{100}\,V$$

That is, the uncorrected value V_1 is approximately 3.5 per cent larger than the correct value.

Worked problems on determining approximate values using the binomial expansion.

Problem 1. Find the value of $(1.003)^{10}$ correct to: (a) three decimal places; and (b) six decimal places, using the binomial expansion.

Writing $(1.003)^{10}$ as $(1 + 0.003)^{10}$ and substituting $x = 0.003$ and $n = 10$ in the general expansion of $(1 + x)^n$ gives:

$$1 + 10\,(0.003) + \frac{10(9)}{(2)(1)}\,(0.003)^2 + \frac{10\,(9)(8)}{(3)(2)(1)}\,(0.003)^3 + \ldots$$

or $(1.003)^{10} = 1 + 0.03 + 0.000\,405 + 0.000\,003\,24 + \ldots$

Hence $(1.003)^{10} =$ **1.030** correct to three decimal places and **1.030 408** correct to six decimal places.

Problem 2. Find the value of $(0.98)^7$ correct to five significant figures by using the binomial expansion.

$(0.98)^7$ is written as $(1 - 0.02)^7$. Using the $(1 + x)^n$-type expansion gives:

$$(1 - 0.02)^7 = 1 + 7\,(-0.02) + \frac{7(6)}{(2)(1)}\,(-0.02)^2 + \frac{(7)(6)(5)}{(3)(2)(1)}\,(-0.02)^3 + \frac{(7)(6)(5)(4)}{(4)(3)(2)(1)}\,(-0.02)^4 + \ldots$$

$$= 1 + 7\,(-0.02) + 21\,(0.000\,4) + 35\,(-0.000\,008) + 35\,(0.000\,000\,16) + \ldots$$
$$= 1 - 0.14 + 0.008\,4 - 0.000\,28 + 0.000\,005\,6 - \ldots$$

$=$ **0.868 13** correct to five significant figures.

Problem 3. Find the value of $(8.016)^4$ correct to six significant figures using the binomial expansion.

$(8.016)^4$ is written in the $(1 + x)^n$ form as follows:

$$(8.016)^4 = [8(1.002)]^4 = 8^4 (1.002)^4 = 8^4 (1 + 0.002)^4$$

$$(1 + 0.002)^4 = 1 + 4(0.002) + \frac{4(3)}{2}(0.002)^2 + \ldots$$

$$= 1 + 0.008 + 0.000\,024 + \ldots$$

$$\simeq 1.008\,024$$

Hence $(8.016)^4 \simeq 8^4 (1.008\,024)$

$\simeq 4\,096\,(1.008\,024)$

$=$ **4 128.87** correct to six significant figures.

Problem 4. Pressure p and volume v are related by the expression

$pv^3 = C$, where C is a constant.

Find the approximate percentage change in C when p is increased by 2 per cent and v decreased by 0.8 per cent.

Let p and v be the original values of pressure and volume.

The new values are $\frac{102}{100}p$ or $(1 + 0.02)p$ and $\frac{99.2}{100}v$ or $(1 - 0.008)v$.

Let the new value of C be C_1, then

$$C_1 = (1 + 0.02)p\,[(1 - 0.008)v]^3 = (1 + 0.02)(1 - 0.008)^3 pv^3$$

$$(1 - 0.008)^3 \simeq 1 - (3)(0.008) + \ldots \simeq 1 - 0.024$$

Hence $C_1 \simeq (1 + 0.02)(1 - 0.024)C$

and neglecting the products of small terms, this becomes

$$C_1 \simeq (1 + 0.02 - 0.024)C \simeq (1 - 0.004)C$$

Hence the value of C **is reduced by approximately 0.4 per cent** when p is increased by 2 per cent and v decreased by 0.8 per cent.

Further problems on determining approximate values, using the binomial expansion, may be found in the following section (4) (Problems 21–30).

4. Further problems

Expansions of the type $(a + b)^n$ where n is a small positive integer using Pascal's triangle

1. Determine the expansion of $(a + b)^8$.
   $[a^8 + 8a^7b + 28a^6b^2 + 56a^5b^3 + 70a^4b^4 + 56a^3b^5 + 28a^2b^6 + 8ab^7 + b^8]$

2. Find the expansion of $(x - y)^5$.
$[x^5 - 5x^4y + 10x^3y^2 - 10x^2y^3 + 5xy^4 - y^5]$
3. Find the expansion of $(2p - 3q)^6$.
$[64p^6 - 576p^5q + 2\,160p^4q^2 - 4\,320p^3q^3 + 4\,860p^2q^4 - 2\,916pq^5 +$
$+ 729q^6]$
4. Expand $(p + 3q)^{11}$ as far as the term containing q^4.
$[p^{11} + 33p^{10}q + 495p^9q^2 + 4\,455p^8q^3 + 26\,730p^7q^4]$
5. Find the expansion of $(x - 2y)^{10}$ as far as the term containing y^5.
$[x^{10} - 20x^9y + 180x^8y^2 - 960x^7y^3 + 3\,360x^6y^4 - 8\,064x^5y^5]$
6. Determine the expansion of $\left(-m - \frac{n}{2}\right)^7$ as far as the term containing n^4.
$$\left[-m^7 - \frac{7}{2}m^6n - \frac{21}{4}m^5n^2 - \frac{35}{8}m^4n^3 - \frac{35}{16}m^3n^4\right]$$
7. Determine the expansion of $\left(\frac{w}{2} - \frac{x}{3}\right)^4$.
$$\left[\frac{1}{16}w^4 - \frac{1}{6}w^3x + \frac{1}{6}w^2x^2 - \frac{2}{27}wx^3 + \frac{1}{81}x^4\right]$$
8. Find the expansion of $(3 + 4y)^6$ and express the result in the form $a + by + cy^2 + \ldots$, where $a, b, c, \ldots$ are constants.
$[729 + 5\,832y + 19\,440y^2 + 34\,560y^3 + 34\,560y^4 + 18\,432y^5 + 4\,096y^6]$
9. Determine the expansion of $\left(\frac{x}{4} - 7\right)^5$ and express the result in the form $ax^5 + bx^4 + cx^3 + \ldots$, where $a, b, c, \ldots$ are constants.
$$\left[\frac{x^5}{1\,024} - \frac{35}{256}x^4 + \frac{245}{32}x^3 - \frac{1\,715}{8}x^2 + \frac{12\,005}{4}x - 16\,807\right]$$
10. Expand $\left(-5 - \frac{p}{3}\right)^8$ as far as the term containing p^4.
$$\left[390\,625 + \frac{625\,000}{3}p + \frac{437\,500}{9}p^2 + \frac{175\,000}{27}p^3 + \frac{43\,750}{81}p^4\right]$$

Expansions of the type $(a + b)^n$ where n is any positive integer

11. Find the expansion of $(a + b)^{12}$ as far as the term containing b^5.
$[a^{12} + 12a^{11}b + 66a^{10}b^2 + 220a^9b^3 + 495a^8b^4 + 792a^7b^5]$
12. Determine the expansion of $\left(x - \frac{y}{2}\right)^{16}$ as far as the term containing y^4.
$$\left[x^{16} - 8x^{15}y + 30x^{14}y^2 - 70x^{13}y^3 + \frac{455}{4}x^{12}y^4\right]$$

In problems 13–15, find the first four terms of the expansions of the expressions given.

13. $\left(m - \frac{n^2}{2}\right)^{13}$ $\left[m^{13} - \frac{13}{2}m^{12}n^2 + \frac{39}{2}m^{11}n^4 - \frac{143}{4}m^{10}n^6\right]$
14. $(-p^2 - 2q)^{17}$ $[-p^{34} - 34\,p^{32}q - 544\,p^{30}q^2 - 5\,440\,p^{28}q^3]$
15. $\left(-\frac{1}{x} + \frac{3}{y}\right)^{19}$ $\left[-\frac{1}{x^{19}} + \frac{57}{x^{18}y} - \frac{1\,539}{x^{17}y^2} + \frac{26\,163}{x^{16}y^3}\right]$

16. Determine the middle term of the expansion of $(x^2 - y^2)^{14}$.
[$-3\,432\,x^{14}y^{14}$]
17. Find the eleventh term of the the expansion of $\left(2p - \frac{q}{2}\right)^{21}$.
[$705\,432\,p^{11}q^{10}$]
18. Find the value of the middle term of the expansion of $\left(2c^2 - \frac{1}{2c^2}\right)^{12}$.
[924]
19. Determine the two middle terms of the expansion of $\left(2p - \frac{1}{3p}\right)^9$.
$\left[\frac{1\,344p}{27}, -\frac{224}{27p}\right]$
20. Find the term involving a^{12} in the expansion of $\left(a^3 - \frac{b}{2}\right)^{14}$.
$\left[\frac{1\,001}{1\,024}a^{12}b^{10}\right]$

Determining the approximate values of expressions

21. Use the binomial expansion to calculate the value of $(0.995)^{12}$ correct to: (a) 4 decimal places; and (b) 6 decimal places. [0.941 6, 0.941 623]
22. Find the value of $(1.05)^3(0.98)^4$ correct to 5 significant figures by using binomial expansions. [1.067 8]
23. Determine the value of $(3.036)^3$ correct to 3 decimal places by using the binomial expansion. [27.984]
24. Use the binomial expansion to find the value of $(2.018)^5$ correct to 6 significant figures. [33.466 2]
25. An error of 3.5 per cent too large was made when measuring the radius of a sphere. Ignoring the products of small quantities, determine the approximate error in calculating: (a) the volume; and (b) the surface area when they are calculated using the correct radius measurement.
[10.5 per cent too large, 7 per cent too large]
26. The area of a triangle is given by $A = \frac{1}{2}ab \sin C$, where C is the angle between the sides a and b of a triangle. Calculate the approximate change in area (ignoring the products of small quantities), when: (a) both a and b are increased by 2 per cent; and (b) when a is increased by 2 per cent and b is reduced by 2 per cent. [4 per cent increase, no change]
27. The moment of inertia of a body about an axis is given by $I = kbd^3$ where k is a constant and b and d are the dimensions of the body. Determine the approximate percentage change in the value of I when b is increased by 5 per cent and d reduced by 1 per cent, if products of small quantities are ignored. [2 per cent increase]
28. The radius of a cone is reduced by 4.5 per cent and its height increased by 1.5 per cent. Determine the approximate percentage change: (a) in its volume; and (b) in its curved surface area, neglecting the products of small quantities. [7.5 per cent reduction, 3 per cent reduction]
29. The power developed by an engine is given by $I = kPLAN$ where k is a

constant. Find the approximate percentage increase in power when P, L, A and N are each increased by 3.5 per cent. [14 per cent]

30. The modules of rigidity G is given by

$$G = \frac{R^4\theta}{L}$$

where R is the radius, θ the angle of twist, and L the length. Find the approximate percentage error in G where R is measured 1.5 per cent too large and θ is measured 5 per cent too small. [1 per cent too large]

Chapter 3

Exponential functions and Naperian logarithms

1. Exponential functions and their power series

The mathematical constant π, which is the ratio of the circumference of any circle to its diameter and has a value of approximately 3.141 6, is accepted and used extensively in calculations involving circular measure. In calculus and more advanced mathematics, another mathematical constant, e, is used frequently. This constant is called the **exponent** and has a value of approximately 2.718 3. A function containing e^x is called an **exponential function** and can also be written as $\exp x$.

The exponent is used mainly in two closely related ways:

(i) the exponential or natural laws of growth or decay; and

(ii) as a base of natural or naperian logarithms.

The natural laws of growth or decay

The natural laws of growth or decay occur frequently in engineering and science and are always of the form

$y = Ae^{kx}$ or $y = A \exp(kx)$ and

$y = A(1 - e^{kx})$ or $y = A(1 - \exp(kx))$

where A and k are constants and can be either positive or negative. The laws relate quantities in which the rate of increase of y is proportional to y itself for the growth law, or in which the rate of decrease of y is proportional to y itself for the decay law. Some of the quantities following these natural laws are given below.

(a) Linear expansion

A rod of length l at temperature θ°C and having a positive coefficient of linear expansion of α will become longer when heated. The natural-growth law is

$$l = l_0\,e^{\alpha\theta} \quad \text{or} \quad l = l_0 \exp(\alpha\theta)$$

where l_0 is the length of the rod at 0°C.

(b) Change of electrical resistance with temperature

A resistor of resistance R_θ at temperature θ°C and having a positive temperature coefficient of resistance of α increases in resistance when heated. The natural-growth law is

$$R_\theta = R_0\,e^{\alpha\theta} \quad \text{or} \quad R_\theta = R_0 \exp(\alpha\theta)$$

where R_0 is the resistance at 0°C.

(c) Tension in belts

A natural-growth law governs the relationship between the tension T_1 in a belt around a pulley wheel and its angle of lap α. It is of the form

$$T_1 = T_0\,e^{\mu\alpha} \quad \text{or} \quad T_1 = T_0 \exp(\mu\alpha)$$

where μ is the coefficient of friction between belt and pulley and T_1 and T_0 are the tensions on the tight and slack sides of the belt respectively.

(d) The growth of current in an inductive circuit

In a circuit of resistance R and inductance L having a final value of steady current I

$$i = I\left(1 - e^{\frac{-Rt}{L}}\right) \quad \text{or} \quad i = I\left(1 - \exp\left(-\frac{Rt}{L}\right)\right)$$

where i is the current flowing at time t. This is an equation which follows the natural-growth law.

(e) Newton's law of cooling

The rate at which a body cools is proportional to the excess of its temperature above that of its surroundings. The law is:

$$\theta = \theta_0\,e^{-kt} \quad \text{or} \quad \theta = \theta_0 \exp(-kt)$$

where the excess of temperature at time $t = 0$ is θ_0 and at time t is θ. The negative power of the exponent indicates a decay curve when k is positive.

(f) Discharge of a capacitor

When a capacitor of capacitance C, having an initial charge of Q, is discharged through a resistor R, then

$$q = Q\,e^{-\frac{t}{CR}} \quad \text{or} \quad q = Q \exp\left(-\frac{t}{CR}\right)$$

where q is the charge after time t.

(g) Atmospheric pressure

The pressure p at height h above ground level is given by

$$p = p_0 \, e^{-\frac{h}{c}} \quad \text{or} \quad p = p_0 \exp\left(-\frac{h}{c}\right)$$

where p_0 is the pressure at ground level and c is a constant.

(h) The decay of current in an inductive circuit

When a circuit having a resistance R, inductance L and initial current I is allowed to decay, it follows a natural law of the form

$$i = I \, e^{-\frac{Rt}{L}} \quad \text{or} \quad i = I \exp\left(-\frac{Rt}{L}\right)$$

where i is the current flowing after time t.

(i) Radioactive decay

The rate of disintegration of a radioactive nucleus having N_0 radioactive atoms present and a decay constant of λ is given by

$$N = N_0 \, e^{-\lambda t} \quad \text{or} \quad N = N_0 \exp(-\lambda t)$$

where N is the number of radioactive atoms present after time t.

(j) Biological growth

The rate of growth of bacteria is proportional to the amount present. When y is the number of bacteria present at time t and y_0 the number present at time $t = 0$ then

$$y = y_0 \, e^{kt} \quad \text{or} \quad y = y_0 \exp(kt)$$

where k is the growth constant.

These are just some of the relationships which exist which follow the natural laws of growth or decay. Both the e^x and $\exp(x)$ notations are in widespread use but to save repetition, in this text the e^x notation will be adopted from here onwards.

The value of e^x

The value of e^x can be calculated to any required degree of accuracy since it is defined in terms of the power series:

$$e^x = 1 + x + \frac{x^2}{2!} + \frac{x^3}{3!} + \ldots$$

This series is said to converge, that is, if all the terms are added the actual value of e^x is obtained, where x is a real number. The more terms that are taken, the closer will be the value of e^x to its actual value. The value of the exponent e, correct to, say, 5 significant figures is found by substituting $x = 1$ in this power series. This gives

$$e = 1 + 1 + \frac{(1)^2}{2!} + \frac{(1)^3}{3!} + \frac{(1)^4}{4!} + \ldots$$

$$= 1 + 1 + \frac{1}{(2)\,(1)} + \frac{1}{(3)\,(2)\,(1)} + \frac{1}{(4)\,(3)\,(2)\,(1)} + \ldots$$

Since $\frac{1}{3!} = \frac{1}{3}\left(\frac{1}{2!}\right)$, $\frac{1}{4!} = \frac{1}{4}\left(\frac{1}{3!}\right)$, and so on, one way of evaluating the decimal fractions is by successive division as shown:

1st term	1.000 00	
2nd term	2)1.000 00	
3rd term	3)0.500 00	$\left(\frac{1}{2!} = \frac{1}{2}\left(\frac{1}{1!}\right)\right)$
4th term	4)0.166 67	$\left(\frac{1}{3!} = \frac{1}{3}\left(\frac{1}{2!}\right)\right)$
5th term	5)0.041 67	$\left(\frac{1}{4!} = \frac{1}{4}\left(\frac{1}{3!}\right)\right)$ and so on
6th term	6)0.008 33	
7th term	7)0.001 39	
8th term	8)0.000 20	
9th term	9)0.000 03	
	0.000 00	
Adding	2.718 29	

Thus e = 2.718 3 correct to 5 significant figures.

The value of $e^{0.01}$, correct to, say, 9 significant figures is found by substituting $x = 0.01$ in the power series for e^x. Thus:

$$e^{0.01} = 1 + 0.01 + \frac{(0.01)^2}{2!} + \frac{(0.01)^3}{3!} + \frac{(0.01)^4}{4!} + \ldots$$

$$= 1 + 0.01 + 0.000\,05 + 0.000\,000\,167 + 0.000\,000\,004$$

and by adding, $e^{0.01} = 1.010\,050\,17$ correct to 9 significant figures. In this example, successive terms in the series grow smaller very rapidly and it is relatively easy to determine the value of $e^{0.01}$ to a high degree of accuracy. However, when x is near to unity or larger than unity, a very large number of terms are required for an accurate result. For many purposes, an accuracy of 4 or 5 significant figures is sufficient and most books of mathematical tables contain tables of the values of exponential functions. One such table is shown in Table 1.

This table enables the values of e^x or e^{-x} to be read over a range of x from 0.02 to 6.0 in discrete steps of 0.01 over most of the range. Some intermediate values can be obtained by using the laws of indices. For example,

$$\begin{aligned} e^{0.74} &= e^{(0.7 + 0.04)} \\ &= (e^{0.7})\,(e^{0.04}) \\ &= (2.013\,8)\,(1.040\,8) \\ &= 2.095\,963 \\ &= 2.096, \text{ allowing for rounding-off errors in the original data.} \end{aligned}$$

Table 1 Exponential functions

x	e^x	e^{-x}	x	e^x	e^{-x}
.02	1.0202	.9802	**1.0**	2.7183	.3679
.04	1.0408	.9608	1.1	3.0042	.3329
.06	1.0618	.9418	1.2	3.3201	.3012
.08	1.0833	.9231	1.3	3.6693	.2725
			1.4	4.0552	.2466
.10	1.1052	.9048			
.11	1.1163	.8958	**1.5**	4.4817	.2231
.12	1.1275	.8869	1.6	4.9530	.2019
.13	1.1388	.8781	1.7	5.4739	.1827
.14	1.1503	.8694	1.8	6.0497	.1653
			1.9	6.6859	.1496
.15	1.1618	.8607			
.16	1.1735	.8521	**2.0**	7.3891	.1353
.17	1.1853	.8437	2.1	8.1662	.1225
.18	1.1972	.8353	2.2	9.0250	.1108
.19	1.2092	.8270	2.3	9.9742	.1003
			2.4	11.023	.0907
.20	1.2214	.8187			
.21	1.2337	.8106	**2.5**	12.182	.0821
22	1.2461	.8025	2.6	13.464	.0743
.23	1.2586	.7945	2.7	14.880	.0672
.24	1.2712	.7866	2.8	16.445	.0608
			2.9	18.174	.0550
.25	1.2840	.7788			
.26	1.2696	.7711	**3.0**	20.085	.0498
.27	1.3100	.7634	3.1	22.198	.0450
.28	1.3231	.7558	3.2	24.532	.0408
.29	1.3364	.7483	3.3	27.113	.0369
			3.4	29.964	.0334
.30	1.3499	.7408			
.31	1.3634	.7335	**3.5**	33.115	.0302
.32	1.3771	.7261	3.6	36.598	.0273
.33	1.3910	.7189	3.7	40.447	.0247
.34	1.4050	.7118	3.8	44.701	.0224
			3.9	49.402	.0202
.35	1.4191	.7047			
.36	1.4333	.6977	**4.0**	54.598	.0183
.37	1.4477	.6907	4.1	60.340	.0166
.38	1.4623	.6839	4.2	66.686	.0150
.39	1.4770	.6771	4.3	73.700	.0136
			4.4	81.451	.0123
.40	1.4918	.6703			
.41	1.5068	.6636	**4.5**	90.017	.0111
.42	1.5220	.6570	4.6	99.484	.0100
.43	1.5373	.6505	4.7	109.95	.00910
.44	1.5527	.6440	4.8	121.51	.00823
			4.9	134.29	.00745
.45	1.5683	.6376			
.46	1.5841	.6313	**5.0**	148.41	.00674
.47	1.6000	.6250	5.1	164.02	.00610
.48	1.6161	.6188	5.2	181.27	.00552
.49	1.6323	.6126	5.3	200.34	.00499
			5.4	221.41	.00452
.50	1.6487	.6065			
.6	1.8221	.5488	**5.5**	244.69	.00409
.7	2.0138	.4966	5.6	270.43	.00370
.8	2.2255	.4493	5.7	298.87	.00335
.9	2.4596	.4066	5.8	330.30	.00303
			5.9	365.04	.00274
			6.0	403.43	.00248

The relationship between the tabular values of e^x is *not* linear and a correct result will **not** be obtained by methods of interpolation between the given values. Other methods of determining the values of e^x are to use an electronic calculator when an $[e^x]$ function is fitted and by using Naperian logarithms, which are introduced in Section 3 of this chapter.

Worked problems on evaluating exponential functions

Problem 1. Determine the value of $2e^{0.3}$ correct to 5 significant figures by using the power series for e^x.

Substituting $x = 0.3$ in the power series

$$e^x = 1 + x + \frac{x^2}{2!} + \frac{x^3}{3!} + \ldots$$

$$e^{0.3} = 1 + 0.3 + \frac{(0.3)^2}{(2)(1)} + \frac{(0.3)^3}{(3)(2)(1)} + \frac{(0.3)^4}{(4)(3)(2)(1)} + \frac{(0.3)^5}{(5)(4)(3)(2)(1)}$$

$$= 1 + 0.3 + 0.045 + 0.004\,5 + 0.000\,338 + 0.000\,020$$

$= 1.349\,86$ correct to 6 significant figures.

Hence $2e^{0.3}$ = **2.699 7** correct to 5 significant figures.

Problem 2. Determine the value of $-4e^{-1}$ correct to 4 decimal places using the power series for e^x.

Substituting $x = -1$ in the power series

$$e^x = 1 + x + \frac{x^2}{2!} + \frac{x^3}{3!} + \ldots$$

$$e^{-1} = 1 + (-1) + \frac{(-1)^2}{(2)(1)} + \frac{(-1)^3}{(3)(2)(1)} + \frac{(-1)^4}{(4)(3)(2)(1)} + \ldots$$

$$= 1 - 1 + 0.5 - 0.166\,667 + 0.041\,667 - 0.008\,333 + 0.001\,389 - 0.000\,198 + \ldots$$

$= 0.367\,858$ correct to 6 decimal places.

Hence $-4e^{-1}$ = **(−4) (0.367 858) = −1.471 4** correct to 4 decimal places.

Problem 3. Use exponential tables to determine the value of: (a) $e^{0.14}$; (b) $e^{-2.6}$; (c) $2e^{-0.66}$; and (d) $4e^{10}$, correct to 4 significant figures.

(a) Using a table of exponential values, when $x = 0.14$
$e^{0.14} = 1.150\,3$ = **1.150** correct to 4 significant figures.

(b) From exponential tables, when $x = 2.6$
$e^{-2.6}$ = **0.074 3** correct to 4 significant figures.

(c) $e^{-0.66} = e^{(-0.7+0.04)}$ and using the laws of indices, $e^{-0.66} = (e^{-0.7})(e^{0.04})$

Using exponential tables:

when $x = 0.7$, $e^{-x} = e^{-0.7} = 0.496\,6$

and when $x = 0.04$, $e^x = e^{0.04} = 1.040\,8$

Hence $e^{-0.66} = (0.496\,6)(1.040\,8) = 0.516\,86$.

Thus $2e^{-0.66} = (2)(0.516\,86) = \mathbf{1.034}$ correct to 4 significant figures.

(d) $e^{10} = e^{(5+5)} = (e^5)(e^5)$

and when $x = 5$, using exponential tables gives

$e^5 = 148.41$

Thus $e^{10} = (148.41)^2$
$= 22\,026$

Then $\mathbf{4e^{10}} = 4\,(22\,026) = \mathbf{88\,100}$ correct to 4 significant figures.

Problem 4. Expand $e^x\,(x^2 + 1)$ as far as the term in x^5.

The power series for e^x is $1 + x + \frac{x^2}{2!} + \frac{x^3}{3!} + \frac{x^4}{4!} + \ldots$

Hence $e^x\,(x^2+1) = \left(1 + x + \frac{x^2}{2!} + \frac{x^3}{3!} + \frac{x^4}{4!} + \ldots\right)(x^2+1)$

$$= \left(x^2 + x^3 + \frac{x^4}{2!} + \frac{x^5}{3!}\right) + \left(1 + x + \frac{x^2}{2!} + \frac{x^3}{3!} + \frac{x^4}{4!} + \frac{x^5}{5!} + \ldots\right)$$

Grouping like terms gives

$$e^x\,(x^2+1) = 1 + x + \left(1 + \frac{1}{2!}\right)x^2 + \left(1 + \frac{1}{3!}\right)x^3 + \left(\frac{1}{2!} + \frac{1}{4!}\right)x^4 + \left(\frac{1}{3!} + \frac{1}{5!}\right)x^5 + \ldots$$

i.e. $\mathbf{e^x\,(x^2+1) = 1 + x + \frac{3}{2}x^2 + \frac{7}{6}x^3 + \frac{13}{24}x^4 + \frac{7}{40}x^5}$ when expanded as far as the term in x^5.

Further problems on exponential functions and their power series may be found in Section 8 (Problems 1–7).

2. Graphs of exponential functions

A graph of the curves of $y = e^x$ and $y = e^{-x}$ over a range $x = -3$ to $x = 3$ is shown in Fig 1. The values of e^x and e^{-x}, correct to 2 decimal places, are obtained from tables of exponential functions and are shown below.

x =	−3.0	−2.5	−2.0	−1.5	−1.0	−0.5	0	0.5	1.0	1.5
e^x =	0.05	0.08	0.14	0.22	0.37	0.61	1	1.65	2.72	4.48
e^{-x} =	20.09	12.18	7.39	4.48	2.72	1.65	1	0.61	0.37	0.22

x =	2.0	2.5	3.0
e^x =	7.39	12.18	20.09
e^{-x} =	0.14	0.08	0.05

For graphs of the form $y = e^{kx}$ where k is any constant, positive or negative, k has the effect of altering the scale of x. For graphs of the form $y = Ae^{kx}$,

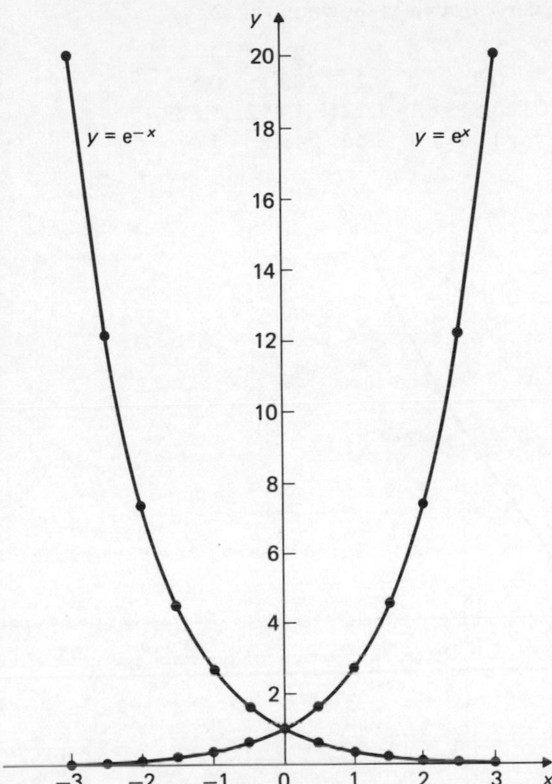

Figure 1 Graphs depicting $y = e^x$ and $y = e^{-x}$

where A is a constant, A has the effect of altering the scale of y. Hence every curve of the form $y = Ae^{kx}$ has the same general shape as shown in Fig. 1 and A and k are called **scale factors** of the graph. Their only function is to alter the values of x and y shown on the axes. Thus similar curves can be obtained for every function of the form $y = Ae^{kx}$ by selecting appropriate scale factors. For example, the curve of $y = 2e^{3x}$ will be identical to the curve of $y = e^x$ in Fig. 1 by making the y-axis markings 4, 8, 12, 16, . . . instead of 2, 4, 6, 8, . . . and the x-axis markings $\frac{1}{3}$, $\frac{2}{3}$, 1, . . . instead of 1, 2, 3, ...

Worked problems on the graphs of exponential functions

Problem 1. Draw a graph of $y = 3e^{0.2x}$ over a range of $x = -3$ to $x = 3$ and hence determine the approximate value of y when $x = 1.7$ and the approximate value of x when $y = 3.3$.

The values of y are calculated for integer values of x over the range required and are shown in the table below.

The values of the exponential functions are obtained using tables. The points are plotted and the curve drawn as shown in Fig. 2.

x	$= -3$	-2	-1	0	1	2	3
$0.2x$	$= -0.6$	-0.4	-0.2	0	0.2	0.4	0.6
$e^{0.2x}$	$= 0.549$	0.670	0.819	1	1.221	1.492	1.822
$3e^{0.2x}$	$= 1.65$	2.01	2.46	3	3.66	4.48	5.47

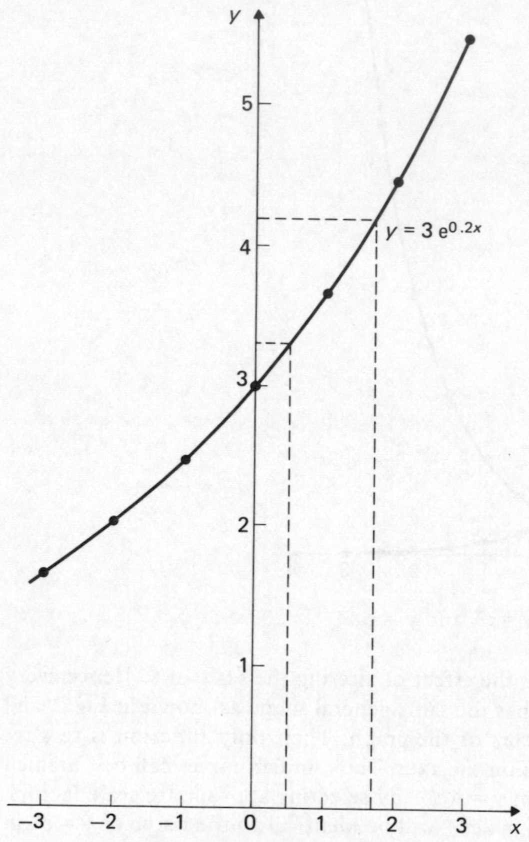

Figure 2 Graph depicting $y = 3e^{0.2x}$ (Problem 1)

From the graph, **when x = 1.7** the corresponding value of **y is 4.2**, and when **y is 3.3** the corresponding value of **x is 0.48.**

Problem 2. Draw a graph of $y = e^{-x^2}$ over a range $x = -2$ to $x = 2$ (this curve is called a **probability curve**).

The values of the coordinates are calculated as shown below:

x	$= -2$	-1.5	-1	-0.5	0	0.5	1.0	1.5	2.0
$-x^2$	$= -4$	-2.25	-1	-0.25	0	-0.25	-1	-2.25	-4
e^{-x^2}	$=$ 0.02	0.11	0.37	0.78	1.0	0.78	0.37	0.11	0.02

Using these values, the graph shown in Fig. 3 is drawn.

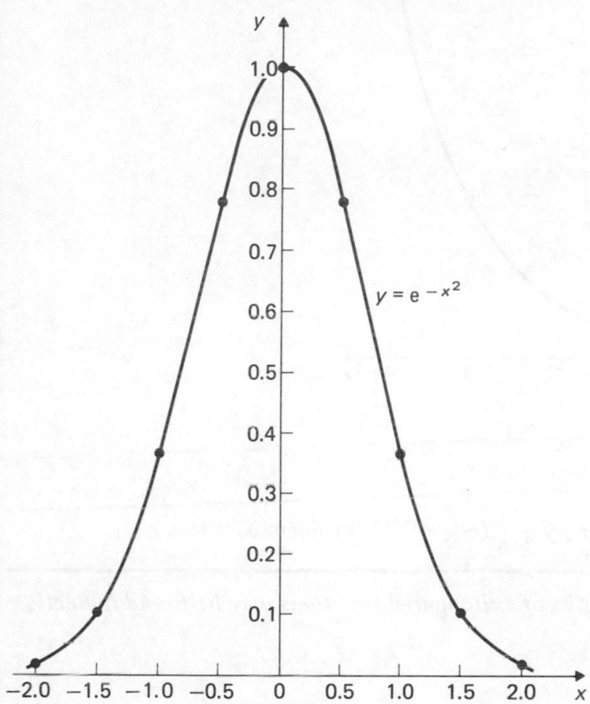

Figure 3 Graph depicting $y = e^{-x^2}$ (Problem 2)

Problem 3. Draw a graph of $y = \frac{1}{5}(e^x - e^{-2x})$ over a range $x = -1$ to $x = 4$.

The values of the coordinates are calculated as shown below. Since the values used to determine $\frac{1}{5}(e^x - e^{-2x})$ range from less than 0.1 to over 50, only 1-decimal-place accuracy is taken.

x	$= -1$	-0.5	0	0.5	1	2	3	4
e^x	$=$ 0.4	0.6	1	1.6	2.7	7.4	20.1	54.6
$-2x$	$=$ 2	1	0	-1	-2	-4	-6	-8
e^{-2x}	$=$ 7.4	2.7	1	0.4	0.1	0.0	0.0	0.0
$e^x - e^{-2x}$	$= -7.0$	-2.1	0	1.2	2.6	7.4	20.1	54.6
$\frac{1}{5}(e^x - e^{-2x})$	$= -1.4$	-0.4	0	0.2	0.5	1.5	4.0	10.9

Using these values, the graph shown in Fig. 4 is drawn.

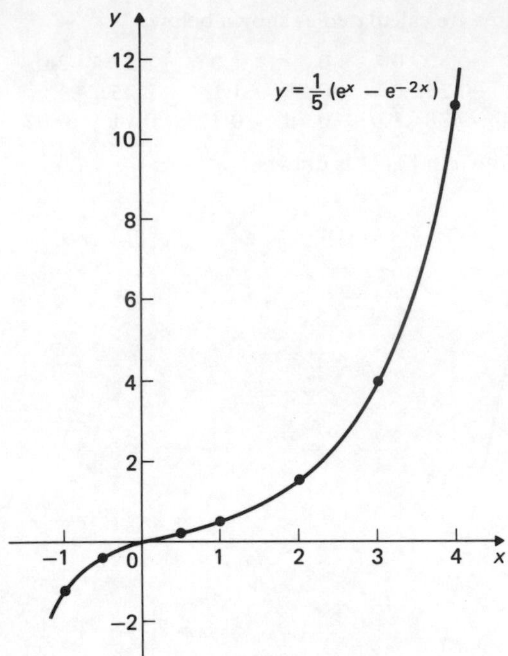

Figure 4 Graph depicting $y = \frac{1}{5}(e^x - e^{-2x})$ (Problem 3)

Further problems on graphs of exponential functions may be found in Section 8 (Problems 8–15).

3. Hyperbolic or Naperian logarithms

A logarithm of a number is defined as the power to which a base has to be raised to be equal to the number. Thus if $y = a^x$, then $x = \log_a y$. When using logarithms as an aid to calculations, a base of 10 is usually selected because the characteristic of the logarithm can be readily obtained. A more convenient base when using calculus and when dealing with problems involving the natural growth or decay laws is the exponent e. Logarithms to a base of e are called hyperbolic, Naperian or natural logarithms and the Naperian logarithm of x is written as $\log_e x$ or $\ln x$, the latter being that recommended by the Système International d'Unités committees, that is, the SI system of units.

One method of determining the value of a Naperian logarithm for values between 0 and 2 is by using the power series

$$\ln(1 + x) = x - \frac{x^2}{2} + \frac{x^3}{3} - \frac{x^4}{4} + \ldots$$

This power series is derived using a technique based on calculus called Maclaurin's series and is valid for $-1 < x \leqslant 1$. This range can be extended by using

$\ln n(1 + x) = \ln n + \ln(1 + x)$ and letting n be equal to 2, 3, 4, . . .

A second way of calculating the value of a Naperian logarithm is by using a conversion factor to convert logarithms which have a base of 10 into logarithms which have a base of e. There is a general rule of logarithms used when changing a base which states:

$$\log_a y = \frac{\log_b y}{\log_b a}$$

Substituting e for a and 10 for b gives

$$\ln y = \frac{\lg y}{\lg e}$$

where lg denotes a logarithm to the base 10.

But $\lg e = \lg 2.718 = 0.434\,3$

Hence $\ln y = \dfrac{\lg y}{0.434\,3}$ or $2.302\,6 \lg y$

Thus, the Naperian logarithm of a number is obtained by multiplying the logarithm of the number which has a base of 10 by 2.302 6.

The processes of determining the values of Naperian logarithms by using a power series or by converting logarithms to a base of 10 are fairly long and tedious and, in practice, either tables of Naperian logarithms are used or an electronic calculating machine is used, where the appropriate function is fitted.

Using Naperian logarithm tables

For numbers from 1 to 10, the tables of Naperian logarithms are used in a similar way as tables of logarithms to a base of 10. With reference to Table 2, care should be taken to see that the correct characteristic has been taken from the first column of the tables, as this is 1 for numbers between 2.718 3 and 7.389 0 and 2 for numbers between 7.389 1 and 10.0.

Numbers larger than 10 are expressed in standard form and the supplementary table of Naperian logarithms of 10^{+n} used. Thus, the Naperian logarithm of, say, 70, is obtained as follows:

$$\begin{aligned}\ln 70 &= \ln(7 \times 10^1)\\ &= \ln 7 + \ln 10^1\\ &= 1.945\,9 + 2.302\,6\\ &= 4.248\,5\end{aligned}$$

Numbers smaller than 1 are also expressed in standard form and the supple-

Table 2 Hyperbolic or Naperian logarithms

	0	1	2	3	4	5	6	7	8	9	Mean Differences								
											1	2	3	4	5	6	7	8	9
1.0	0.0000	0099	0198	0296	0392	0488	0583	0677	0770	0862	10	19	29	38	48	57	67	76	86
1.1	.0953	1044	1133	1222	1310	1398	1484	1570	1655	1740	9	17	26	35	44	52	61	70	78
1.2	.1823	1906	1989	2070	2151	2231	2311	2390	2469	2546	8	16	24	32	40	48	56	64	72
1.3	.2624	2700	2776	2852	2927	3001	3075	3148	3221	3293	7	15	22	30	37	44	52	59	67
1.4	.3365	3436	3507	3577	3646	3716	3784	3853	3920	3988	7	14	21	28	35	41	48	55	62
1.5	.4055	4121	4187	4253	4318	4383	4447	4511	4574	4637	6	13	19	26	32	39	45	52	58
1.6	.4700	4762	4824	4886	4947	5008	5068	5128	5188	5247	6	12	18	24	30	36	42	48	55
1.7	.5306	5365	5423	5481	5539	5596	5653	5710	5766	5822	6	11	17	24	29	34	40	46	51
1.8	.5878	5933	5988	6043	6098	6152	6206	6259	6313	6366	5	11	16	22	27	32	38	43	49
1.9	.6419	6471	6523	6575	6627	6678	6729	6780	6831	6881	5	10	15	20	26	31	36	41	46
2.0	.6931	6981	7031	7080	7129	7178	7227	7275	7324	7372	5	10	15	20	24	29	34	39	44
2.1	.7419	7467	7514	7561	7608	7655	7701	7747	7793	7839	5	9	14	19	23	28	33	37	42
2.2	.7885	7930	7975	8020	8065	8109	8154	8198	8242	8286	4	9	13	18	22	27	31	36	40
2.3	.8329	8372	8416	8459	8502	8544	8587	8629	8671	8713	4	9	13	17	21	26	30	34	38
2.4	.8755	8796	8838	8879	8920	8961	9002	9042	9083	9123	4	8	12	16	20	24	29	33	37
2.5	.9163	9203	9243	9282	9322	9361	9400	9439	9478	9517	4	8	12	16	20	24	27	31	35
2.6	.9555	9594	9632	9670	9708	9746	9783	9821	9858	9895	4	8	11	15	19	23	26	30	34
2.7	.9933	9969	1.0006	0043	0080	0116	0152	0188	0225	0260	4	7	11	15	18	22	25	29	33
2.8	1.0296	0332	0367	0403	0438	0473	0508	0543	0578	0613	4	7	11	14	18	21	25	28	32
2.9	1.0647	0682	0716	0750	0784	0818	0852	0886	0919	0953	3	7	10	14	17	20	24	27	31
3.0	1.0986	1019	1053	1086	1119	1151	1184	1217	1249	1282	3	7	10	13	16	20	23	26	30
3.1	1.1314	1346	1378	1410	1442	1474	1506	1537	1569	1600	3	6	10	13	16	19	22	25	29
3.2	1.1632	1663	1694	1725	1756	1787	1817	1848	1878	1909	3	6	9	12	15	18	22	25	28
3.3	1.1939	1969	1.2000	2030	2060	2090	2119	2149	2179	2208	3	6	9	12	15	18	21	24	27
3.4	1.2238	2267	2296	2326	2355	2384	2413	2442	2470	2499	3	6	9	12	15	17	20	23	26
3.5	1.2528	2556	2585	2613	2641	2669	2698	2726	2754	2782	3	6	8	11	14	17	20	23	25
3.6	1.2809	2837	2865	2892	2920	2947	2975	3002	3029	3056	3	5	8	11	14	16	19	22	25
3.7	1.3083	3110	3137	3164	3191	3218	3244	3271	3297	3324	3	5	8	11	13	16	19	21	24
3.8	1.3350	3376	3403	3429	3455	3481	3507	3533	3558	3584	3	5	8	10	13	16	18	21	23
3.9	1.3610	3635	3661	3686	3712	3737	3762	3788	3813	3838	3	5	8	10	13	15	18	20	23
4.0	1.3863	3888	3913	3938	3962	3987	4012	4036	4061	4085	2	5	7	10	12	15	17	20	22
4.1	1.4110	4134	4159	4183	4207	4231	4255	4279	4303	4327	2	5	7	10	12	14	17	19	22
4.2	1.4351	4375	4398	4422	4446	4469	4493	4516	4540	4563	2	5	7	9	12	14	16	19	21
4.3	1.4586	4609	4633	4656	4679	4702	4725	4748	4770	4793	2	5	7	9	12	14	16	18	21
4.4	1.4816	4839	4861	4884	4907	4929	4951	4974	4996	5019	2	5	7	9	11	14	16	18	20
4.5	1.5041	5063	5085	5107	5129	5151	5173	5195	5217	5239	2	4	7	9	11	13	15	18	20
4.6	1.5261	5282	5304	5326	5347	5369	5390	5412	5433	5454	2	4	6	9	11	13	15	17	19
4.7	1.5476	5497	5518	5539	5560	5581	5602	5623	5644	5665	2	4	6	8	11	13	15	17	19
4.8	1.5686	5707	5728	5748	5769	5790	5810	5831	5851	5872	2	4	6	8	10	12	14	16	19
4.9	1.5892	5913	5933	5953	5974	5994	6014	6034	6054	6074	2	4	6	8	10	12	14	16	18
5.0	1.6094	6114	6134	6154	6174	6194	6214	6233	6253	6273	2	4	6	8	10	12	14	16	18
5.1	1.6292	6312	6332	6351	6371	6390	6409	6429	6448	6467	2	4	6	8	10	12	14	16	18
5.2	1.6487	6506	6525	6544	6563	6582	6601	6620	6639	6658	2	4	6	8	10	11	13	15	17
5.3	1.6677	6696	6715	6734	6752	6771	6790	6808	6827	6845	2	4	6	7	9	11	13	15	17
5.4	1.6864	6882	6901	6919	6938	6956	6974	6993	7011	7029	2	4	5	7	9	11	13	15	17

Hyperbolic or Naperian logarithms of 10^{+n}

n	1	2	3	4	5	6	7	8	9
$\log_e 10^n$	2.3026	4.6052	6.9078	9.2103	11.5129	13.8155	16.1181	18.4207	20.7233

	0	1	2	3	4	5	6	7	8	9	Mean Differences								
											1	2	3	4	5	6	7	8	9
5.5	1.7047	7066	7084	7102	7120	7138	7156	7174	7192	7210	2	4	5	7	9	11	13	14	16
5.6	1.7228	7246	7263	7281	7299	7317	7334	7352	7370	7387	2	4	5	7	9	11	12	14	16
5.7	1.7405	7422	7440	7457	7475	7492	7509	7527	7544	7561	2	3	5	7	9	10	12	14	16
5.8	1.7579	7596	7613	7630	7647	7664	7681	7699	7716	7733	2	3	5	7	9	10	12	14	15
5.9	1.7750	7766	7783	7800	7817	7834	7851	7867	7884	7901	2	3	5	7	8	10	12	13	15
6.0	1.7918	7934	7951	7967	7984	8001	8017	8034	8050	8066	2	3	5	7	8	10	12	13	15
6.1	1.8083	8099	8116	8132	8148	8165	8181	8197	8213	8229	2	3	5	6	8	10	11	13	15
6.2	1.8245	8262	8278	8294	8310	8326	8342	8358	8374	8390	2	3	5	6	8	10	11	13	14
6.3	1.8405	8421	8437	8453	8469	8485	8500	8516	8532	8547	2	3	5	6	8	9	11	13	14
6.4	1.8563	8579	8594	8610	8625	8641	8656	8672	8687	8703	2	3	5	6	8	9	11	12	14
6.5	1.8718	8733	8749	8764	8779	8795	8810	8825	8840	8856	2	3	5	6	8	9	11	12	14
6.6	1.8871	8886	8901	8916	8931	8946	8961	8976	8991	9006	2	3	5	6	8	9	11	12	14
6.7	1.9021	9036	9051	9066	9081	9095	9110	9125	9140	9155	1	3	4	6	7	9	10	12	13
6.8	1.9169	9184	9199	9213	9228	9242	9257	9272	9286	9301	1	3	4	6	7	9	10	12	13
6.9	1.9315	9330	9344	9359	9373	9387	9402	9416	9430	9445	1	3	4	6	7	9	10	12	13
7.0	1.9459	9473	9488	9502	9516	9530	9544	9559	9573	9587	1	3	4	6	7	9	10	11	13
7.1	1.9601	9615	9629	9643	9657	9671	9685	9699	9713	9727	1	3	4	6	7	8	10	11	13
7.2	1.9741	9755	9769	9782	9796	9810	9824	9838	9851	9865	1	3	4	6	7	8	10	11	12
7.3	1.9879	9892	9906	9920	9933	9947	9961	9974	9988	2.0001	1	3	4	5	7	8	10	11	12
7.4	2.0015	0028	0042	0055	0069	0082	0096	0109	0122	0136	1	3	4	5	7	8	9	11	12
7.5	2.0149	0162	0176	0180	0202	0215	0229	0242	0255	0268	1	3	4	5	7	8	9	11	12
7.6	2.0281	0259	0308	0321	0334	0347	0360	0373	0386	0399	1	3	4	5	7	8	9	10	12
7.7	2.0412	0425	0438	0451	0464	0477	0490	0503	0516	0528	1	3	4	5	6	8	9	10	12
7.8	2.0541	0554	0567	0580	0592	0605	0618	0631	0643	0656	1	3	4	5	6	8	9	10	11
7.9	2.0669	0681	0694	0707	0719	0732	0744	0757	0769	0782	1	3	4	5	6	8	9	10	11
8.0	2.0794	0807	0819	0832	0844	0857	0869	0882	0894	0906	1	3	4	5	6	7	9	10	11
8.1	2.0919	0931	0943	0956	0968	0980	0992	1005	1017	1029	1	2	4	5	6	7	9	10	11
8.2	2.1041	1054	1066	1078	1090	1102	1114	1126	1138	1150	1	2	4	5	6	7	9	10	11
8.3	2.1163	1175	1187	1199	1211	1223	1235	1247	1258	1270	1	2	4	5	6	7	8	10	11
8.4	2.1282	1294	1306	1318	1330	1342	1353	1365	1377	1389	1	2	4	5	6	7	8	9	11
8.5	2.1401	1412	1424	1436	1448	1459	1471	1483	1494	1506	1	2	4	5	6	7	8	9	11
8.6	2.1518	1529	1541	1552	1564	1576	1587	1599	1610	1622	1	2	3	5	6	7	8	9	10
8.7	2.1633	1645	1656	1668	1679	1691	1702	1713	1725	1736	1	2	3	5	6	7	8	9	10
8.8	2.1748	1759	1770	1782	1793	1804	1815	1827	1838	1849	1	2	3	5	6	7	8	9	10
8.9	2.1861	1872	1883	1894	1905	1917	1928	1939	1950	1961	1	2	3	4	6	7	8	9	10
9.0	2.1972	1983	1994	2006	2017	2028	2039	2050	2061	2072	1	2	3	4	6	7	8	9	10
9.1	2.2083	2094	2105	2116	2127	2138	2148	2159	2170	2181	1	2	3	4	5	7	8	9	10
9.2	2.2192	2203	2214	2225	2235	2246	2257	2268	2279	2289	1	2	3	4	5	6	8	9	10
9.3	2.2300	2311	2322	2332	2343	2354	2364	2375	2386	2396	1	2	3	4	5	6	7	9	10
9.4	2.2407	2418	2428	2439	2450	2460	2471	2481	2492	2502	1	2	3	4	5	6	7	8	10
9.5	2.2513	2523	2534	2544	2555	2565	2576	2586	2597	2607	1	2	3	4	5	6	7	8	9
9.6	2.2618	2628	2638	2649	2659	2670	2680	2690	2701	2711	1	2	3	4	5	6	7	8	9
9.7	2.2721	2732	2742	2752	2762	2773	2783	2793	2803	2814	1	2	3	4	5	6	7	8	9
9.8	2.2824	2834	2844	2854	2865	2875	2885	2895	2905	2915	1	2	3	4	5	6	7	8	9
9.9	2.2925	2935	2946	2956	2966	2976	2986	2996	3006	3016	1	2	3	4	5	6	7	8	9
10.0	2.3026																		

Hyperbolic or Naperian logarithms of 10^{-n}

n	1	2	3	4	5	6	7	8	9
$\log_e 10^{-n}$	$\bar{3}.6974$	$\bar{5}.3948$	$\bar{7}.0922$	$\overline{10}.7897$	$\overline{12}.4871$	$\overline{14}.1845$	$\overline{17}.8819$	$\overline{19}.5793$	$\overline{21}.2767$

mentary table of Naperian logarithms of 10^{-n} used. The Naperian logarithm of 0.07 is obtained as follows:

$$\begin{aligned}\ln 0.07 &= \ln (7 \times 10^{-2})\\ &= \ln 7 + \ln 10^{-2}\\ &= 1.945\,9 + \bar{5}.394\,8\\ &= \bar{3}.340\,7 = -3 + 0.340\,7 \text{ or } -2.659\,3\end{aligned}$$

Tables of antilogarithms are not normally given for Naperian logarithms. The antilogarithm of a number between 0 and 2.302 6 is obtained by finding the value of the logarithm within the table and the antilogarithm is then obtained from the corresponding row and columns. For example, when $\ln x$ = 1.234 6, the nearest value lower than the logarithm within the table is 1.232 6 and corresponds to the row 3.4 and column 3. The mean difference is 1.234 6 − 1.232 6, that is, 20 when these numbers are treated as integer values, and gives a mean-difference-column value of 7. Hence, when $\ln x$ = 1.234 6, then x = 3.437.

For logarithms larger than 2.302 6, the value of the logarithm is expressed in the form

$A + B$ = the value of the logarithm,

where A is the nearest *lower* number to $A + B$ in the supplementary table of Naperian logarithms of 10^{+n} and B is the number required to make the equation correct. Thus, the value of x, when $\ln x = 6.774\,3$, is obtained as follows:

$$6.774\,3 = 4.605\,2 + (6.774\,3 - 4.605\,2)$$

The $\ln 10^{+n}$ value (4.605 2) — The value to make the equation correct (6.774 3 − 4.605 2)

$$\begin{aligned}\text{Thus, } 6.774\,3 &= 4.605\,2 + 2.169\,1\\ \text{i.e.}\quad \ln x &= 4.605\,2 + 2.169\,1\end{aligned}$$

(If, when determining the antilogarithm of a number in this way, the B number, i.e. the value to make the equation correct, is not between 0 and 2.302 6, then the wrong supplementary-table value of A has been chosen. A check should be made to make sure that the value of A is the nearest lower number to $A + B$.)

When determining an antilogarithm, addition becomes multiplication.

$$\begin{aligned}\text{Hence, } x &= 10^2 \times 8.750\\ &= 875\end{aligned}$$

When the logarithm has a negative characteristic, it is expressed in the form: $A + B$ = the value of the logarithm, where A is the nearest *more negative* number to $A + B$ in the supplementary tables of $\ln 10^{-n}$ and B is the number required to make the equation correct. The value of x, when $\ln x$ = $\bar{4}.567\,1$, is obtained as follows:

$$\bar{4}.567\,1 = \bar{5}.394\,8 + (\bar{4}.567\,1 - \bar{5}.394\,8)$$

The ln 10^{-n} value	The value to make the equation correct

Thus, $\bar{4}.567\,1 = \bar{5}.394\,8 + 1.172\,3$

i.e. $\ln x = \bar{5}.394\,8 + 1.172\,3$

(If, when determining the antilogarithm of a number in this way, the *B* number, i.e. the value to make the equation correct, is not between 0 and 2.302 6, then the wrong supplementary-table value of *A* has been chosen. A check should be made to make sure that the value of *A* is the nearest more negative number to *A* + *B*.)

When determining an antilogarithm, addition becomes multiplication.

Hence, $x = 10^{-2} \times 3.229$
$= 0.032\,29$

The method of determining the value of e^x using tables of exponential functions is shown in Section 1. However, only certain discrete values of e^x can be found by this method. Using tables of Naperian logarithms, values of e^x for all values of x can be determined. The way of doing this is to express e^x in logarithmic form. Let e^x be equal to y, then by taking Naperian logarithms of each side of the equation

$$\ln e^x = \ln y$$

or $$x \ln e = \ln y$$

But ln e is $\log_e e$ or unity, since if $\log_e e = x$, then $e = e^x$ or $x = 1$, by the definition of a logarithm.

Hence $x = \ln y$

Thus, to determine the value of y when $y = e^{1.254\,6}$, take Naperian logarithms of each side.

Hence, $\ln y = \ln e^{1.234\,6}$

or, $\ln y = 1.234\,6 \ln e = 1.234\,6$ since $\ln e = 1$.

Determining the antilogarithm gives

$$y = 3.437, \text{ i.e. } e^{1.234\,6} = 3.437$$

Worked problems on Naperian logarithms

Problem 1. Determine the value of ln x when x is: (a) 2.019; (b) 371.4; and (c) 0.000 837.

(a) Since 2.019 lies between 0 and 2.302 6, its value can be obtained directly from tables of Naperian logarithms. The row is 2.0 and column 1, giving a matrix number of 0.698 1. The mean-difference column headed 9 gives a mean difference of 44 and adding this to the matrix number, treating both numbers as integers, gives 6 981 + 44 or 7 025. Thus **when x is 2.019, ln x is 0.702 5.**

(b) Expressing 371.4 in standard form gives:

$$371.4 = 3.714 \times 10^2$$

Hence, $\ln 371.4 = \ln (3.714 \times 10^2)$

$$= \ln 3.714 + \ln 10^2$$

Using the main tables to find the value of ln 3.3714 and the supplementary table of $\ln 10^{+n}$ to find $\ln 10^2$ gives:

$$\ln 371.4 = 1.312\,1 + 4.605\,2$$
$$= 5.917\,3$$

i.e. **when x is 371.4, ln x is 5.917 3.**

(c) Expressing 0.000 837 in standard form gives:

$$0.000\,837 = 8.37 \times 10^{-4}$$

Hence, $\ln 0.000\,837 = \ln (8.37 \times 10^{-4})$

$$= \ln 8.37 + \ln 10^{-4}$$

The main table is used to evaluate ln 8.37 and the supplementary table of $\ln 10^{-n}$ to find the value of 10^{-4}. This gives:

$$\ln 0.000\,837 = 2.124\,7 + \overline{10}.789\,7$$
$$= \bar{8}.914\,4$$

i.e. **when x = 0.000 837, In x = $\bar{8}$.914 4 or −7.085 6.**

Problem 2. Determine the value of x when ln x is (a) 2.172 0; (b) 9.571 3; and (c) $\bar{7}$.371 4.

(a) Since ln x lies between 0 and 2.302 6, its antilogarithm is found by using the main tables of Naperian logarithms only. The row 8.7 and column 7 give a matrix number of 2.171 3. The mean difference is 21 720 − 21 713, treating the matrix number as integers, giving a mean difference of 7. Thus the fourth significant figure in the result is 6, the number at the top of the appropriate mean-difference column.

Thus, **when ln x = 2.172 0, x = 8.776.**

(b) Expressing 9.571 3 in the form: 9.571 3 = A + B, where A is the nearest smaller number to 9.571 3 in the supplementary table of $\ln 10^{+n}$ and B is the number needed to make the equation correct, gives:

$$9.571\,3 = 9.210\,3 + (9.571\,3 - 9.210\,3)$$

i.e. $\quad \ln x = 9.571\,3 = 9.210\,3 + 0.361\,0$

Finding the antilogarithms and changing the addition to multiplication gives

$$x = 10^4 \times 1.435$$
$$= 14\,350$$

i.e. **when ln x = 9.571 3, x = 14 350.**

(c) As for part (b), expressing $\bar{7}$.371 4 in the form: $\bar{7}$.371 4 = A + B, where A is the nearest more negative number in the supplementary table of $\ln 10^{-n}$ and B is the number to make the equation correct, gives:

$$\bar{7}.371\,4 = \bar{7}.092\,2 + (\bar{7}.371\,4 - \bar{7}.092\,2)$$

(At first sight, $\bar{7}.092\,2$ does not appear to be more negative than $\bar{7}.371\,4$. However, since $\bar{7}.092\,2 = -7 + 0.092\,2 = -6.901\,8$ and $\bar{7}.371\,4 = -7 + 0.371\,4 = -6.628\,6$, it can be seen that $\bar{7}.092\,2$ is more negative than $\bar{7}.371\,4$.)
Then, $\ln x = \bar{7}.371\,4 = \bar{7}.092\,2 + 0.279\,2$
and taking the antilogarithms of each side gives

$$x = 10^{-3} \times 1.322$$

i.e. **when $\ln x = \bar{7}.371\,4$, $x = 0.001\,322$.**

Problem 3. Determine the value of y, when: (a) $y = e^{3.173}$; (b) $y = -7.6\,e^{0.017\,3}$; and (c) $y = \frac{7}{4}e^{-8.721}$.

(a) $y = e^{3.173}$, and taking Naperian logarithms of each side of the equation gives

$$\ln y = \ln e^{3.173}$$
$$= 3.173 \ln e$$
$$= 3.173, \text{ since } \ln e = 1$$

Then, $\ln y = 2.302\,6 + (3.173 - 2.302\,6)$
i.e. $\ln y = 2.302\,6 + 0.870\,4$
Determining the antilogarithms of each side of the equation gives

$$y = 10 \times 2.388$$
$$\mathbf{y = 23.88}$$

(b) Let, u, say, $= e^{0.017\,3}$. Then $\ln u = 0.017\,3$.
Since $0.017\,3$ is a number between 0 and $2.302\,6$, its antilogarithm is found directly from the tables of Naperian logarithms. Determining the antilogarithms of each side of the equation gives

$$u = 1.018$$

Hence $y = -7.6 \times 1.018$
$= \mathbf{-7.74}$

(c) As for part (b), let $u = e^{-8.721}$

$$\ln u = -8.721$$

To determine the antilogarithm of a negative number, it is expressed as having a negative characteristic and a positive mantissa.

Thus $-8.721 = \bar{9}.279\,0$

Now, $\bar{9}.279\,0 = \overline{10}.789\,7 + (\bar{9}.279\,0 - \overline{10}.789\,7)$

i.e. $\ln u = \overline{10}.789\,7 + 0.489\,3$

i.e. $u = 10^{-4} \times 1.631$,

and $y = \dfrac{7}{4} \times 10^{-4} \times 1.631$

$= 2.854 \times 10^{-4}$

Further problems on Naperian logarithms may be found in Section 8 (Problems 16–30).

4. Solving problems involving the natural laws of growth and decay

The previous three sections of this chapter have introduced the natural laws of growth and decay, exponential functions and Naperian logarithms. A knowledge of these three topics enables problems on the natural laws of growth and decay to be solved.

Worked problems on the natural laws of growth and decay

Problem 1. A belt is in contact with a pulley for a sector of $\theta = 1.073$ radians and the coefficient of friction between these two surfaces is $\mu = 0.27$. Determine the tension on the taut side of the belt, T newtons, when the tension on the slack side is given by $T_0 = 23.8$ newtons, given that these quantities are related by the law

$$T = T_0 \, e^{\mu\theta}$$

If we require the transmitted force $(T - T_0)$ to be increased to 25.0 newtons, assuming that T_0 remains at 23.8 newtons and θ at 1.073 radians, determine the coefficient of friction.

$$T = T_0 \, e^{\mu\theta} = 23.8 \, e^{(0.27 \times 1.073)}$$
$$= 23.8 \, e^{0.290}$$

Let, u, say $= e^{0.290}$, then $\ln u = 0.290$

and using tables of Naperian logarithms to find the antilogarithm, gives

$$u = 1.336$$

Hence $\quad T = 23.8 \times 1.336$

$$= \mathbf{31.80 \text{ newtons}}$$

For the transmitted force to be 25 newtons, T becomes $23.8 + 25$, or 48.8 newtons.

Then $48.8 = 23.8 \, e^{\mu \times 1.073}$

$$\frac{48.8}{23.8} = e^{1.073\mu}$$

or $\quad 2.050 = e^{1.073\mu}$

Taking Naperian logarithms of each side of this equation gives

$$\ln 2.050 = 1.073\mu$$
$$0.7178 = 1.073\mu$$

i.e. **the coefficient of friction, $\mu = 0.669\,0$**

Problem 2. The instantaneous current, i amperes at time t seconds is given by:

$$i = 6.0 \, e^{-\frac{t}{CR}},$$

when a capacitor is being charged. The capacitance C is 8.3×10^{-6} farads and the resistance R has a value of 0.24 megohms. Determine the instantaneous current when t is 3.0 seconds. Also determine the time for the instantaneous current to fall to 4.2 amperes.

$$i = 6.0\,e^{\left(-\frac{3}{8.3 \times 10^{-6} \times 0.24 \times 10^{6}}\right)}$$

$$= 6.0\,e^{\left(-\frac{3}{8.3 \times 0.24}\right)}$$

$$= 6.0\,e^{-1.506\,0}$$

Letting u, say, equal $e^{-1.506\,0}$ and taking Naperian logarithms of each side of this equation, gives

$$\ln u = -1.506\,0$$

or $\ln u = \bar{2}.494\,0$

Taking antilogarithms gives, $u = 0.221\,8$

Hence $i = 6 \times 0.221\,8$

$= 1.33$ amperes

That is, **the current flowing when t is 3.0 seconds is 1.33 amperes.**

It is usually easier to transpose and make t the subject of the formula before evaluation. Thus, since $i = 6.0\,e^{-\frac{t}{CR}}$:

$$\frac{i}{6.0} = e^{-\frac{t}{CR}}$$

$$e^{\frac{t}{CR}} = \frac{6.0}{i}$$

$$\ln e^{\frac{t}{CR}} = \ln\left(\frac{6.0}{i}\right)$$

$$\frac{t}{CR} = \ln\left(\frac{6.0}{i}\right)$$

or $$t = CR \ln\left(\frac{6.0}{i}\right)$$

Substituting the values of C, R and i gives

$$t = 8.3 \times 10^{-6} \times 0.24 \times 10^{6} \ln \frac{6.0}{4.2}$$

$$= 1.992 \ln 1.428\,6$$

$$= 1.992 \times 0.356\,7$$

$= 0.710\,5$ seconds

i.e. **the time for the current to fall to 4.2 amperes is 0.711 seconds.**

Problem 3. The temperature θ_2 degress Celsius of a winding which is being heated electrically, at time t seconds is given by

$$\theta_2 = \theta_1 (1 - e^{-\frac{t}{T}}),$$

where θ_1 is the temperature at time $t = 0$ seconds and T seconds is a constant. Calculate: (a) θ_1 in degrees Celsius when θ_2 is 45°C, t is 28 s and T is 73 s; and (b) the time t seconds for θ_2 to fall to half the value of θ_1.

(a) Transposing the formula to make θ_1 the subject gives

$$\theta_1 = \frac{\theta_2}{1 - e^{-\frac{t}{T}}}$$

 Substituting the values of θ_2, t and T gives

$$\theta_1 = \frac{45}{1 - e^{-\frac{28}{73}}} = \frac{45}{1 - e^{-0.3836}}$$

i.e. $$\theta_1 = \frac{45}{1 - 0.6814} = \frac{45}{0.3186}$$

$$\theta_1 = 141.24°\text{C}$$

That is, **the initial temperature is 141°C.**

(b) Transposing to make t the subject of the formula:

$$\frac{\theta_2}{\theta_1} = 1 - e^{-\frac{t}{T}}$$

$$\frac{\theta_2}{\theta_1} - 1 = -e^{-\frac{t}{T}}$$

or $$e^{-\frac{t}{T}} = 1 - \frac{\theta_2}{\theta_1}$$

Hence $$-\frac{t}{T} = \ln\left(1 - \frac{\theta_2}{\theta_1}\right)$$

i.e. $$t = -T \ln\left(1 - \frac{\theta_2}{\theta_1}\right)$$

Substituting the values of T, θ_1 and θ_2 gives

$$t = -73 \ln\left(1 - \frac{1}{2}\right)$$

$$= -73 \ln 0.5$$

$$= -73 \times (-0.6931)$$

$$= 50.6 \text{ seconds}$$

i.e. **the time for the temperature to fall to half its original value is 50.6 seconds.**

Further problems involving the natural laws of growth and decay may be found in Section 8 (Problems 31–40).

5. Reducing equations of the form $y = ax^n$ to linear form using log-log graph paper

It has been shown in earlier work that equations which are not linear, when values of y are plotted against corresponding values of x, have to be altered in some way to get them into the general straight-line form of $Y = mX + c$. In this section, the methods of reducing to linear form three types of equations, namely, $y = ax^n$, $y = ab^x$ and $y = ae^{kx}$, by using logarithmic graph paper are dealt with.

Equations of the form $y = ax^n$ are reduced to linear form by taking logarithms to a base of 10 of each side of the equation.

Thus, when $y = ax^n$

$$\lg y = \lg (ax^n)$$
$$= \lg a + \lg x^n$$

i.e., $\lg y = n \lg x + \lg a$

Comparing this with the straight-line equation $Y = mX + c$ gives:

$$\boxed{\lg y} = n \boxed{\lg x} + \lg a$$
$$Y = m \; X + c$$

Thus by plotting $\lg y$ vertically and $\lg x$ horizontally, the equation $y = ax^n$ is reduced to linear form. Special graph paper called **log-log** graph paper can be used for this purpose, on which the distances along the x- and y-axes are proportional to the logarithms of numbers (see Fig. 5). Using this graph paper, values of x and y can be plotted directly, without first having to determine their logarithms. The values of the y-axis intercept, c, can be obtained directly from the graph without having to use antilogarithms. However, determining the slope of the graphs and the constant n is not so straightforward. Hence laws of the form $y = ax^n$ can generally be more readily verified than by using normal graph paper, and there is less likelihood of mistakes being made.

Examination of the scale markings on either of the axes of a sheet of log-log graph paper shows that the scales do not have equal divisions. They are marked from 1 to 9 and this pattern of marking can be repeated several times. The number of times the pattern of markings is repeated on an axis signifies the number of **cycles**. When the y-axis has, say, two sets of values from 1 to 9 and the x-axis has 3 sets of values from 1 to 9, then this particular log-log graph paper is called 'log 2 cycle × 3 cycle'. Many different arrangements are produced, but those in most common use vary from 'log 1 cycle × 1 cycle' through the various permutations to 'log 5 cycle × 5 cycle'.

One cycle of log-log graph paper is used to signify the values of x or y between 10^n and 10^{n+1}, where n is an integer.

Since $\lg 1 - \lg 0.1 = 0 - \bar{1} = 0 - (-1) = 1$,
$\lg 10 - \lg 1 = 1 - 0 = 1$,
$\lg 100 - \lg 10 = 2 - 1 = 1$ and so on, the distance each cycle occupies on a logarithmic scale is the same. Thus, one cycle can be used to signify values from 0.1 to 1, or from 1 to 10, or from 10 to 100, and so on.

To depict a set of numbers from 0.6 to 174, say, on one axis of log-log graph paper, 4 cycles will be required (0.1 to 1, 1 to 10, 10 to 100 and 100 to 1000) and the start of each cycle is marked 0.1, 1, 10 and 100, or 10^{-1}, 10^0, 10^1 and 10^2. The divisions within a cycle are proportional to the logarithms of the numbers 1 to 10 and the distance from, say, the 1 to 2 marks is 0.301 0 (i.e. lg 2.000 0 − lg 1.000 0), of the total distance in the cycle. Similarly, the distance from the 9 to 10 marks is 0.457 6 of the total distance in the cycle, i.e. the distance between marks decreases logarithmically within the cycle.

The worked problems at the end of this section show the method of verifying laws of the form $y = ax^n$, and determining the values of a and n.

 When the value of y at $x = 1$ (since $\lg 1 = 0.000\,0$) can be read directly from the graph, i.e. the point at which the graph cuts the ordinate $x = 1$, the y-value gives the constant a in the equation $y = ax^n$. When the selection of scales is such that $x = 1$ does not appear on the graph, values of x and y from the graph can be used to determine a from the equation (see Problem 2). Since distances are directly proportional to the logarithms of the quantities, the value of n is found by measuring directly (say in centimetres) the distances of the slope of the graph (see Problem 1). An alternative method of determining the constants a and n is by selecting any two points on the graph, reading the values of x and y at these points and forming simultaneous equations to solve for a and n (see Problem 1).

Worked problems using log-log graph paper to reduce equations of the form $y = ax^n$ to linear form

---

Problem 1. Two quantities are believed to be related by a law of the form $y = ax^b$, where a and b are constants. Measured values of x and y are taken and the results obtained are:

y	0.25	0.59	1.67	4.62	12.70	42.60
x	0.34	0.52	0.87	1.45	2.40	4.40

Prove that the law relating these quantities is as stated and determine the approximate values of a and b.

Taking logarithms to a base of 10 of each side of the equation gives:

$$\lg y = \lg (ax^b)$$
$$= \lg a + \lg x^b$$
$$= \lg a + b \lg x$$

i.e $\lg y = b \lg x + \lg a$

Comparing this equation with the straight-line equation, $Y = mX + c$ gives:

$$\boxed{\lg y} = b\ \boxed{\lg x} + \lg a$$
$$\boxed{Y} = m\ \boxed{X} + c$$

This shows that it is necessary to plot $\lg y$ against $\lg x$ to verify the law. By using log-log graph paper, the values of y and x can be plotted directly without first having to determine their logarithms. The values of y range from 0.25 to 42.60 and 3 cycles are required for the y-axis (0.1 to 1.0, 1.0 to 10.0 and 10.0 to 100.0). The values of x range from 0.34 to 4.40 and two cycles are therefore required (0.1 to 1.0 and 1.0 to 10.0). Hence 'log 3 cycle × 2 cycle' graph paper is selected. The axes are marked as shown in Fig. 5, the values of x and y plotted, and a straight line is drawn through the points. Since the distance between the logarithmic markings on log-log graph paper is usually small, no attempt is made to subdivide the distance contained within the small squares logarithmically when plotting points which fall within squares. Any two points on the line are selected, say A and B, and by measuring the

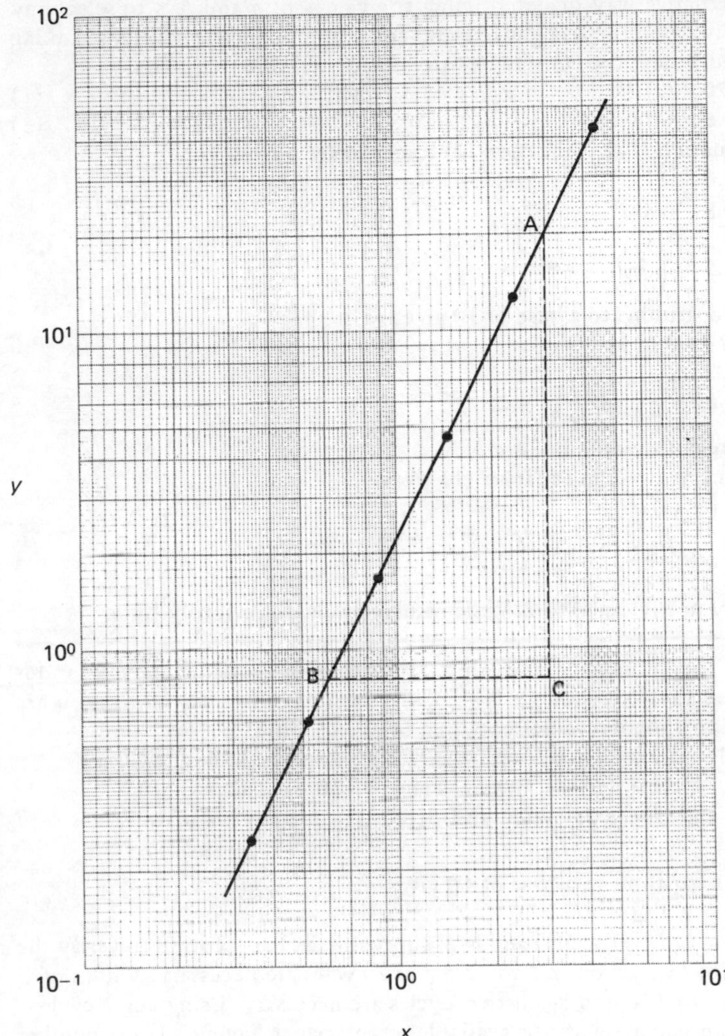

Figure 5 Graph to verify a law of the form $y = ax^b$ (Problem 1)

distances AC and BC (say in centimetres), the slope of the line, i.e. $\frac{\text{AC}}{\text{BC}}$ is obtained. This is found to be 2, i.e. the value of b (corresponding to m in the straight-line equation) is 2. The graph crosses the ordinate $x = 1$ at $y = 2.2$, hence the value of a is 2.2 (lg a corresponds to c in the straight-line equation, but since the distances along the axes are proportional to the logarithms of the values shown, there is no need to find the antilogarithm). Thus the values of a and b are 2.2 and 2 respectively, and since the points lie on a straight line, the law is verified.

An alternative way of determining the values of a and b is to select any two points lying on the straight line and form simultaneous equations. Taking points A and B, say, the coordinates are (3, 20) and (0.6, 0.8). Hence

$$20 = (a)\,3^b \qquad \ldots (1)$$

$$0.8 = (a)\,0.6^b \qquad \ldots (2)$$

Dividing equation (1) by equation (2) to eliminate a gives

$$\frac{2.0}{0.8} = \frac{3^b}{0.6^b}$$

$$\text{Thus, } 25 = \left(\frac{3}{0.6}\right)^b = 5^b$$

and taking logarithms to a base of 10 gives

$$\lg 25 = b \lg 5$$

$$b = \frac{\lg 25}{\lg 5} = \frac{2 \lg 5}{\lg 5} = 2,$$

the result previously obtained for b.

Substituting $b = 2$ in equation (1) gives

$$20 = (a)3^2$$

$$a = \frac{20}{9},$$

which is very nearly equal to 2.2, the result obtained previously for a.

Problem 2. The luminosity of a lamp, I, varies with applied voltage V, and it is anticipated that it follows a law of the form $I = kV^n$, where k and n are constants. Experimental values obtained for I and V are:

I candela	2.5	4.9	8.1	12.1	16.9	22.5	44.0
V volts	50	70	90	110	130	150	210

Verify that the law is as stated and determine the approximate values of k and n.

Taking logarithms to a base of 10 gives

$$\lg I = n \lg V + \lg k$$

Thus values of $\lg I$ (vertically) and $\lg V$ (horizontally) are plotted to verify the law. Values of I vary from 2.5 to 22.5 and 2 cycles are necessary. Values of V vary from 50 to 150 and again two cycles are necessary. Using 'log 2 cycle × 2 cycle' graph paper (or if not available, graph paper having a larger number of cycles per axis), the values are plotted and a straight line drawn through the points. This is shown in Fig. 6.

The points plotted do lie on a straight line, hence the law $I = kV^n$ does relate to the given values. The slope of the graph, n, can be found by selecting any two points on the graph, say points A and B shown in Fig. 6, and by measuring the distances AC and BC. Then

$$\text{slope of graph} = n = \frac{\text{distance AC}}{\text{distance BC}} = 2$$

The 'x-axis' of the graph starts at $I = 10$ and since the 'y-axis' intercept method can only be used when the graph contains the ordinate '$x = 1$', an equation is formed from which the value of k can be obtained. Selecting any

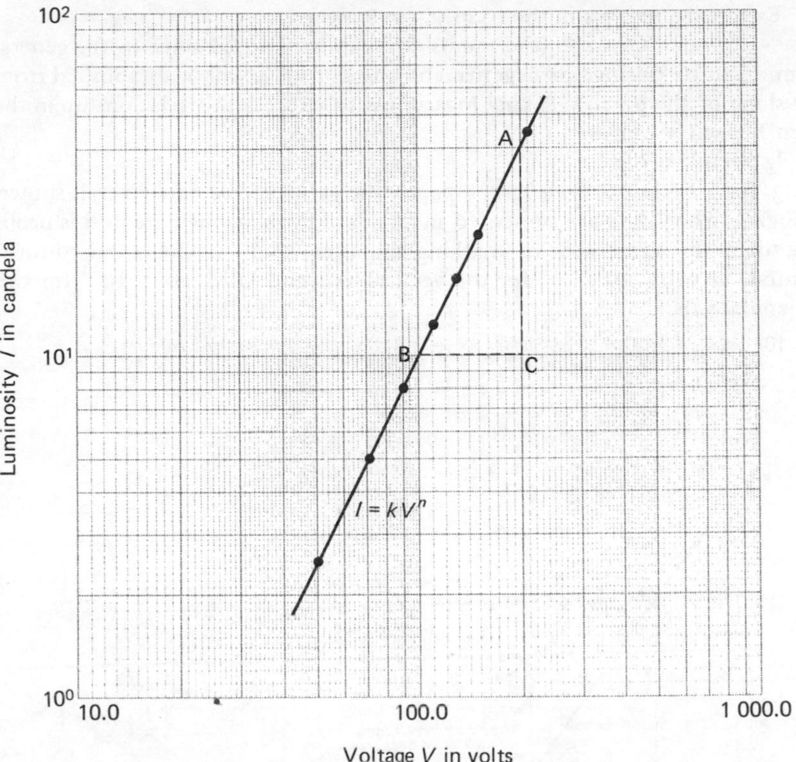

Figure 6 Variation of luminosity with applied voltage (Problem 2)

point on the graph, say, $V = 100$ at $I = 10$, and substituting these valucs in the equation $I = kV^n$ gives:

$$10 = (k)100^2 \text{ (since } n = 2)$$

$$\text{or} \quad k = \frac{10}{100^2} = \frac{1}{1\,000} = 0.001$$

Thus, **the values of k and n are 2 and 0.001 respectively and the law is $I = 0.001\ V^2$**. The values of k and n could have been determined by selecting any two points on the graph, obtaining the coordinates for these points, and forming simultaneous equations using $\lg I = n \lg V + \lg k$.

Problem 3. The pressure p and volume v of a gas are believed to be related by a law of the form $pv^n = c$, where c is a constant. Values of p and corresponding values of v obtained in a laboratory are:

p pascals	5.0×10^5	1.45×10^6	3.8×10^6	9.8×10^6	2.35×10^7
v cubic metres	2.35×10^{-2}	1.1×10^{-2}	5.5×10^{-3}	2.8×10^{-3}	1.5×10^{-3}

Verify that the law is of the form $pv^n = c$ and determine the approximate values of n and c.

Expressing $pv^n = c$ in the form $y = ax^n$ gives

$p = cv^n$ (the minus sign can be ignored since in the general form n can be positive or negative and the sign of n will be established from the slope of the graph). Taking logarithms to express this relationship in the form $Y = mX + c$ gives

$\lg p = n \lg v + \lg c$

Using 'log 3 cycle × 2 cycle' graph paper, the graph of the data given is shown in Fig. 7. When data are presented in standard form, as they are in this problem, then the ideal way of marking the axes is to use the standard-form method, i.e. 10^5, 10^6, 10^7 for the vertical axis and 10^{-3}, 10^{-2}, 10^{-1} for the horizontal axis.

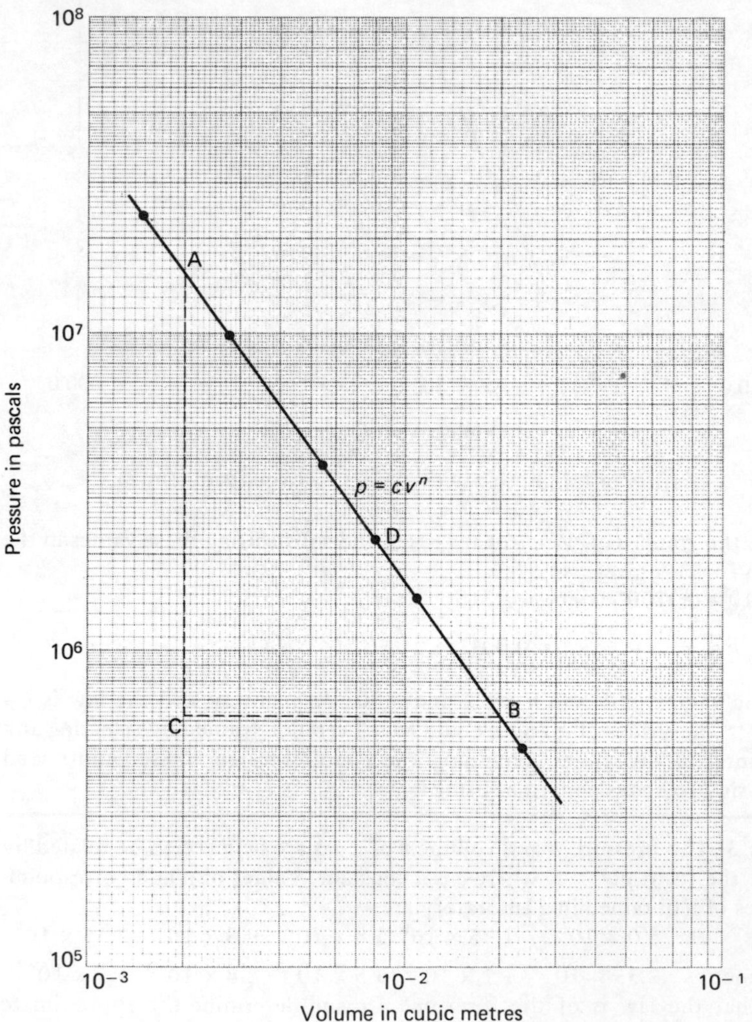

Figure 7 Variation of pressure with volume (Problem 3)

Since a straight line can be drawn through the points, the law is verified. Selecting any two points on the graph, say A and B, and finding the ratio of the distances $\frac{AC}{BC}$ gives a slope of 1.4. Since the values of p are decreasing as v increases, the slope of the graph is negative and hence $n = -1.4$. Selecting any point on the graph, say D, having coordinates $(8 \times 10^{-3}, 2.2 \times 10^6)$ and substituting these values in the equation $p = cv^n$ gives

$$2.2 \times 10^6 = c\,(8 \times 10^{-3})^{-1.4}$$
$$= c\,(0.008)^{-1.4}$$

or
$$c = \frac{2.2 \times 10^6}{(0.008)^{-1.4}}$$
$$= \frac{2.2 \times 10^6}{862.3}$$
$$= 2\,550, \text{ correct to 3 significant figures.}$$

Thus the law is $p = 2\,550\,v^{-1.4}$ or $pv^{1.4} = 2\,550$, i.e. **$n = 1.4$ and $c = 2\,550$.**

Further problems on reducing equations of the form y = axⁿ *to linear form may be found in Section 8 (Problems 41–46).*

6. Reducing equations of the form $y = ab^x$ to linear form using log-linear graph paper

Taking logarithms to a base of 10 of each side of the equation $y = ab^x$ gives

$$\lg y = \lg (ab^x)$$
$$= \lg a + \lg b^x$$
$$= x \lg b + \lg a, \text{ where } a \text{ and } b \text{ are constants.}$$

Comparing this equation with the straight-line equation $Y = mX + c$ gives:

$\lg y$	$= (\lg b)$	x	$+ \lg a$
Y	$= m$	X	$+ c$

This shows that by plotting $\lg y$ against x, the slope of the resultant straight-line graph is $\lg b$ and the y-axis intercept value is $\lg a$. In this case, graph paper which has a linear scale on one axis (x) and a logarithmic scale on the other axis ($\lg y$) can be used. This type of graph paper is called **log-linear** graph paper and is specified by the number of cycles on the logarithmic scale. For example, paper having three cycles on the logarithmic scale is called 'log 3 cycle × linear' graph paper (shown in Fig. 8). The method of determining the constants a and b in the equation $y = ab^n$ is shown in the worked problem following.

Worked problem using log-linear graph paper to reduce an equation of the form $y = ab^x$ to linear form

---

102 **Problem.** Quantities x and y are believed to be related by a law of the form $y = ab^x$. The values of x and corresponding values of y are:

x	0	0.5	1.0	1.5	2.0	2.5	3.0
y	1	3.2	10	31.6	100	316.2	1 000

Verify the law is as stated, find the approximate values of a and b, and comment on the significance of the graph drawn.

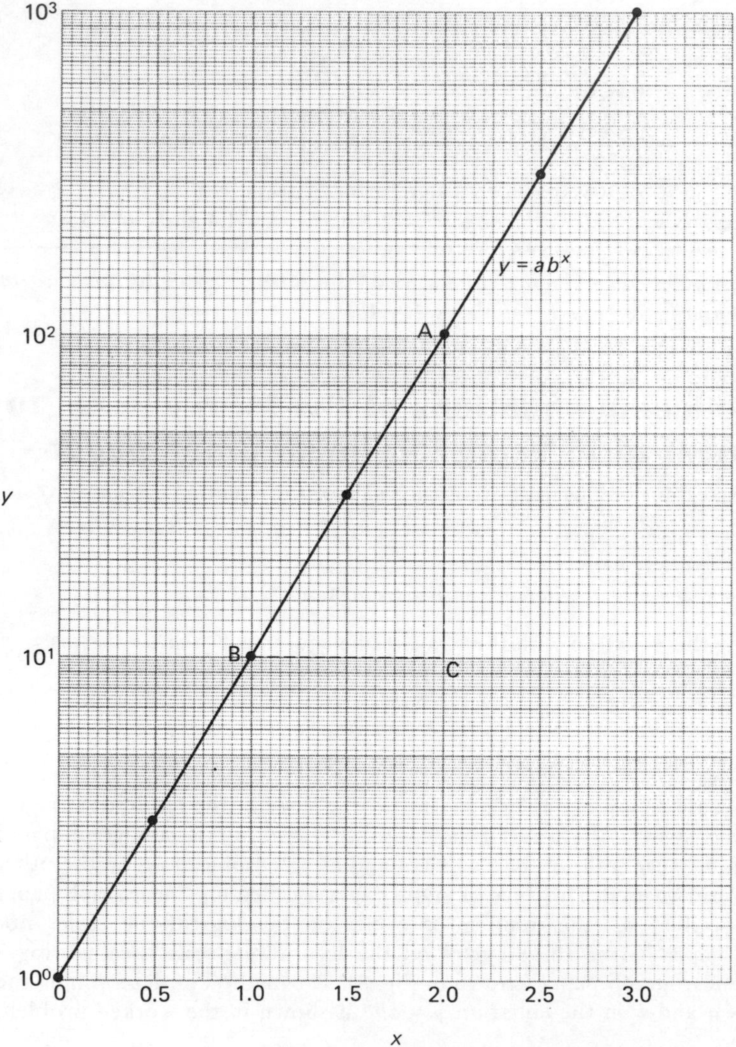

Figure 8 Graph to verify a law of the form $y = ab^x$

Since $y = ab^x$, then $\lg y = (\lg b)\,x + \lg a$. Comparing this equation with the straight-line equation $Y = mX + c$ gives

$$\boxed{\lg y} = \lg b \; \boxed{x} + \lg a$$
$$\boxed{Y} = m \; \boxed{X} + c$$

Thus $\lg y$ is plotted against x to verify that the law is as stated. Using log-linear graph paper, values of x are selected on the linear scale over a range 0 to 3. Values of y have a range from 1 to 1 000 and 3 cycles are needed to span this range. The graph is shown in Fig. 8.

A straight line can be drawn through the points so the law is verified. Since there is now a mixture of linear values on the x-axis and logarithmic values on the y-axis, direct measurement of the slope of the graph is not possible to determine $\lg b$. Selecting any two points on the graph, say A and B, having coordinates $(2, 10^2)$ and $(1, 10^1)$ gives

$$\text{slope} = \frac{AC}{BC} = \frac{\lg 10^2 - \lg 10^1}{2 - 1} = \frac{2-1}{2-1} = 1.$$

Hence, slope $= \lg b = 1$, giving $b = 10$.
Also, when $x = 0$, $y = 10^0 = 1$
i.e. $a = 1$.
That is, the constants a and b have values of 1 and 10 respectively.

The significance of the graph is that the relationship is $y = 10^x$ and by the definition of a logarithm, when $y = 10^x$, $x = \lg y$. Hence the graph may be used to determine the approximate value of any logarithm between $y = 1$ and $y = 1\ 000$. Also the approximate value of any antilogarithm between $x = 0$ and $x = 3$ can be found by finding the corresponding value of y. For example, when $y = 500$, from the graph, $x = 2.7$, i.e. the value of lg 500 is approximately 2.7.

By drawing a graph of $y = a^x$, where a is any positive number, the values of logarithms to a base of a can be obtained, and this is one of the principal uses of equations of the form $y = ab^x$.

Further problems on reducing equations of the form $y = ab^x$ *to linear form may be found in Section 8 (Problems 47 and 48).*

7. Reducing equations of the form $y = ae^{kx}$ to linear form using log-linear graph paper

Taking logarithms to a base of e of each side of the equation $y = ae^{kx}$ gives

$$\begin{aligned} \ln y &= \ln (ae^{kx}) \\ &= \ln a + \ln e^{kx} \\ &= \ln a + kx \ln e \end{aligned}$$

 However, by the basic definition of a logarithm, when

$$y = a^x \qquad \ldots (1)$$

then

$$x = \log_a y \qquad \ldots (2)$$

If $y = a$, then from equation (1), $a = a^x$ or $x = 1$. Also from equation (2), $\log_a a = 1$. It follows that $\log_e e = 1$ or $\ln e = 1$. Hence, $\ln y = kx + \ln a$, where a and k are constants. Comparing this equation with the straight-line equation $Y = mX + c$ gives:

$$\boxed{\ln y} = k \boxed{x} + \ln a$$
$$\boxed{Y} = m \boxed{X} + c$$

This shows that by plotting $\ln y$ against x we will obtain a straight-line graph (and verify a relationship of the form $y = ae^{kx}$ where it exists).

The same log-linear graph paper can be used as for logarithms to a base of 10. It is shown in Section 3 that

$$\ln x = 2.302\,6 \lg x$$
i.e. $\ln x = (\text{a constant})(\lg x)$

Thus, when using logarithmic graph paper to depict Naperian logarithms, the distances along an axis representing ln 100–ln 10, ln 10–ln 1, and so on are uniform, as they were for logarithms to a base of 10, and the distances within cycles alter logarithmically as they did for logarithms to a base of 10. Thus the effect of using Naperian logarithms instead of logarithms to a base of 10 is to introduce a scale factor, which is automatically allowed for in subsequent calculations and measurements, as shown in the worked problems. When log-linear graph paper is used, it is not necessary to determine the values of $\ln y$, since values of y are plotted directly on the logarithmic axis. The graphs in the worked problems following show that the slope of the graph, k, is given by $\frac{\ln y}{x}$ and is obtained by selecting two points on the graph and determining the values of y (and hence $\ln y$) from the vertical axis and the values of x from the horizontal axis. When the straight-line graph cuts the ordinate $x = 0$, the intercept value gives the constant a (as shown in Problem 1). However, when the range of values is such that the straight-line graph does not cut the ordinate $x = 0$ within the scales selected, a point is selected on the straight-line graph and values of x, y and k substituted in the equation $y = ae^{kx}$, to determine the value of a (as shown in Problem 2).

Worked problems to reduce equations of the form $y = ae^{kx}$ to linear form using log-linear graph paper.

Problem 1. It is believed that x and y are related by a law of the form $y = ae^{kx}$ where a and k are constants. Values of x and y are measured and the results are:

x	−0.9	0.25	0.9	2.1	2.8	3.7	4.8
y	2.5	6.0	10.0	25.0	42.5	85.0	198.0

Verify that the law stated does relate these quantities and determine the approximate values of a and x.

Taking Naperian logarithms of each side of the equation $y = ae^{kx}$ gives
$\ln y = kx + \ln a$

The values of y vary from 2.5 to 198, hence 'log 3 cycle × linear' graph paper is required. The graph is shown in Fig. 9.

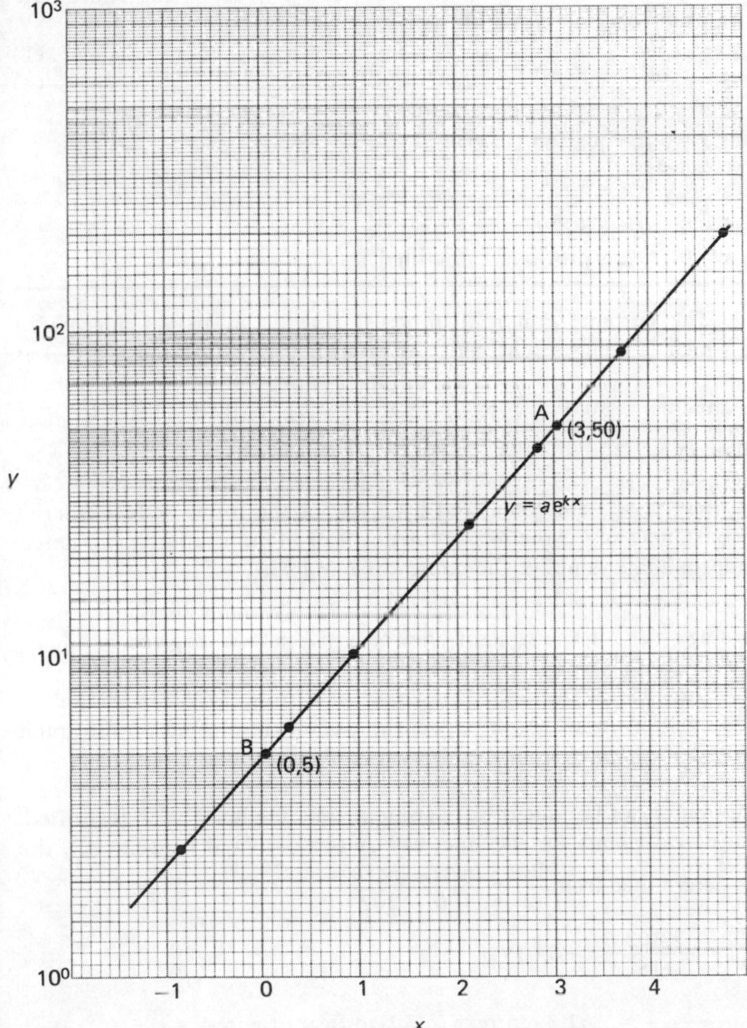

Figure 9 Graph to verify a law of the form $y = ae^{kx}$ (Problem 1)

Since the points can be joined by a straight line, **the law $y = ae^{kx}$ for the values given is verified.** Selecting any two points on the line, say A having coordinates (3, 50) and B having coordinates (0, 5), the slope k is determined from:

$$\text{slope} = k = \frac{\ln y}{x} = \frac{\ln 50 - \ln 5}{3 - 0} = \frac{2.302\,6}{3} = 0.768$$

The y-axis intercept value at $x = 0$ is $y = 5$, hence $a = 5$. Therefore **the law is $y = 5e^{0.768x}$.**

Alternatively, the values of a and k can be determined by solving simultaneous equations. Substituting in the equation $y = ae^{kx}$, the coordinate values of x and y from points A and B, gives

$$50 = ae^{3k} \quad \ldots (1)$$

$$5 = ae^{0k} \quad \ldots (2)$$

Since $e^{0k} = e^0 = 1$, $a = 5$ from equation (2).
Substituting $a = 5$ in equation (1) gives

$$50 = 5e^{3k}$$

or $e^{3k} = 10$

$$3k = \ln 10 = 2.302\,6$$

$$k = 0.768, \text{ as previously obtained.}$$

Problem 2. The current i (in milliamperes) flowing in an 8.3 microfarad capacitor, which is being discharged, varies with time t (in milliseconds) as shown:

i milliamperes	50.0	17.0	5.8	1.7	0.58	0.24
t milliseconds	200	255	310	375	425	475

Show that these results are connected by a law of the form $i = Ie^{\frac{t}{T}}$, where I and T are constants and I is the initial current flow in milliamperes, and determine the approximate values of the constants I and T.

Expressing $i = Ie^{\frac{t}{T}}$ in the straight-line form of $Y = mX + c$ gives:

$$\ln i = \frac{1}{T}t + \ln I$$

Using 'log 3 cycle × linear' graph paper, the points are plotted on the graph shown in Fig. 10.

Since these points can be joined by a straight line, **the law $i = Ie^{\frac{t}{T}}$ is verified.** Selecting any two points on the line, say A (400, 1) and B (282, 10), the slope, $\frac{1}{T}$, is determined from $\frac{\ln i}{t} = \frac{\ln 1 - \ln 10}{400 - 282}$

$$\text{i.e.} \quad \frac{1}{T} = \frac{-2.302\,6}{118} = -0.019\,5$$

Hence $T = \dfrac{1}{-0.019\,5} = -51.3$, correct to 3 significant figures.

Since the line does not cross the i-axis at $t = 0$, the value of I is found by

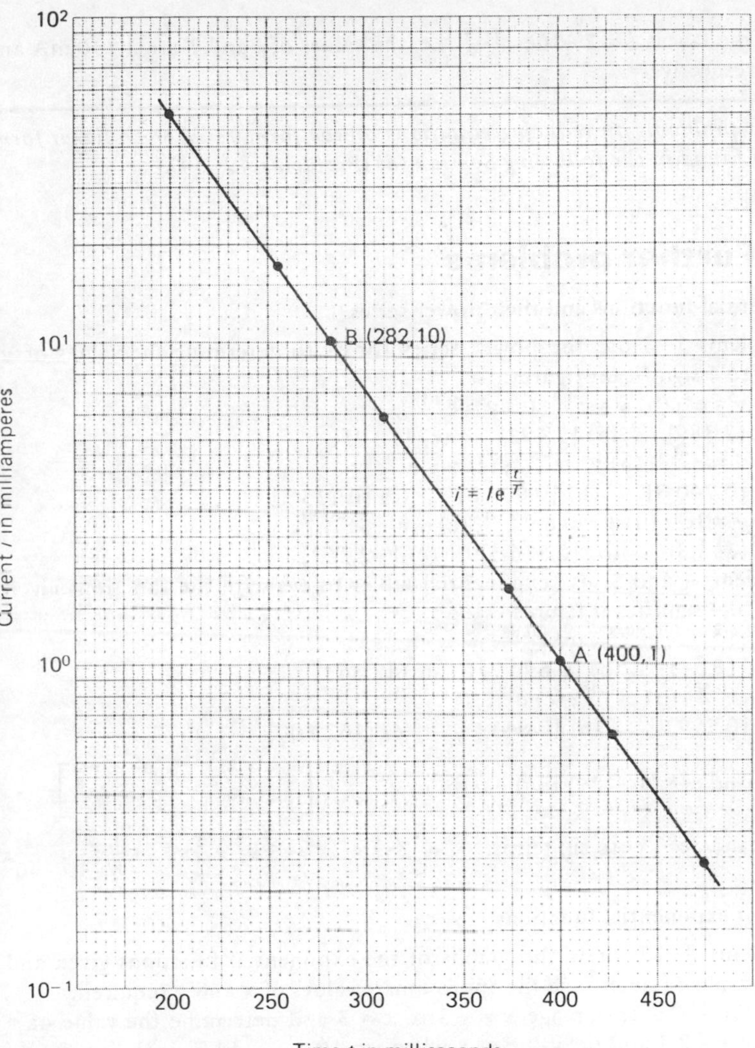

Figure 10 Variation of current with time (Problem 2)

selecting a point on the line and using the coordinates of this point, together with the value of T in the equation $i = Ie^{\frac{t}{T}}$. Using the coordinates of point A gives

$$1 = Ie^{-\frac{400}{51 \cdot 3}}$$

$$I = e^{\frac{400}{51 \cdot 3}}$$

$$= 2\,430 \text{ mA}$$

Hence the law is $i = 2\,430\mathrm{e}^{-\frac{t}{51.3}}$, i.e. the values of I and T are 2 430 mA and -51.3 respectively.

Further problems on reducing equations on the form $y = a\mathrm{e}^{kx}$ *to linear form may be found in the following Section (8) (Problems 49–54).*

8. Further problems

Exponential functions and their power series

In Problems 1–3 use the power series for e^x to determine the values of y, correct to 4 significant figures.

1. (a) $y = \mathrm{e}^2$; (b) $y = \mathrm{e}^{0.4}$; (c) $y = \mathrm{e}^{0.1}$.
   (a) [7.389] (b) [1.492] (c) [1.105]
2. (a) $y = \mathrm{e}^{-0.3}$; (b) $y = \mathrm{e}^{-0.1}$; (c) $y = \mathrm{e}^{-2}$.
   (a) [0.740 8] (b) [0.904 8] (c) [0.135 3]
3. (a) $y = 3\mathrm{e}^4$; (b) $y = 0.6\mathrm{e}^{-0.04}$; (c) $y = -5\mathrm{e}^{-0.75}$.
   (a) [163.8] (b) [0.576 5] (c) [−2.362]

In Problems 4 and 5, use exponential tables to evaluate the functions given, correct to 4 significant figures.

4. (a) $\mathrm{e}^{5.2}$; (b) $\mathrm{e}^{-0.37}$; (c) $\mathrm{e}^{0.86}$.
   (a) [181.3] (b) [0.690 7] (c) [2.363]
5. (a) $\mathrm{e}^{-0.58}$; (b) $\mathrm{e}^{-0.17}$; (c) e^{12}.
   (a) [0.559 9] (b) [0.843 7] (c) [162 800]
6. Expand $\mathrm{e}^{2x}(1-2x)$ to the term in x^4. $\left[1 - 4x^2 - \frac{8x^3}{3} - 2x^4\right]$
7. Expand $3\mathrm{e}^{x^2}x^{\frac{1}{2}}$ to six terms. $\left[3x^{\frac{1}{2}} + 3x^{\frac{5}{2}} + \frac{3}{2}x^{\frac{9}{2}} + \frac{1}{2}x^{\frac{13}{2}} + \frac{1}{8}x^{\frac{17}{2}} + \frac{1}{40}x^{\frac{21}{2}}\right]$

Graphs of exponential functions

In Problems 8–15, draw the graphs of the exponential functions given and use the graphs to determine the approximate values of x and y required.

8. $y = 5\mathrm{e}^{0.4x}$ over a range $x = -3$ to $x = 3$ and determine the value of y when $x = 2.7$ and the value of x when $y = 10$. [14.7, 1.7]
9. $y = 0.35\mathrm{e}^{2.5x}$ over a range $x = -2$ to $x = 2$ and determine the value of y when $x = 1.8$ and the value of x when $y = 40$. [31.5, 1.9]
10. $y = \frac{1}{3}\mathrm{e}^{-2x}$ over a range $x = -2$ to $x = 2$ and determine the value of y when $x = -1.75$ and the value of x when $y = 5$. [11, −1.35]
11. $y = 0.46\mathrm{e}^{-0.27x}$ over a range $x = -10$ to $x = 10$ and determine the value of y when $x = -8.5$ and the value of x when $y = 6.1$. [4.6, −9.6]
12. $y = 4\mathrm{e}^{-2x^2}$ over a range $x = -1.5$ to $x = 1.5$ and determine the value of y when $x = -1.2$ and the value of x when $y = 2.9$. [0.22, ±0.4]
13. $y = 100\mathrm{e}^{\frac{x^2}{3}}$ over a range $x = -3$ to $x = 3$ and determine the value of y when $x = \frac{7}{3}$ and the value of x when $y = 40$. [614, ±1.66]

14. $y = \frac{1}{2}(e^{x} - e^{-x})$ over a range $x = -3$ to $x = 3$ and determine the value of y when $x = -2.3$ and the value of x when $y = 5$. [−4.94, 2.31]
15. $y = 3(e^{2x} - 4e^{-x})$ over a range $x = 2$ to $x = -2$ and determine the value of y when $x = -1.6$ and the value of x when $y = 35$. [−59, 1.28]

Naperian logarithms

In Problems 16−21, determine the values of $\ln x$ when x has the values shown.

16. (a) 2.614; (b) 6.775; (c) 9.213.
    (a) [0.960 9] (b) [1.913 2] (c) [2.220 6]
17. (a) $3\frac{9}{17}$; (b) $\frac{127}{19}$; (c) $8\frac{4}{7}$.
    (a) [1.261 1] (b) [1.899 7] (c) [2.148 4]
18. (a) 77.34; (b) 190; (c) 2 377.
    (a) [4.348 2] (b) [5.247 0] (c) [7.773 6]
19. (a) 23.2; (b) 17 140; (c) 7 601.
    (a) [3.144 2] (b) [9.749 2] (c) [8.936 0]
20. (a) 0.171; (b) 0.005 35; (c) 0.877 4.
    (a) $\bar{2}.233\,9$ or −1.766 1
    (b) $\bar{6}.769\,3$ or −5.270 7
    (c) $\bar{1}.869\,2$ or −0.130 8
21. (a) 3.74×10^{-3}; (b) 7.818×10^{-2}; (c) 9.671×10^{-4}.
    (a) $\bar{6}.411\,3$ or −5.588 7
    (b) $\bar{3}.451\,3$ or −2.548 7
    (c) $\bar{7}.058\,8$ or −6.841 2

In Problems 22−27, determine the values of x when $\ln x$ has the values shown.

22. (a) 0.277 4; (b) 1.308 9; (c) 2.101 3.
    (a) [1.320] (b) [3.702] (c) [8.177]
23. (a) 0.871 5; (b) 1.001 7; (c) 2.277 3.
    (a) [2.390] (b) [2.723] (c) [9.750]
24. (a) 2.917 4; (b) 7.173 8; (c) 14.207 4.
    (a) [18.49] (b) [1 305] (c) [1.480×10^{6}]
25. (a) 4.61; (b) 10.371; (c) 17.214.
    (a) [100.5] (b) [31 920] (c) [2.992×10^{7}]
26. (a) $\bar{1}.172\,9$; (b) $\bar{5}.217\,3$; (c) $\overline{10}.0271$.
    (a) [0.437 3] (b) [8.373×10^{-3}] (c) [4.665×10^{-5}]
27. (a) $\bar{2}.367\,5$; (b) $\bar{6}.007\,7$; (c) $\overline{15}.381\,4$.
    (a) [0.195 4] (b) [2.498×10^{-3}] (c) [4.479×10^{-7}]
28. Find the value of: (a) $e^{2.173}$; (b) $e^{4.179}$; (c) $e^{6.71}$.
    (a) [8.785] (b) [65.30] (c) [820.6]
29. Calculate the value of $y = 2.7e^{0.1x}$ when the value of x is: (a) 0.0174; (b) 14; and (c) $\frac{7}{16}$. (a) [2.705] (b) [10.95] (c) [2.821]
30. Determine the value of $y = -0.371e^{-\frac{2x}{3}}$ when the value of x is: (a) −1.84; (b) 3.716; and (c) 12.
    (a) [−1.265] (b) [-3.115×10^{-2}] (c) [-1.245×10^{-4}]

The natural laws of growth and decay

In Problems 31−35, the natural laws of growth or decay are given. Determine in each case the quantities specified.

31. The instantaneous voltage v in a capacitance circuit is related to time t by the equation $v = Ve^{-\frac{t}{CR}}$, where V, C and R are constants. Determine v when $t = 27 \times 10^{-3}$, $C = 8.0 \times 10^{-6}$, $R = 57 \times 10^3$ and $V = 100$. Also determine R when $v = 83$, $t = 9.0 \times 10^{-3}$, $C = 8.0 \times 10^{-6}$ and $V = 100$. [94.3, 6.04×10^3]
32. The length of a bar, l, at temperature θ is given by $l = l_0 e^{\alpha\theta}$, where l_0 and α are constants. Determine α when $l = 2.738$, $l_0 = 2.631$ and $\theta = 315.7$. Also determine l_0 when l and θ are as for the first part of the problem but $\alpha = 1.771 \times 10^{-4}$. [$1.263 \times 10^{-4}$, 2.589]
33. Two quantities x and y are found to be related by the equation $y = ae^{-kx}$, where a and k are constants.
(a) Determine y when $a = 1.671 \times 10^4$, $k = -4.60$ and $x = 1.537$.
(b) Determine x when $y = 76.31$, $a = 15.3$ and $k = 4.77$.
(a) [1.966×10^7] (b) [-0.3369]
34. Quantities p and q are related by the equation $p = 7.413\,(1 - e^{\frac{kq}{t}})$, where k and t are constants. Determine p when $k = 3.7 \times 10^{-2}$, $q = 712.8$ and $t = 5.747$. Also determine t when $p = -98.3$ and q and k are as for the first part of the problem. [-722, 9.92]
35. When quantities I and C are related by the equation $I = BT^2 e^{-\frac{C}{T}}$, and B and T are constants, determine I when $B = 14.3$, $T = 1.27$ and $C = 8.15$. Also determine C when $I = 7.47 \times 10^{-2}$ and B and T are as for the first part of the problem. [0.037 66, 7.280]
36. In an experiment involving Newton's law of cooling, the temperature θ after a time t of 73.0 seconds is found to be 51.8°C. Using the relationship $\theta = \theta_0 e^{-kt}$, determine k when $\theta_0 = 15.0$°C. [$-0.016\,98$]
37. The pressure p at height h above ground level is given by $p = p_0 e^{-\frac{h}{c}}$, where p_0 is the pressure at ground level and c is a constant. When p_0 is 1.013×10^5 pascals and the pressure at a height of 1 570 metres is 9.871×10^4 pascals, determine the value of c. [60 620]
38. The current i amperes flowing in a capacitor at time t seconds is given by $i = 7.51\,(1 - e^{-\frac{t}{CR}})$, where the circuit resistance R is 27.4 kilohms and the capacitance C is 14.71 microfarads. Determine: (a) the time for the current to reach 6.37 amperes; and (b) the current flow after 0.458 seconds. (a) [0.759 8 s] (b) [5.099 A]
39. The voltage drop, V volts, across an inductor of L henrys at time t seconds is given by $V = 125\,e^{-\frac{Rt}{L}}$. Determine: (a) the time for the voltage to reach 98.0 volts; and (b) the voltage when t is 14.7 microseconds, given that the circuit resistance R is 128 ohms and its inductance is 10.3 millihenrys. (a) [0.019 58 ms] (b) [104.1 V]
40. The resistance R_t of an electrical conductor at temperature t degrees Celsius is given by $R_t = R_0 e^{\alpha t}$, where α is a constant and R_0 is 3.41

kilohms. Calculate the value of α when R_t is 3.72 kilohms, and t = 1 710 degrees Celsius. For this conductor, at what temperature will the resistance be 3.50 kilohms? [5.088×10^{-5}, 512°C]

Log-log and log-linear graph paper

In Problems 41–54, use log-log or log-linear graph paper to obtain the required solutions.

41. The periodic time T of oscillation of a pendulum is believed to be related to its length l by a law of the form $T = k\,l^n$, where k and n are constants. Values of T were measured for various lengths and the results are:

T (s)	1.0	1.2	1.4	1.6	1.8	2.0	2.4
l (m)	2.4	3.5	4.8	6.3	7.9	9.8	14.1

Verify that the law stated is correct and determine the approximate values of k and n. [0.64, $\frac{1}{2}$]

42. Two quantities x and y are measured and the results obtained are:

x	3.7	5.9	7.4	9.1	11.6	13.8
y	6.2	3.9	3.1	2.5	2.0	1.7

Show that x and y are related by a law of the form $y = ax^n$ and determine the approximate values of a and n. [23, −1]

43. The variation of the voltage V and admittance Y of an electrical circuit are:

Voltage, V volts	0.37	0.51	0.74	0.98	1.13	1.34
Admittance, Y siemens	0.446	0.614	0.892	1.181	1.361	1.614

Show that the law connecting V and Y is of the form $V = RY^n$ and determine the approximate values of R and n. [0.83, −1]

44. Two quantities x and y are believed to be related by a law of the form $y = A\,x^n$. The values of y and corresponding values of x are:

y	1.3×10^3	4.3×10^3	1.44×10^4	7.6×10^4	3.8×10^5
x	3.8	2.1	1.1	0.46	0.20

Show that x and y are related by this law and determine the approximate values of A and n. [1.7×10^4, −1.93]

45. The activity of a mixture of radioactive isotopes is believed to vary according to the law $R = R_0 t^{-c}$ where R_0 and c are constants. If R varies with t as shown below verify that this is true and determine R_0 and C.

R/m C $\text{Kg}^{-1}\,\text{hr}^{-1}$	5.93	2.94	2.30	0.67	0.44	0.31
t (h)	3.0	5.0	6.0	15.0	20.0	27.0

[R_0 = 25.8, C = 1.35]

46. The periods and mean distances of the six innermost planets are given in the following table:

Period, P days	88.0	224.7	365.3	687.0	4 333	10 760
Mean distance, s miles $\times 10^6$	36	67	93	142	483	886

Show that these values are related by the Keplerian equation $P = ks^n$ and find approximate values of k and n. [n = 1.50, k = 0.41]

47. Values of p and q are believed to be related by a law of the form $p = ab^q$

where a and b are constants. The values of p and corresponding values of q are:

p	4.5	7.4	11.2	15.8	39.0	68.0	271.5
q	0.6	1.3	1.9	2.4	3.7	4.5	6.5

Verify that the law relating p and q is correct and determine the approximate values of a and b. [3,2]

48. The values of k and corresponding values of l are:

l	0.2	0.7	1.3	1.9	2.4	3.6
k	6.0	14.2	39.4	109.7	257.2	1 990.0

The law relating these quantities is of the form $k = mn^l$. Determine the approximate values of m and n. [4.3, 5.5]

49. Determine the law of the form $y = ae^{kx}$ which relates the following values:

y	0.015	0.08	0.17	0.30	0.96	3.0
x	−6.0	8.5	15.0	20.0	30.0	40.0

[$y = 0.03\ e^{0.115x}$]

50. The voltage drop across an inductor, v volts, is believed to be related to time, t milliseconds, by the relationship $v = V\ e^{\frac{t}{T}}$, where V and T are constants. The variation of voltage with time for this inductor is:

v volts	700	400	190	100	50	17
t milliseconds	27.3	31.7	37.5	42.5	47.8	56.3

Show that the law relating these quantities is as stated and determine the approximate values of V and T. [24.1 kV, −7.75 ms]

51. The tension in two sides of a belt, T and T_0 newtons, passing round a pulley wheel and in contact with the pulley over an angle of θ radians, is given by $T = T_0\ e^{\mu\theta}$ where T_0 and μ are constants. Determine the approximate values of T_0 and the coefficient of friction μ, from the following observations:

T newtons	67.3	76.4	90.0	107.4	117.6	125.5
θ radians	1.13	1.54	2.07	2.64	2.93	3.14

[47.4, 0.31]

52. A liquid which is cooling is believed to follow a law of the form $\theta = \theta_0\ e^{kt}$ where θ_0 and k are constants and θ is the temperature of the body at time t. Measurements are made of the temperature and time and the results are:

θ °C	83	58	41.5	32	26
t minutes	16.7	25	32.5	37	43.5

Show that these quantities are related by this law and determine the approximate values of θ_0 and k. [174°C, −0.044]

53. The mass (m) of a given substance is believed to dissolve in one litre of water at temperature (t °C) according to the law

$m = ae^{kt}$

m was measured at various temperatures and the following results obtained:

t °C	10	20	30	40	50
m Kg	35.5	39.1	43.2	47.7	52.8

Show that the law is true and find approximate values for a and k.
[$a = 32.1$, $k = 0.009\ 90$]

54. The following results were obtained when measuring the growth of duckweed:

Weeks (t)	0	1	2	3	4	5
No. of fronds (n)	20	30	52	77	135	211

It is thought the measurements are connected by the equation $n = Ae^{kt}$.
Verify that this is so and obtain approximate values for A and k.
[$A = 18.4$, $k = 0.49$]

Chapter 4

Methods of differentiation

1. Introduction to differentiation

Let P and Q be two points close together on a curve as shown in Fig. 1.

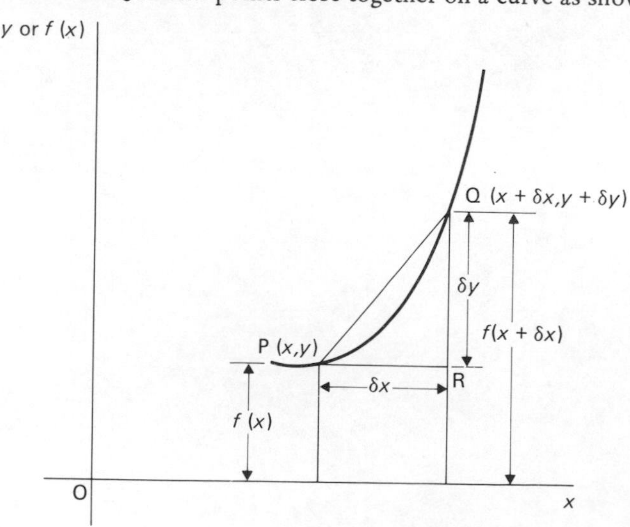

Figure 1

Let the length PR be δx, representing a small increment in x, and the length QR be δy, the corresponding increase in y.

Let P be any point on the curve with coordinates (x, y). Then Q has the coordinates $(x + \delta x, y + \delta y)$.

The slope of the chord PQ $= \dfrac{\delta y}{\delta x}$

But from Fig. 1, $\delta y = (y + \delta y) - y = f(x + \delta x) - f(x)$

$$\text{Hence } \frac{\delta y}{\delta x} = \frac{f(x + \delta x) - f(x)}{\delta x}$$

The smaller δx becomes, the closer the gradient of the chord PQ approaches the gradient of the tangent at P. That is, as $\delta x \to 0$, the gradient of the chord $\to$ the gradient of the tangent. As δx approaches zero, the value of $\dfrac{\delta y}{\delta x}$ approaches what is called a **limiting value.** There are two notations commonly used when finding the gradient of the tangent drawn to a curve.

(i) The gradient of the curve at P is represented as

$$\lim_{\delta x \to 0} \frac{\delta y}{\delta x}. \text{ This is written as } \frac{dy}{dx}$$

$$\text{i.e. } \frac{dy}{dx} = \lim_{\delta x \to 0} \frac{\delta y}{\delta x}$$

This way of stating the gradient of a curve is called **Leibniz notation.**

(ii) The gradient of the curve at P $= \lim\limits_{\delta x \to 0} \left\{ \dfrac{f(x + \delta x) - f(x)}{\delta x} \right\}$

This is written as $f'(x)$.

$$\text{i.e. } f'(x) = \lim_{\delta x \to 0} \left\{ \frac{f(x + \delta x) - f(x)}{\delta x} \right\}$$

This way of stating the gradient of a curve is called **functional notation.**

$\dfrac{dy}{dx}$ is the same as $f'(x)$ and is called the **differential coefficient** of y with respect to x, or the **derivative** of y with respect to x.

The process of finding a differential coefficient is called **differentiation.**

2. Differential coefficients of common functions

(i) Differential coefficient of ax^n

Let $f(x) = ax^n$

then $f(x + \delta x) = a(x + \delta x)^n$

By definition, $f'(x) = \lim_{\delta x \to 0} \left\{ \frac{f(x + \delta x) - f(x)}{\delta x} \right\}$

$$= \lim_{\delta x \to 0} \left\{ \frac{a(x + \delta x)^n - ax^n}{\delta x} \right\}$$

$a(x + \delta x)^n$ may be expanded using the binomial expansion.

$$a(x + \delta x)^n = a \left[x^n + nx^{n-1}\,\delta x + \frac{n(n-1)}{(1)(2)} x^{n-2} (\delta x)^2 + \dots \right]$$

$$= ax^n + an\,x^{n-1}\,\delta x + \frac{a\,n(n-1)}{(1)(2)} x^{n-2} (\delta x)^2 + \dots$$

$$a(x + \delta x)^n - ax^n = an\,x^{n-1}\,\delta x + \frac{a\,n(n-1)}{(1)(2)} x^{n-2} (\delta x)^2 + \dots$$

$$\frac{a(x + \delta x)^n - ax^n}{\delta x} = an\,x^{n-1} + \frac{a\,n(n-1)}{(1)(2)} x^{n-2}\,\delta x + \dots$$

$$f'(x) = \lim_{\delta x \to 0} \left\{ an\,x^{n-1} + \frac{a\,n(n-1)}{(1)(2)} x^{n-2}\,\delta x + \dots \right\}$$

i.e. $\boldsymbol{f'(x) = an\,x^{n-1}}$, since all subsequent terms in the bracket will contain δx raised to some power and will become zero when a limiting value is taken. This result is true for all values of n, whether they are positive, negative or fractional.

Hence when $\boldsymbol{f(x) = ax^n, f'(x) = an\,x^{n-1}}$

or when $\boldsymbol{y = ax^n, \frac{dy}{dx} = an\,x^{n-1}}$

(ii) Differential coefficient of sin *x*

Let $f(x) = \sin x$
then $f(x + \delta x) = \sin(x + \delta x)$

By definition, $f'(x) = \lim_{\delta x \to 0} \left\{ \frac{f(x + \delta x) - f(x)}{\delta x} \right\}$

$$= \lim_{\delta x \to 0} \left\{ \frac{\sin(x + \delta x) - \sin x}{\delta x} \right\}$$

It was shown in Chapter 1, Section 12 that:

$$\sin x - \sin y = 2 \cos \left[\frac{x + y}{2} \right] \sin \left[\frac{x - y}{2} \right]$$

Hence $\sin(x + \delta x) - \sin x = 2 \cos \left[\frac{(x + \delta x) + x}{2} \right] \sin \left[\frac{(x + \delta x) - x}{2} \right]$

$$= 2 \cos \left[x + \frac{\delta x}{2} \right] \sin \left[\frac{\delta x}{2} \right]$$

$$\frac{\sin(x+\delta x)-\sin x}{\delta x}=\frac{2\cos\left[x+\frac{\delta x}{2}\right]\sin\left[\frac{\delta x}{2}\right]}{\delta x}$$

$$=\cos\left[x+\frac{\delta x}{2}\right]\frac{\sin\left[\frac{\delta x}{2}\right]}{\left[\frac{\delta x}{2}\right]}$$

When δx is small, $\sin \delta x \simeq \delta x$

(For example, if $\delta x = 1^\circ$, $\sin 1^\circ = 0.017\,5$ and 1° in radians $= 0.017\,5$.)

Thus when δx is small, say, less than 2°, $\sin\left[\frac{\delta x}{2}\right]=\left[\frac{\delta x}{2}\right]$, correct to 3 significant figures

i.e. $\dfrac{\sin\left[\frac{\delta x}{2}\right]}{\left[\frac{\delta x}{2}\right]}=1$

Hence $f'(x)=\lim\limits_{\delta x\to 0}\left\{\cos\left[x+\frac{\delta x}{2}\right]\frac{\sin\left[\frac{\delta x}{2}\right]}{\left[\frac{\delta x}{2}\right]}\right\}$

i.e. $\quad \boldsymbol{f'(x) = \cos x}$

Hence when $\boldsymbol{f(x) = \sin x, f'(x) = \cos x}$

or when $\quad \boldsymbol{y = \sin x, \dfrac{dy}{dx} = \cos x}$

(iii) Differential coefficient of cos *x*

Let $f(x) = \cos x$

then $f(x + \delta x) = \cos (x + \delta x)$

By definition, $f'(x)=\lim\limits_{\delta x\to 0}\left\{\dfrac{f(x+\delta x)-f(x)}{\delta x}\right\}$

$$=\lim_{\delta x\to 0}\left\{\frac{\cos(x+\delta x)-\cos x}{\delta x}\right\}$$

It was shown in Chapter 1, Section 12 that:

$$\cos x-\cos y=-2\sin\left[\frac{x+y}{2}\right]\sin\left[\frac{x-y}{2}\right]$$

Hence $\cos(x+\delta x)-\cos x=-2\sin\left[\dfrac{(x+\delta x)+x}{2}\right]\sin\left[\dfrac{(x+\delta x)-x}{2}\right]$

$$=-2\sin\left[x+\frac{\delta x}{2}\right]\sin\left[\frac{\delta x}{2}\right]$$

$$\frac{\cos(x+\delta x)-\cos x}{\delta x}=\frac{-2\sin\left[x+\frac{\delta x}{2}\right]\sin\left[\frac{\delta x}{2}\right]}{\delta x}$$

$$=-\sin\left[x+\frac{\delta x}{2}\right]\frac{\sin\left[\frac{\delta x}{2}\right]}{\left[\frac{\delta x}{2}\right]}$$

Hence $f'(x)=\lim_{\delta x\to 0}\left\{-\sin\left[x+\frac{\delta x}{2}\right]\frac{\sin\left[\frac{\delta x}{2}\right]}{\left[\frac{\delta x}{2}\right]}\right\}$

$= -(\sin x)$, since in the limit, $\frac{\sin\left[\frac{\delta x}{2}\right]}{\left[\frac{\delta x}{2}\right]}=1$

i.e. $\quad f'(x)=-\sin x$

Hence when $f(x)=\cos x$, $f'(x)=-\sin x$

or when $\quad y=\cos x, \frac{dy}{dx}=-\sin x$

(iv) Differential coefficient of e^{ax}

Let $f(x)=e^{ax}$

then $f(x+\delta x)=e^{a(x+\delta x)}$

By definition, $f'(x)=\lim_{\delta x\to 0}\left\{\frac{f(x+\delta x)-f(x)}{\delta x}\right\}$

$$=\lim_{\delta x\to 0}\left\{\frac{e^{a(x+\delta x)}-e^{ax}}{\delta x}\right\}$$

$$e^{a(x+\delta x)}-e^{ax}=e^{ax}(e^{a\delta x}-1)$$

Since $e^x=1+x+\frac{x^2}{2!}+\frac{x^3}{3!}+\ldots$ (see p.76)

then $e^{a\delta x}=1+(a\delta x)+\frac{(a\delta x)^2}{2!}+\frac{(a\delta x)^3}{3!}+\ldots$

Therefore $e^{ax}(e^{a\delta x}-1)=e^{ax}\left(1+a\delta x+\frac{(a\delta x)^2}{2!}+\ldots-1\right)$

$$=e^{ax}\left(a\delta x+\frac{(a\delta x)^2}{2!}+\ldots\right)$$

$$\frac{e^{a(x+\delta x)}-e^{ax}}{\delta x}=e^{ax}\left(a+\frac{a^2\delta x}{2!}+\ldots\right)$$

$$\text{Hence } f'(x) = \lim_{\delta x \to 0} \left\{ e^{ax}\left(a + \frac{a^2\delta x}{2!} + \ldots \right) \right\}$$

$$= (e^{ax})(a)$$

i.e. $\quad f'(x) = ae^{ax}$

Hence when $f(x) = e^{ax}, f'(x) = ae^{ax}$

or when $\qquad y = e^{ax}, \dfrac{dy}{dx} = ae^{ax}$

(v) Differential coefficient of ln *ax*

Let $f(x) = \ln ax$
then $f(x + \delta x) = \ln a(x + \delta x)$

$$\text{By definition, } f'(x) = \lim_{\delta x \to 0} \left\{ \frac{f(x + \delta x) - f(x)}{\delta x} \right\}$$

$$= \lim_{\delta x \to 0} \left\{ \frac{\ln a(x + \delta x) - \ln ax}{\delta x} \right\}$$

$$\ln a(x + \delta x) - \ln ax = \ln ax \left(1 + \frac{\delta x}{x}\right) - \ln ax$$

$$= \ln ax + \ln\left(1 + \frac{\delta x}{x}\right) - \ln ax$$

$$= \ln\left(1 + \frac{\delta x}{x}\right)$$

Now $\ln(1 + x) = x - \dfrac{x^2}{2} + \dfrac{x^3}{3} - \ldots$ (see p. 84)

$$\text{Therefore } \ln\left(1 + \frac{\delta x}{x}\right) = \frac{\delta x}{x} - \frac{\left(\frac{\delta x}{x}\right)^2}{2} + \frac{\left(\frac{\delta x}{x}\right)^3}{3} - \ldots$$

$$\frac{\ln a(x + \delta x) - \ln ax}{\delta x} = \frac{1}{x} - \frac{\delta x}{2x^2} + \frac{\delta x^2}{3x^3} - \ldots$$

$$\text{Hence } f'(x) = \lim_{\delta x \to 0} \left\{ \frac{\ln a(x + \delta x) - \ln ax}{\delta x} \right\}$$

$$= \lim_{\delta x \to 0} \left\{ \frac{1}{x} - \frac{\delta x}{2x^2} + \frac{\delta x^2}{3x^3} - \ldots \right\}$$

i.e. $\quad f'(x) = \dfrac{1}{x}$

Hence when $f(x) = \ln ax, f'(x) = \dfrac{1}{x}$

or when $\qquad y = \ln ax, \dfrac{dy}{dx} = \dfrac{1}{x}$

	y or $f(x)$	$\frac{dy}{dx}$ or $f'(x)$
(i)	ax^n	$an\,x^{n-1}$
(ii)	$\sin x$	$\cos x$
(iii)	$\cos x$	$-\sin x$
(iv)	e^{ax}	$a\,e^{ax}$
(v)	$\ln ax$	$\frac{1}{x}$

If $y = ax^n$ and $n = 0$

then $y = ax^0$ and $\frac{dy}{dx} = (a)\,(0)\,x^{0-1} = 0$

Hence the differential coefficient of a constant is zero.

If $f(x) = c$, where c is a constant, $f'(x) = 0$.

The differential coefficient of a sum or difference is the sum or difference of the differential coefficients of the separate terms.

i.e. when $f(x) = g(x) + h(x) - j(x)$

then $f'(x) = g'(x) + h'(x) - j'(x)$

The differential coefficients obtained in the following worked problems are deduced using only the above summary.

Worked problems on differential coefficients of common functions

Problem 1. Find the differential coefficient of: (a) $5x^4$; (b) $\frac{3}{x^2}$; (c) $4\sqrt{x}$.

When $f(x) = ax^n$, $f'(x) = an\,x^{n-1}$

(a) $f(x) = 5x^4$

$f'(x) = (5)(4)\,x^{4-1} = \mathbf{20x^3}$

(b) $f(x) = \frac{3}{x^2} = 3x^{-2}$

$f'(x) = (3)(-2)\,x^{-2-1} = -6x^{-3}$

i.e. $f'(x) = \frac{-6}{x^3}$

(c) $f(x) = 4\sqrt{x} = 4x^{\frac{1}{2}}$

$f'(x) = (4)(\tfrac{1}{2})\,x^{\frac{1}{2}-1} = 2x^{-\frac{1}{2}}$

i.e. $f'(x) = \frac{2}{\sqrt{x}}$

Problem 2. Differentiate $2x^3 + 7x + \frac{1}{3x^2} - \frac{4}{x^3} + \frac{4}{3}\sqrt{x^3} - 8$ with respect to x.

$$f(x) = 2x^3 + 7x + \frac{1}{3x^2} - \frac{4}{x^3} + \frac{4}{3}\sqrt{x^3} - 8$$

$$= 2x^3 + 7x + \frac{x^{-2}}{3} - 4x^{-3} + \frac{4}{3}x^{\frac{3}{2}} - 8$$

$$f'(x) = (2)(3)\,x^{3-1} + (7)\,x^{1-1} + \frac{(-2)}{3}x^{-2-1} - (4)(-3)\,x^{-3-1} + \left(\frac{4}{3}\right)\left(\frac{3}{2}\right)x^{\frac{3}{2}-1} - 0$$

$$= 6x^2 + 7 - \frac{2}{3}x^{-3} + 12\,x^{-4} + 2x^{\frac{1}{2}}$$

i.e. $\boldsymbol{f'(x) = 6x^2 + 7 - \frac{2}{3x^3} + \frac{12}{x^4} + 2\sqrt{x}}$

Problem 3. (a) If $f(x) = 2 \sin x$ find $f'(x)$;
(b) If $y = 5 \cos x$ find $\frac{dy}{dx}$.

(a) When $f(x) = \sin x$, $f'(x) = \cos x$
When $f(x) = 2 \sin x$
then $\boldsymbol{f'(x) = 2 \cos x}$

(b) When $y = \cos x$, $\frac{dy}{dx} = -\sin x$

When $y = 5 \cos x$

then $\boldsymbol{\frac{dy}{dx} = -5 \sin x}$

Problem 4. Differentiate: (a) e^{6t}; (b) $5e^{-3t}$ with respect to t.

When $f(t) = e^{at}$, $f'(t) = a\,e^{at}$
(a) $f(t) = e^{6t}$
$\boldsymbol{f'(t) = 6\,e^{6t}}$
(b) $f(t) = 5e^{-3t}$
$f'(t) = (5)(-3)\,e^{-3t} = \boldsymbol{-15\,e^{-3t}}$

Problem 5. Find the differential coefficient of: (a) ln $4x$; (b) 3 ln $2x$.

When $f(x) = \ln ax$, $f'(x) = \frac{1}{x}$

(a) $f(x) = \ln 4x$, $\boldsymbol{f'(x) = \frac{1}{x}}$

(b) $f(x) = 3 \ln 2x$, $\boldsymbol{f'(x) = \frac{3}{x}}$

Problem 6. If $g = 3.2\,h^5 - 3\sin h - 5e^{7h} + \sqrt[3]{h^4} + 6$, find $\dfrac{dg}{dh}$.

$$g = 3.2\,h^5 - 3\sin h - 5e^{7h} + h^{\frac{4}{3}} + 6$$

$$\frac{dg}{dh} = (3.2)(5)\,h^4 - 3\cos h - (5)(7)\,e^{7h} + \left(\frac{4}{3}\right) h^{\frac{1}{3}} + 0$$

$$= 16\,h^4 - 3\cos h - 35\,e^{7h} + \frac{4}{3}\,h^{\frac{1}{3}}$$

i.e. $$\frac{dg}{dh} = 16\,h^4 - 3\cos h - 35\,e^{7h} + \frac{4}{3}\sqrt[3]{h}$$

Problem 7. $f(x) = 4\ln(2.6x) - \dfrac{3}{\sqrt[3]{x^2}} + \dfrac{4}{e^{3x}} + \dfrac{1}{5} - 2\cos x$. Find $f'(x)$.

$$f(x) = 4\ln(2.6x) - 3x^{-\frac{2}{3}} + 4e^{-3x} + \frac{1}{5} - 2\cos x$$

$$f'(x) = \frac{4}{x} - (3)\left(-\frac{2}{3}\right)x^{-\frac{5}{3}} + (4)(-3)\,e^{-3x} + 0 - (-2\sin x)$$

i.e. $$f'(x) = \frac{4}{x} + \frac{2}{\sqrt[3]{x^5}} - \frac{12}{e^{3x}} + 2\sin x$$

Problem 8. Find the gradient of the curve $y = \dfrac{3}{2\sqrt{x}}$ at the point $\left(4, \dfrac{3}{4}\right)$.

$$y = \frac{3}{2\sqrt{x}} = \frac{3}{2}\,x^{-\frac{1}{2}}$$

$$\text{Gradient} = \frac{dy}{dx} = \left(\frac{3}{2}\right)\left(-\frac{1}{2}\right)x^{-\frac{3}{2}} = -\frac{3}{4\sqrt{x^3}}$$

When $x = 4$, gradient $= -\dfrac{3}{4\sqrt{4^3}} = -\dfrac{3}{4(8)}$

i.e. **Gradient** $= -\dfrac{3}{32}$

Problem 9. Find the coordinates of the point on the curve $y = 3\sqrt[3]{x^2}$ where the gradient is 1.

$$y = 3\sqrt[3]{x^2} = 3x^{\frac{2}{3}}$$

$$\text{Gradient} = \frac{dy}{dx} = (3)\left(\tfrac{2}{3}\right)x^{-\frac{1}{3}} = \frac{2}{\sqrt[3]{x}}$$

If the gradient is equal to 1, then $1 = \dfrac{2}{\sqrt[3]{x}}$

i.e. $\sqrt[3]{x} = 2$

$x = 2^3 = 8$

When $x = 8$, $y = 3\sqrt[3]{8^2} = 3(4) = 12$

Hence the gradient is 1 at the point (8, 12)

Problem 10. If $f(x) = \dfrac{4x^3 - 8x^2 + 6x}{2x}$ find the coordinates of the point at which the gradient is: (a) zero; and (b) 4.

$$f(x) = \frac{4x^3 - 8x^2 + 6x}{2x} = \frac{4x^3}{2x} - \frac{8x^2}{2x} + \frac{6x}{2x}$$
$$= 2x^2 - 4x + 3$$

The derivative, $f'(x)$, gives the gradient of the curve.
Hence $f'(x) = (2)(2)\,x^1 - 4$
$= 4x - 4$

(a) When $f'(x)$ is zero
$4x - 4 = 0$
i.e. $x = 1$
When $x = 1, y = 2x^2 - 4x + 3 = 2\,(1)^2 - 4\,(1) + 3$
i.e. $y = 1$
Hence the coordinates of the point where the gradient is zero are (1, 1)

(b) When $f'(x)$ is 4
$4x - 4 = 4$
i.e. $x = 2$
When $x = 2, y = 2\,(2)^2 - 4\,(2) + 3 = 3$
Hence the coordinates of the point where the gradient is 4 are (2, 3)

Further problems on differential coefficients of common functions may be found in Section 6 (Problems 1–30).

3. Differentiation of products and quotients of two functions

(i) Differentiation of a product

The function $y = 3x^2 \sin x$ is a product of two terms in x, i.e. $3x^2$ and $\sin x$.
Let $u = 3x^2$ and $v = \sin x$.
Let x increase by a small increment δx, causing incremental changes in u, v and y of δu, δv and δy respectively.

Then
$$y = (3x^2)(\sin x)$$
$$= (u)(v)$$
$$y + \delta y = (u + \delta u)(v + \delta v)$$
$$= uv + v\delta u + u\delta v + \delta u\delta v$$
$$(y + \delta y) - (y) = uv + v\delta u + u\delta v + \delta u\delta v - uv$$
$$\delta y = v\delta u + u\delta v + \delta u\delta v$$

Dividing both sides by δx gives:

$$\frac{\delta y}{\delta x} = v\frac{\delta u}{\delta x} + u\frac{\delta v}{\delta x} + \frac{\delta u}{\delta x}\delta v$$

As $\delta x \to 0$ then $\delta u \to 0$, $\delta v \to 0$ and $\delta y \to 0$

 However, the fact that δu and δx, for example, both approach zero does not mean that $\frac{\delta u}{\delta x}$ will approach zero.

Ratios of small quantities, such as $\frac{\delta u}{\delta x}, \frac{\delta v}{\delta x}$ or $\frac{\delta y}{\delta x}$ can be significant.

Consider two lines AB and AC meeting at A and whose intersecting angle (i.e. $\angle$ BAC) is any value.

If AB = δy = 1 unit, say, and AC = δx = 2 units, then the ratio

$$\frac{\delta y}{\delta x} = \frac{1}{2}.$$

This ratio of $\frac{1}{2}$ is still the same whether the unit of δy and δx is in, say, kilometres or millimetres. No matter how small δy or δx is made, the ratio is still $\frac{1}{2}$. Thus when $\delta y \to 0$ and when $\delta x \to 0$, the ratio $\frac{\delta y}{\delta x}$ is still a significant value.

As $\delta x \to 0$, $\frac{\delta u}{\delta x} \to \frac{du}{dx}, \frac{\delta v}{\delta x} \to \frac{dv}{dx}, \frac{\delta y}{\delta x} \to \frac{dy}{dx}$ and $\delta v \to 0$

Hence $\frac{dy}{dx} = v\frac{du}{dx} + u\frac{dv}{dx}$

This is known as the **product rule.**

Summary

When $y = uv$ and u and v are functions of x, then

$$\frac{dy}{dx} = v\frac{du}{dx} + u\frac{dv}{dx}$$

Using functional notation: When $F(x) = f(x)\,g(x)$ then:

$$\mathbf{F'(x) = f(x)\,g'(x) + g(x)\,f'(x)}$$

Applying the product rule to $y = 3x^2 \sin x$:

let $u = 3x^2$ and $v = \sin x$

Then $\frac{dy}{dx} = (\sin x)\frac{d}{dx}(3x^2) + (3x^2)\frac{d}{dx}(\sin x)$

$$= (\sin x)(6x) + (3x^2)(\cos x)$$

$$= 6x \sin x + 3x^2 \cos x$$

i.e. $\mathbf{\frac{dy}{dx} = 3x\,(2 \sin x + x \cos x)}$

From the above it should be noted that the differential coefficient of a product **cannot** be obtained merely by differentiating each term and multiplying the two answers together. The above formula **must** be used whenever differentiating products.

(ii) Differentiation of a quotient

The function $y = \dfrac{3 \cos x}{5x^3}$ is a quotient of two terms in x, i.e. $3 \cos x$ and $5x^3$.

Let $u = 3 \cos x$ and $v = 5x^3$.

Let x increase by a small increment δx causing incremental changes in u, v and y of δu, δv and δy respectively.

Then $$y = \frac{3 \cos x}{5x^3}$$

$$= \frac{u}{v}$$

$$y + \delta y = \frac{u + \delta u}{v + \delta v}$$

$$(y + \delta y) - (y) = \frac{u + \delta u}{v + \delta v} - \frac{u}{v}$$

$$= \frac{uv + v\delta u - uv - u\delta v}{v(v + \delta v)}$$

i.e. $$\delta y = \frac{v\delta u - u\delta v}{v^2 + v\delta v}$$

Dividing both sides by δx gives:

$$\frac{\delta y}{\delta x} = \frac{v\dfrac{\delta u}{\delta x} - u\dfrac{\delta v}{\delta x}}{v^2 + v\delta v}$$

As $\delta x \to 0$, $\dfrac{\delta u}{\delta x} \to \dfrac{du}{dx}$, $\dfrac{\delta v}{\delta x} \to \dfrac{dv}{dx}$, $\dfrac{\delta y}{\delta x} \to \dfrac{dy}{dx}$ and $\delta v \to 0$

Hence $$\frac{dy}{dx} = \frac{v\dfrac{du}{dx} - u\dfrac{dv}{dx}}{v^2}$$

This is known as the **quotient rule.**

Summary

When $y = \dfrac{u}{v}$ and u and v are functions of x, then

$$\frac{dy}{dx} = \frac{v\dfrac{du}{dx} - u\dfrac{dv}{dx}}{v^2}$$

Using functional notation:

When $F(x) = \dfrac{f(x)}{g(x)}$, then $F'(x) = \dfrac{g(x)\, f'(x) - f(x)\, g'(x)}{[g(x)]^2}$

Applying the quotient rule to $y = \dfrac{3 \cos x}{5x^3}$:

Let $u = 3 \cos x$ and $v = 5x^3$

$$\text{Then } \frac{dy}{dx} = \frac{(5x^3)\frac{d}{dx}(3 \cos x) - (3 \cos x)\frac{d}{dx}(5x^3)}{(5x^3)^2}$$

$$= \frac{(5x^3)(-3 \sin x) - (3 \cos x)(15x^2)}{25x^6}$$

$$= \frac{-15x^2 (x \sin x + 3 \cos x)}{25x^6}$$

$$\text{i.e. } \mathbf{\frac{dy}{dx} = \frac{-3 (x \sin x + 3 \cos x)}{5x^4}}$$

From above it should be noted that the differential coefficient of a quotient **cannot** be obtained by merely differentiating each term and dividing the numerator by the denominator. The above formula **must** be used when differentiating quotients.

The first step when differentiating a product such as $y = uv$ or a quotient such as $y = \dfrac{u}{v}$ is to decide clearly which is the u part and which is the v part. When this has been decided differentiation involves substitution into the appropriate formula.

Worked Problems on differentiating products and quotients

Problem 1. Find the differential coefficient of $5x^2 \cos x$.

Let $y = 5x^2 \cos x$

Also, let $u = 5x^2$ and $v = \cos x$

Then $\dfrac{du}{dx} = 10x$ and $\dfrac{dv}{dx} = -\sin x$

$$\text{Then } \frac{dy}{dx} = v\frac{du}{dx} + u\frac{dv}{dx}$$

$$= (\cos x)(10x) + (5x^2)(-\sin x)$$

$$= 10x \cos x - 5x^2 \sin x$$

$$\text{i.e. } \mathbf{\frac{dy}{dx} = 5x (2 \cos x - x \sin x)}$$

Problem 2. Differentiate $3e^{2h} \sin h$ with respect to h.

Let $F(h) = 3e^{2h} \sin h$

Let $f(h) = 3e^{2h}$ and $g(h) = \sin h$

then $f'(h) = 6e^{2h}$ and $g'(h) = \cos h$

Then $F'(h) = g(h)\,f'(h) + f(h)\,g'(h)$
$= (\sin h)(6e^{2h}) + (3e^{2h})(\cos h)$
$= 6e^{2h} \sin h + 3e^{2h} \cos h$

i.e. $\mathbf{F'(h) = 3e^{2h}\,[2 \sin h + \cos h]}$

Problem 3. If $y = 7\sqrt{x} \ln 4x$ find $\dfrac{dy}{dx}$.

$$y = 7x^{\frac{1}{2}} \ln 4x$$

Let $u = 7x^{\frac{1}{2}}$ and $v = \ln 4x$

then $\dfrac{du}{dx} = \dfrac{7}{2}x^{-\frac{1}{2}}$ and $\dfrac{dv}{dx} = \dfrac{1}{x}$

Then $\dfrac{dy}{dx} = v\dfrac{du}{dx} + u\dfrac{dv}{dx}$

$$= (\ln 4x)\left[\frac{7}{2}x^{-\frac{1}{2}}\right] + [7x^{\frac{1}{2}}]\left[\frac{1}{x}\right]$$

$$= \frac{7}{2\sqrt{x}} \ln 4x + \frac{7}{\sqrt{x}}$$

i.e. $\mathbf{\dfrac{dy}{dx} = \dfrac{7}{2\sqrt{x}}(\ln 4x + 2)}$

Problem 4. Find the differential coefficients of: (a) $\tan x$; (b) $\cot x$; (c) $\sec x$; (d) $\operatorname{cosec} x$.

(a) Let $y = \tan x = \dfrac{\sin x}{\cos x}$

Differentiation of $\tan x$ is treated as a quotient with $u = \sin x$ and $v = \cos x$.

Then $\dfrac{du}{dx} = \cos x$ and $\dfrac{dv}{dx} = -\sin x$

$$\frac{dy}{dx} = \frac{v\dfrac{du}{dx} - u\dfrac{dv}{dx}}{v^2}$$

$$= \frac{(\cos x)(\cos x) - (\sin x)(-\sin x)}{(\cos x)^2}$$

$$= \frac{(\cos^2 x + \sin^2 x)}{(\cos x)^2}$$

$$= \frac{1}{\cos^2 x} \quad (\text{since } \cos^2 x + \sin^2 x = 1)$$

i.e. $\dfrac{dy}{dx} = \sec^2 x$

Hence, when $y = \tan x$, $\dfrac{dy}{dx} = \sec^2 x$

or, when $f(x) = \tan x$, $f'(x) = \sec^2 x$

(b) Let $y = \cot x = \dfrac{\cos x}{\sin x}$

Differentiation of $\cot x$ is treated as a quotient with $u = \cos x$ and $v = \sin x$.

Then $\dfrac{du}{dx} = -\sin x$ and $\dfrac{dv}{dx} = \cos x$

$$\frac{dy}{dx} = \frac{v\dfrac{du}{dx} - u\dfrac{dv}{dx}}{v^2}$$

$$= \frac{(\sin x)(-\sin x) - (\cos x)(\cos x)}{(\sin x)^2}$$

$$= \frac{-(\sin^2 x + \cos^2 x)}{\sin^2 x}$$

$$= \frac{-1}{\sin^2 x}$$

i.e. $\dfrac{dy}{dx} = -\operatorname{cosec}^2 x$

Hence when $y = \cot x$, $\dfrac{dy}{dx} = -\operatorname{cosec}^2 x$

or, when $f(x) = \cot x$, $f'(x) = -\operatorname{cosec}^2 x$

(c) Let $y = \sec x = \dfrac{1}{\cos x}$

Differentiation of $\sec x$ is treated as a quotient with $u = 1$ and $v = \cos x$.

Then $\dfrac{du}{dx} = 0$ and $\dfrac{dv}{dx} = -\sin x$

$$\frac{dy}{dx} = \frac{v\dfrac{du}{dx} - u\dfrac{dv}{dx}}{v^2}$$

$$= \frac{(\cos x)(0) - (1)(-\sin x)}{(\cos x)^2}$$

$$= \frac{\sin x}{\cos^2 x}$$

$$= \left[\frac{1}{\cos x}\right]\left[\frac{\sin x}{\cos x}\right]$$

i.e. $\dfrac{dy}{dx} = \sec x \tan x$

Hence when $y = \sec x$, $\dfrac{dy}{dx} = \sec x \tan x$

or, when $f(x) = \sec x$, $f'(x) = \sec x \tan x$

(d) Let $y = \operatorname{cosec} x = \dfrac{1}{\sin x}$

Differentiation of $\operatorname{cosec} x$ is treated as a quotient with $u = 1$ and $v = \sin x$.

Then $\frac{du}{dx} = 0$ and $\frac{dv}{dx} = \cos x$

$$\frac{dy}{dx} = \frac{v\frac{du}{dx} - u\frac{dv}{dx}}{v^2}$$

$$= \frac{(\sin x)(0) - (1)(\cos x)}{(\sin x)^2}$$

$$= \frac{-\cos x}{\sin^2 x}$$

$$= -\left[\frac{1}{\sin x}\right]\left[\frac{\cos x}{\sin x}\right]$$

i.e. $\frac{dy}{dx} = -\operatorname{cosec} x \cot x$

Hence when $y = \operatorname{cosec} x$, $\frac{dy}{dx} = -\operatorname{cosec} x \cot x$

or, when $f(x) = \operatorname{cosec} x$, $f'(x) = -\operatorname{cosec} x \cot x$

The differential coefficients of the six trigonometrical ratios may thus be summarised as below:

	y or $f(x)$	$\frac{dy}{dx}$ or $f'(x)$
1.	$\sin x$	$\cos x$
2.	$\cos x$	$-\sin x$
3.	$\tan x$	$\sec^2 x$
4.	$\sec x$	$\sec x \tan x$
5.	$\operatorname{cosec} x$	$-\operatorname{cosec} x \cot x$
6.	$\cot x$	$-\operatorname{cosec}^2 x$

Problem 5. If $f(t) = \frac{4e^{7t}}{\sqrt[3]{t^2}}$ find $f'(t)$

$$f(t) = \frac{4e^{7t}}{t^{\frac{2}{3}}}$$

Let $g(t) = 4e^{7t}$ and $h(t) = t^{\frac{2}{3}}$

then $g'(t) = 28e^{7t}$ and $h'(t) = \frac{2}{3}t^{-\frac{1}{3}}$

$$f'(t) = \frac{h(t)\,g'(t) - g(t)\,h'(t)}{[h(t)]^2} = \frac{(t^{\frac{2}{3}})(28e^{7t}) - (4e^{7t})(\frac{2}{3}t^{-\frac{1}{3}})}{(t^{\frac{2}{3}})^2}$$

$$= \frac{28t^{\frac{2}{3}}e^{7t} - \frac{8}{3}t^{-\frac{1}{3}}e^{7t}}{t^{\frac{4}{3}}} = \frac{28t^{\frac{2}{3}}e^{7t}}{t^{\frac{4}{3}}} - \frac{8t^{-\frac{1}{3}}e^{7t}}{3t^{\frac{4}{3}}}$$

$$= 28t^{-\frac{2}{3}}e^{7t} - \frac{8}{3}t^{-\frac{5}{3}}e^{7t}$$

$$= \frac{4}{3}e^{7t}\,t^{-\frac{5}{3}}(21t - 2)$$

i.e. $f'(t) = \dfrac{4e^{7t}}{3\sqrt[3]{t^5}}(21t - 2)$

(Note that initially, $f(t) = \dfrac{4e^{7t}}{t^{\frac{2}{3}}}$ could have been treated as a product $f(t) =$ $4e^{7t}\, t^{-\frac{2}{3}}$.)

Problem 6. Find the coordinates of the points on the curve $y = \dfrac{\frac{1}{3}(5 - 6x)}{3x^2 + 2}$ where the gradient is zero.

$$y = \frac{\frac{1}{3}(5 - 6x)}{3x^2 + 2}$$

Let $u = \frac{1}{3}(5 - 6x)$ and $v = 3x^2 + 2$

then $\dfrac{du}{dx} = -2$ and $\dfrac{dv}{dx} = 6x$

$$\frac{dy}{dx} = \frac{v\dfrac{du}{dx} - u\dfrac{dv}{dx}}{v^2} = \frac{(3x^2 + 2)(-2) - \frac{1}{3}(5 - 6x)(6x)}{(3x^2 + 2)^2}$$

$$= \frac{-6x^2 - 4 - 10x + 12x^2}{(3x^2 + 2)^2} = \frac{6x^2 - 10x - 4}{(3x^2 + 2)^2}$$

When the gradient is zero, $\dfrac{dy}{dx} = 0$

Hence
$$6x^2 - 10x - 4 = 0$$
$$2(3x^2 - 5x - 2) = 0$$
$$2(3x + 1)(x - 2) = 0$$

i.e. $x = -\frac{1}{3}$ or $x = 2$

Substituting in the original equation for y:

When $x = -\frac{1}{3}$, $y = \dfrac{\frac{1}{3}[5 - 6(-\frac{1}{3})]}{3(-\frac{1}{3})^2 + 2} = \dfrac{\frac{7}{3}}{\frac{7}{3}} = 1$

When $x = 2$, $y = \dfrac{\frac{1}{3}[5 - 6(2)]}{3(2)^2 + 2} = \dfrac{-\frac{7}{3}}{14} = -\frac{1}{6}$

Hence the coordinates of the points on the curve $y = \dfrac{\frac{1}{3}(5 - 6x)}{3x^2 + 2}$ where the gradient is zero are $(-\frac{1}{3}, 1)$ and $(2, -\frac{1}{6})$.

Problem 7. Differentiate $\dfrac{\sqrt{x}\sin x}{2e^{4x}}$ with respect to x.

The function $\dfrac{\sqrt{x}\sin x}{2e^{4x}}$ is a quotient, although the numerator (i.e. $\sqrt{x}\sin x$) is a product.

Let $\quad y = \dfrac{x^{\frac{1}{2}} \sin x}{2e^{4x}}$

Let $\quad u = x^{\frac{1}{2}} \sin x$ and $v = 2e^{4x}$

then $\quad \dfrac{du}{dx} = (x^{\frac{1}{2}})(\cos x) + (\sin x)(\frac{1}{2}x^{-\frac{1}{2}})$

and $\quad \dfrac{dv}{dx} = 8e^{4x}$

$$\frac{dy}{dx} = \frac{v\dfrac{du}{dx} - u\dfrac{dv}{dx}}{v^2}$$

$$= \frac{(2e^{4x})(x^{\frac{1}{2}}\cos x + \frac{1}{2}x^{-\frac{1}{2}}\sin x) - (x^{\frac{1}{2}}\sin x)(8e^{4x})}{(2e^{4x})^2}$$

Dividing throughout by $2e^{4x}$ gives:

$$\frac{dy}{dx} = \frac{x^{\frac{1}{2}}\cos x + \frac{1}{2}x^{-\frac{1}{2}}\sin x - 4x^{\frac{1}{2}}\sin x}{2e^{4x}}$$

Hence $\quad \dfrac{dy}{dx} = \dfrac{\sqrt{x}\cos x + \sin x\left(\dfrac{1}{2\sqrt{x}} - 4\sqrt{x}\right)}{2e^{4x}}$

or $\quad \dfrac{dy}{dx} = \dfrac{\sqrt{x}\cos x + \left(\dfrac{1-8x}{2\sqrt{x}}\right)\sin x}{2e^{4x}}$

Further problems on differentiating products and quotients may be found in Section 6 (Problems 31–64).

4. Differentiation by substitution

The function $y = (4x - 3)^7$ can be differentiated by firstly multiplying $(4x - 3)$ by itself seven times, and then differentiating each term produced in turn. This would be a long process. In this type of function a substitution is made.

Let $u = 4x - 3$, then instead of $y = (4x - 3)^7$ we have $y = u^7$.

An important rule that is used when differentiating by substitution is:

$$\frac{dy}{dx} = \frac{dy}{du}\frac{du}{dx}$$

This is often known as the **chain rule.**

From above, $y = (4x - 3)^7$
If $u = 4x - 3$ then $y = u^7$

Thus $\frac{dy}{du} = 7u^6$ and $\frac{du}{dx} = 4$

Hence since $\frac{dy}{dx} = \frac{dy}{du}\frac{du}{dx}$

$$\frac{dy}{dx} = (7u^6)(4) = 28u^6$$

Rewriting $u = 4x - 3$, $\mathbf{\frac{dy}{dx} = 28\,(4x - 3)^6}$

Since y is a function of u, and u is a function of x, then y is a 'function of a function' of x. The method of obtaining differential coefficients by making substitutions is often called the 'function of a function process'.

Worked problems on differentiation by substitution

Problem 1. Differentiate sin $(6x + 1)$.

Let $y = \sin(6x + 1)$
and $u = 6x + 1$

Then $y = \sin u$, giving $\frac{dy}{du} = \cos u$

and $\frac{du}{dx} = 6$

Using the 'differentiation by substitution' formula: $\frac{dy}{dx} = \frac{dy}{du}\frac{du}{dx}$ gives $\frac{dy}{dx} = (\cos u)(6) = 6\cos u$

Rewriting $u = 6x + 1$ gives:

$$\mathbf{\frac{dy}{dx} = 6\cos(6x + 1)}$$

Note that this result could have been obtained by firstly differentiating the trigonometric function (i.e. differentiating sin $f(x)$) giving cos $f(x)$ and then multiplying by the differential coefficient of $f(x)$, i.e. 6.

Problem 2. Find the differential coefficient of $(3t^4 - 2t)^5$.

Let $y = (3t^4 - 2t)^5$
and $u = (3t^4 - 2t)$

Then $y = u^5$, giving $\frac{dy}{du} = 5u^4$

and $\frac{du}{dt} = 12t^3 - 2$

Using the 'chain rule': $\frac{dy}{dt} = \frac{dy}{du}\frac{du}{dt}$ gives $\frac{dy}{dt} = (5u^4)(12t^3 - 2)$

Rewriting $u = 3t^4 - 2t$ gives:

$$\mathbf{\frac{dy}{dt} = 5(3t^4 - 2t)^4 (12t^3 - 2)}$$

Note that this result could have been obtained by firstly differentiating the bracket, giving $5[f(x)]^4$ and then multiplying this result by the differential coefficient of $f(x)$ (i.e. $(12t^3 - 2)$).

Problem 3. If $y = 5 \operatorname{cosec}(3\sqrt{x} + 2x)$ find $\dfrac{dy}{dx}$.

$$y = 5 \operatorname{cosec}(3\sqrt{x} + 2x)$$

Let $u = (3\sqrt{x} + 2x)$ then $\dfrac{du}{dx} = \dfrac{3}{2\sqrt{x}} + 2$

Thus $y = 5 \operatorname{cosec} u$ and $\dfrac{dy}{du} = -5 \operatorname{cosec} u \cot u$

Now $\dfrac{dy}{dx} = \dfrac{dy}{du}\dfrac{du}{dx} = (-5 \operatorname{cosec} u \cot u)\left(\dfrac{3}{2\sqrt{x}} + 2\right)$

Rewriting $u = 3\sqrt{x} + 2x$ gives:

$$\mathbf{\frac{dy}{dx} = -5\left(\frac{3}{2\sqrt{x}} + 2\right) \operatorname{cosec}(3\sqrt{x} + 2x) \cot(3\sqrt{x} + 2x)}$$

In a similar way to Problem 1, this result could have been obtained by firstly differentiating $5 \operatorname{cosec} f(x)$ giving $-5 \operatorname{cosec} f(x) \cot f(x)$ and then multiplying this result by the differential coefficient of $f(x)$.

Problem 4. If $p = 2\tan^5 v$ find $\dfrac{dp}{dv}$.

$$p = 2\tan^5 v$$

Let $u = \tan v$ then $\dfrac{du}{dv} = \sec^2 v$

Then $p = 2u^5$ and $\dfrac{dp}{du} = 10u^4$

Now $\dfrac{dp}{dv} = \dfrac{dp}{du}\dfrac{du}{dv} = (10u^4)(\sec^2 v)$

Rewriting $u = \tan v$ gives:

$$\frac{dp}{dv} = 10(\tan v)^4 \sec^2 v$$

$$\mathbf{\frac{dp}{dv} = 10\tan^4 v \sec^2 v}$$

In a similar way to Problem 2, this result could have been obtained by firstly differentiating the bracket (i.e. differentiating $2[f(v)]^5$) giving $10[f(v)]^4$ and then multiplying this result by the differential coefficient of $f(v)$.

Problem 5. Write down the differential coefficients of the following: (a) $\sqrt{(4x^2 + x - 3)}$; (b) $2\sec^3 t$; (c) $4\cot(5g^2 + 2)$; (d) $\sqrt{(4x^3 + 2)^3} \cos(3x^2 + 2)$.

(a) $f(x) = \sqrt{(4x^2 + x - 3)} = (4x^2 + x - 3)^{\frac{1}{2}}$

$$f'(x) = \tfrac{1}{2}(4x^2 + x - 3)^{-\frac{1}{2}}(8x + 1)$$

$$= \mathbf{\frac{8x + 1}{2\sqrt{(4x^2 + x - 3)}}}$$

(b) $f(t) = 2\sec^3 t = 2(\sec t)^3$

$f'(t) = 6(\sec t)^2 (\sec t \tan t)$

$= \mathbf{6\sec^3 t \tan t}$

(c) $f(g) = 4\cot(5g^2 + 2)$

$f'(g) = 4[-\operatorname{cosec}^2(5g^2 + 2)]\,(10g)$

$= \mathbf{-40g\operatorname{cosec}^2(5g^2 + 2)}$

(d) $f(x) = \sqrt{(4x^3 + 2)^3}\cos(3x^2 + 2)$

$= (4x^3 + 2)^{\frac{3}{2}}\cos(3x^2 + 2)$ (i.e. a product)

$f'(x) = [\cos(3x^2 + 2)]\,[\tfrac{3}{2}(4x^3 + 2)^{\frac{1}{2}}(12x^2)] + [(4x^3 + 2)^{\frac{3}{2}}]$
$[(-\sin(3x^2 + 2))(6x)]$

$= \mathbf{6x\sqrt{(4x^3 + 2)}\,[3x\cos(3x^2 + 2) - (4x^3 + 2)\sin(3x^2 + 2)]}$

Use of functional notation in differentiation by substitution

An alternative method of dealing with a differentiation-by-substitution problem is by using functional notation. Before attempting to differentiate such a problem it is important to realise how a 'function of a function' is built up, and this is shown below.

'Function of a function'

Let $F(x) = \sin(x^2)$. For the input of x to the 'function boxes' shown in Fig. 2 the expression $F(x) = \sin(x^2)$ is obtained by operating firstly on x with the square function, which we will call f, and then on x^2 with the sine function, which we will call g.

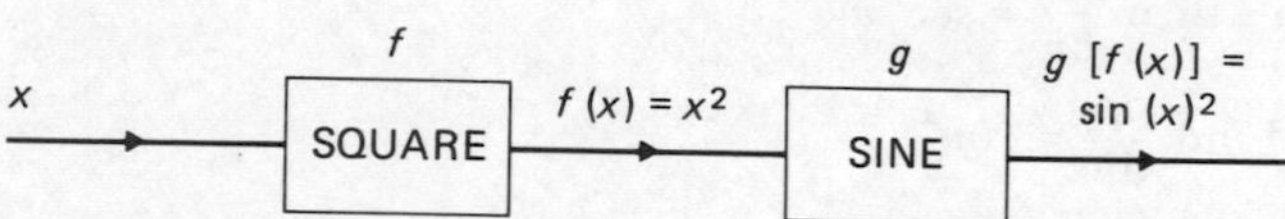

Figure 2

Hence if $F(x) = \sin(x^2)$
then from Fig. 2, $\mathbf{F(x) = g[f(x)]}$ ... (1)

Similarly, the function $G(x) = \cos^2(3x + 4)$ is built up as shown in Fig. 3, the three operations being labelled f, g and h.

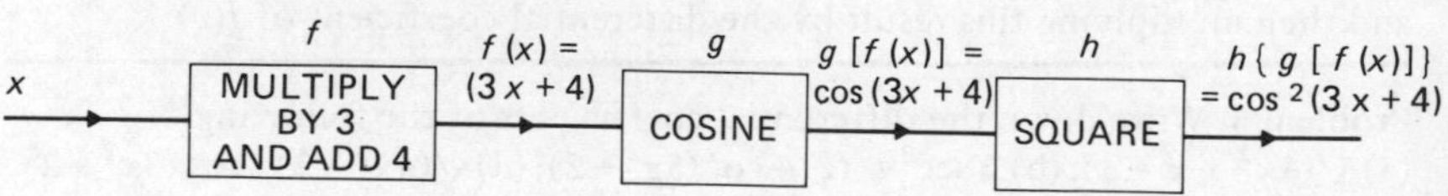

Figure 3

Hence if $G(x) = \cos^2(3x + 4)$
then from Fig. 3, $\mathbf{G(x) = h\{g[f(x)]\}}$ ... (2)

To differentiate such functions as shown in equations (1) and (2) above using functional notation, the relationships used are given below.

If $\quad F(x) = g[f(x)]$

then $\mathbf{F'(x) = g'[f(x)]\, f'(x)}$

If $\quad G(x) = h\,\{g[f(x)]\}$

then $\mathbf{G'(x) = h'\{g[f(x)]\}\, g'[f(x)]\, f'(x)}$

Similarly, if a function is built up by four operations then the differential coefficient will be composed of the product of four separate differentials. Such a 'chain-rule operation' is easy to apply as long as the components of the function can be identified.

For example, when $F(x) = \sin(x^2)$

$$f(x) = x^2, \text{ giving } f'(x) = 2x$$
$$g[f(x)] = \sin(x^2), \text{ giving } g'[f(x)] = \cos(x^2)$$

Hence $\mathbf{F'(x)} = g'[f(x)]\, f'(x)$
$= [\cos(x^2)]\,[2x] = 2x\cos(x^2)$

Similarly, when $G(x) = \cos^2(3x + 4)$

$$f(x) = (3x + 4), \text{ giving } f'(x) = 3$$
$$g[f(x)] = \cos(3x + 4), \text{ giving } g'[f(x)] = -\sin(3x + 4)$$
$$h\{g[f(x)]\} = [\cos(3x + 4)]^2, \text{ giving } h'\{g[f(x)]\} = 2[\cos(3x + 4)]$$

Hence $\mathbf{G'(x)} = h'\{g[f(x)]\}\, g'[f(x)]\, f'(x)$
$= [2\cos(3x + 4)]\,[-\sin(3x + 4)]\,[3]$
$= \mathbf{-6\cos(3x + 4)\sin(3x + 4)}$

Worked problems on differentiation by substitution using functional notation

In Problems 1–4 a full explanation is given on how the particular function is built up and then differentiated. When conversant with such problems it is unnecessary to write down such detail. Problem 5 shows how the differential coefficient may be written down on sight.

Problem 1. Find the differential coefficient of $\tan(2x^3 + 7)$.

Let $F(x) = \tan(2x^3 + 7)$

The function $F(x)$ is built up as shown in Fig. 4.

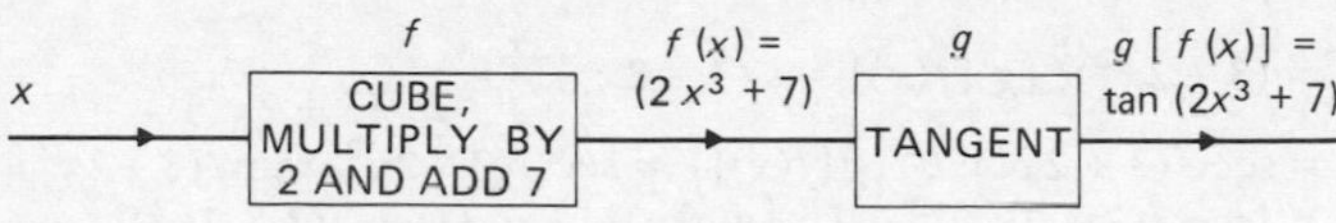

Figure 4

$$f(x) = (2x^3 + 7); f'(x) = 6x^2$$
$$g[f(x)] = \tan(2x^3 + 7); g'[f(x)] = \sec^2(2x^3 + 7)$$
$$F'(x) = g'[f(x)]\, f'(x)$$
$$= [\sec^2(2x^3 + 7)]\ [6x^2]$$
i.e. $$\mathbf{F'(x) = 6x^2 \sec^2(2x^3 + 7)}$$

Problem 2. If $F(\theta) = \cot^2(8\theta^2 + 5)$ find $F'(\theta)$.

The function $F(\theta)$ is built up as shown in Fig. 5.

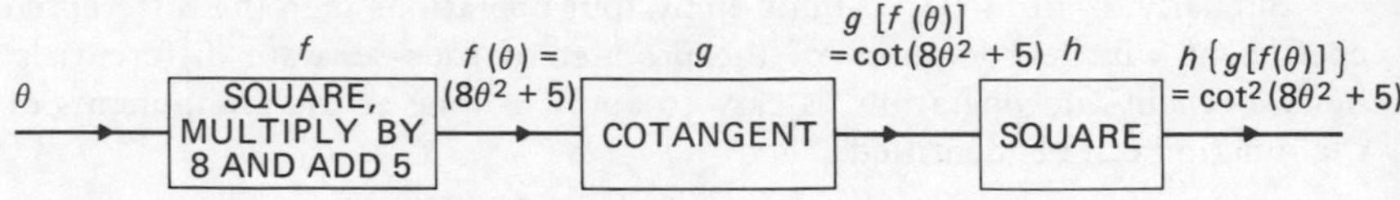

Figure 5

$$f(\theta) = (8\theta^2 + 5); f'(\theta) = 16\theta$$
$$g[f(\theta)] = \cot(8\theta^2 + 5); g'[f(\theta)] = -\operatorname{cosec}^2(8\theta^2 + 5)$$
$$h\left\{g[f(\theta)]\right\} = [\cot(8\theta^2 + 5)]^2; h'\left\{g[f(\theta)]\right\} = 2[\cot(8\theta^2 + 5)]$$
$$F'(\theta) = h'\left\{g[f(\theta)]\right\}\ g'[f(\theta)]\ f'(\theta)$$
$$= [2\cot(8\theta^2 + 5)]\ [-\operatorname{cosec}^2(8\theta^2 + 5)]\ [16\theta]$$
i.e. $$\mathbf{F'(\theta) = -32\theta \cot(8\theta^2 + 5) \operatorname{cosec}^2(8\theta^2 + 5)}$$

Problem 3. Differentiate $\sec^3\sqrt{(3 + 2x^2)}$.

Let $F(x) = \sec^3\sqrt{(3 + 2x^2)} = \sec^3(3 + 2x^2)^{\frac{1}{2}} = [\sec(3 + 2x^2)^{\frac{1}{2}}]^3$
The function $F(x)$ is built up as shown in Fig. 6.

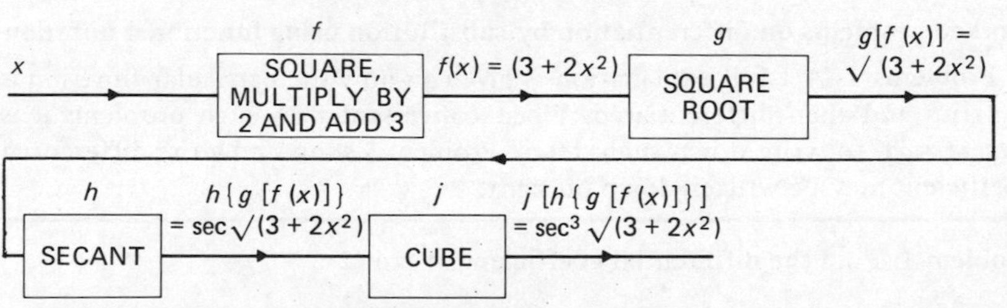

Figure 6

$$f(x) = (3 + 2x^2); f'(x) = 4x$$
$$g[f(x)] = (3 + 2x^2)^{\frac{1}{2}}; g'[f(x)] = \tfrac{1}{2}(3 + 2x^2)^{-\frac{1}{2}} = \frac{1}{2\sqrt{(3 + 2x^2)}}$$
$$h\left\{g[f(x)]\right\} = \sec\sqrt{(3 + 2x^2)}; h'\left\{g[f(x)]\right\} = \sec\sqrt{(3 + 2x^2)}\tan\sqrt{(3 + 2x^2)}$$
$$j[h\left\{g[f(x)]\right\}] = [\sec\sqrt{(3 + 2x^2)}]^3; j'[h\left\{g[f(x)]\right\}] = 3[\sec\sqrt{(3 + 2x^2)}]^2$$
$$= 3\sec^2\sqrt{(3 + 2x^2)}$$

Hence $F'(x) = j'[h\{g[f(x)]\}]\ h'\{g[f(x)]\}\ g'[f(x)]\ f'(x)$

$$= [3\sec^2\sqrt{(3+2x^2)}][\sec\sqrt{(3+2x^2)}\tan\sqrt{(3+2x^2)}]\left[\frac{1}{2\sqrt{(3+2x^2)}}\right][4x]$$

i.e. $$\mathbf{F'(x) = \frac{6x}{\sqrt{(3+2x^2)}}\sec^3\sqrt{(3+2x^2)}\tan\sqrt{(3+2x^2)}}$$

Problem 4. If $F(x) = \ln[\sin^2(3x-2)]$ find $F'(x)$.

The function $F(x)$ is built up as shown in Fig. 7.

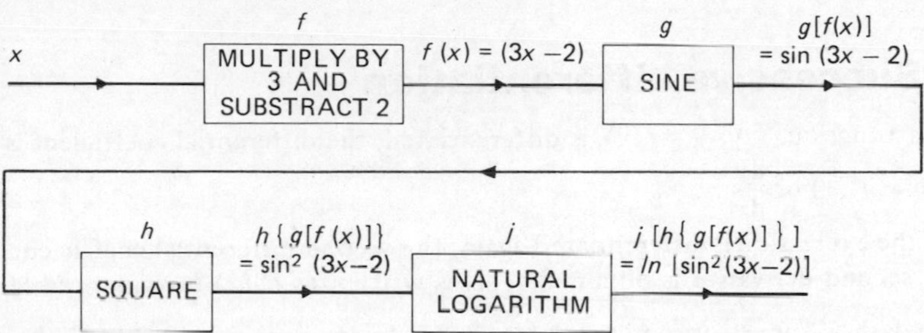

Figure 7

$$f(x) = (3x-2);\ f'(x) = 3$$
$$g[f(x)] = \sin(3x-2);\ g'[f(x)] = \cos(3x-2)$$
$$h\{g[f(x)]\} = [\sin(3x-2)]^2;\ h'\{g[f(x)]\} = 2[\sin(3x-2)]$$
$$j[h\{g[f(x)]\}] = \ln[\sin^2(3x-2)];\ j'[h\{g[f(x)]\}] = \frac{1}{\sin^2(3x-2)}$$

Hence $F'(x) = j'[h\{g[f(x)]\}]\ h'\{g[f(x)]\}\ g'[f(x)]\ f'(x)$

$$= \left[\frac{1}{\sin^2(3x-2)}\right][2\sin(3x-2)][\cos(3x-2)][3]$$

i.e. $$\mathbf{F'(x)} = \frac{6\cos(3x-2)}{\sin(3x-2)} = \mathbf{6\cot(3x-2)}$$

Problem 5. Write down the differential coefficients of the following functions: (a) $3\cos^4(5x+6)$; (b) $5\,e^{\tan 4t}$; (c) $5\ln[\mathrm{cosec}^3\sqrt{(5x^2-2)}]$.

(a) $F(x) = 3\cos^4(5x+6)$
$F'(x) = 3[4\cos^3(5x+6)][-\sin(5x+6)][5]$
$= \mathbf{-60\cos^3(5x+6)\sin(5x+6)}$

(b) $F(t) = 5\,e^{\tan 4t}$
$F'(t) = 5[e^{\tan 4t}][\sec^2 4t][4]$
$= \mathbf{20\,e^{\tan 4t}\sec^2 4t}$

(c) $F(x) = 5 \ln[\text{cosec}^3 \sqrt{(5x^2 - 2)}]$

$$F'(x) = 5 \left[\frac{1}{\text{cosec}^3 \sqrt{(5x^2 - 2)}}\right] [3\ \text{cosec}^2 \sqrt{(5x^2 - 2)}]$$
$$[-\text{cosec}\sqrt{(5x^2 - 2)}\ \cot\sqrt{(5x^2 - 2)}]\ [\tfrac{1}{2}(5x^2 - 2)^{-\frac{1}{2}}]\ [10x]$$

i.e. $$F'(x) = \frac{-75x \cot\sqrt{(5x^2 - 2)}}{\sqrt{(5x^2 - 2)}}$$

Further problems on differentiation by substitution may be found in Section 6 (Problems 65–131).

5. Successive differentiation

When a function, say, $y = f(x)$, is differentiated, the differential coefficient is written as $f'(x)$ or $\frac{dy}{dx}$.

If the expression is differentiated again, the second differential coefficient or the second derivative is obtained. This is written as $f''(x)$ (pronounced 'f double-dash x') or $\frac{d^2y}{dx^2}$ (pronounced 'dee two y by dee x squared'). (Similarly, if differentiated again the third differential coefficient or third derivative is obtained, and is written as $f'''(x)$ or $\frac{d^3y}{dx^3}$, and so on.)

Worked problems on successive differentiation

Problem 1. If $f(x) = 3x^4 + 2x^3 + x - 1$, find $f'(x)$ and $f''(x)$.

$$f(x) = 3x^4 + 2x^3 + x - 1$$
$$f'(x) = (3)(4)x^3 + (2)(3)x^2 + 1 - 0$$
$$= 12x^3 + 6x^2 + 1$$
$$f''(x) = (12)(3)x^2 + (6)(2)x + 0$$
$$= 36x^2 + 12x$$

Problem 2. If $y = \frac{4}{3}x^3 - \frac{2}{x^2} + \frac{1}{3x} - \sqrt{x}$ find $\frac{d^2y}{dx^2}$ and $\frac{d^3y}{dx^3}$.

$$y = \frac{4}{3}x^3 - \frac{2}{x^2} + \frac{1}{3x} - \sqrt{x}$$
$$= \frac{4}{3}x^2 - 2x^{-2} + \frac{1}{3}x^{-1} - x^{\frac{1}{2}}$$

$$\frac{dy}{dx} = [\tfrac{4}{3}]\ (3)\ x^2 - (2)(-2)x^{-3} + \tfrac{1}{3}(-1)x^{-2} - [\tfrac{1}{2}]\ x^{-\frac{1}{2}}$$
$$= 4x^2 + 4x^{-3} - \tfrac{1}{3}x^{-2} - \tfrac{1}{2}x^{-\frac{1}{2}}$$

$$\frac{d^2y}{dx^2} = (4)(2)x + (4)(-3)x^{-4} - [\tfrac{1}{3}](-2)x^{-3} - [\tfrac{1}{2}][-\tfrac{1}{2}]x^{-\frac{3}{2}}$$

$$= 8x - 12x^{-4} + \tfrac{2}{3}x^{-3} + \tfrac{1}{4}x^{-\frac{3}{2}}$$

i.e. $$\frac{d^2y}{dx^2} = 8x - \frac{12}{x^4} + \frac{2}{3x^3} + \frac{1}{4\sqrt{x^3}}$$

$$\frac{d^3y}{dx^3} = 8 - (12)(-4)x^{-5} + [\tfrac{2}{3}](-3)x^{-4} + [\tfrac{1}{4}][-\tfrac{3}{2}]x^{-\frac{5}{2}}$$

$$= 8 + 48x^{-5} - 2x^{-4} - \tfrac{3}{8}x^{-\frac{5}{2}}$$

i.e. $$\frac{d^3y}{dx^3} = 8 + \frac{48}{x^5} - \frac{2}{x^4} - \frac{3}{8\sqrt{x^5}}$$

Problem 3. If $f(t) = 3 \ln [\cos^3(4t^2 - 1)]$, evaluate $f''(t)$ when $t = \frac{1}{2}$.

$$f(t) = 3 \ln [\cos^3(4t^2 - 1)]$$

$$f'(t) = 3\left[\frac{1}{\cos^3(4t^2-1)}\right][3\cos^2(4t^2-1)][-\sin(4t^2-1)][8t]$$

$$= -\frac{72t\cos^2(4t^2-1)\sin(4t^2-1)}{\cos^3(4t^2-1)}$$

$$= -72t\,\frac{\sin(4t^2-1)}{\cos(4t^2-1)}$$

$$= -72t\tan(4t^2-1)$$

Thus $f'(t)$ is a product of the two terms, $(-72t)$ and $\tan(4t^2 - 1)$.

Hence $f''(t) = [-72t][\sec^2(4t^2-1)][8t] + [\tan(4t^2-1)][-72]$

$= -72[8t^2\sec^2(4t^2-1) + \tan(4t^2-1)]$

When $t = \frac{1}{2}$, $f''(t) = -72[8(\frac{1}{2})^2\sec^2\{4(\frac{1}{2})^2 - 1\} + \tan\{4(\frac{1}{2})^2 - 1\}]$

$= -72[2\sec^2 0 + \tan 0]$

$= -72[2(1)^2 + 0]$

i.e. $\quad$ **$f''(t) = -144$**

Problem 4. If $y = Ae^{2x} + Be^{-3x}$ prove that $\dfrac{d^2y}{dx^2} + \dfrac{dy}{dx} - 6y = 0$.

$$y = Ae^{2x} + Be^{-3x}$$

$$\frac{dy}{dx} = 2Ae^{2x} - 3Be^{-3x}$$

$$\frac{d^2y}{dx^2} = 4Ae^{2x} + 9Be^{-3x}$$

$$6y = 6(Ae^{2x} + Be^{-3x}) = 6Ae^{2x} + 6Be^{-3x}$$

Substituting into $\dfrac{d^2y}{dx^2} + \dfrac{dy}{dx} - 6y$ gives:

$(4Ae^{2x} + 9Be^{-3x}) + (2Ae^{2x} - 3Be^{-3x}) - (6Ae^{2x} + 6Be^{-3x})$

$= 4Ae^{2x} + 9Be^{-3x} + 2Ae^{2x} - 3Be^{-3x} - 6Ae^{2x} - 6Be^{-3x} = 0$

Thus $\frac{d^2y}{dx^2} + \frac{dy}{dx} - 6y = 0$

(Note that an equation of the form $\frac{d^2y}{dx^2} + \frac{dy}{dx} - 6y = 0$ is known as a 'differential equation' and such equations are discussed in Chapter 9.)

Further problems on successive differentiation may be found in the following section (6) (Problems 132–156).

6. Further problems

Differentiation of common functions

Find the differential coefficients with respect to x of the functions in Problems 1–6.

1. (a) x^4 (b) x^6 (c) x^9 (d) $x^{3.2}$ (e) $x^{4.7}$
   (a) $[4x^3]$ (b) $[6x^5]$ (c) $[9x^8]$ (d) $[3.2x^{2.2}]$ (e) $[4.7x^{3.7}]$
2. (a) $3x^3$ (b) $4x^7$ (c) $2x^{10}$ (d) $4.6x^{1.5}$ (e) $6x^{5.4}$
   (a) $[9x^2]$ (b) $[28x^6]$ (c) $[20x^9]$ (d) $[6.9x^{0.5}]$
   (e) $[32.4x^{4.4}]$
3. (a) x^{-2} (b) x^{-3} (c) x^{-5} (d) $\frac{1}{x}$ (e) $-\frac{1}{x^3}$ (f) $\frac{1}{x^{10}}$
   (a) $[-2x^{-3}]$ (b) $[-3x^{-4}]$ (c) $[-5x^{-6}]$ (d) $\left[-\frac{1}{x^2}\right]$
   (e) $\left[\frac{3}{x^4}\right]$ (f) $\left[-\frac{10}{x^{11}}\right]$
4. (a) $4x^{-1}$ (b) $-5x^{-4}$ (c) $3x^{-7}$ (d) $-\frac{6}{x^2}$ (e) $\frac{4}{3x^5}$ (f) $\frac{2}{5x^{1.4}}$
   (a) $[-4x^{-2}]$ (b) $[20x^{-5}]$ (c) $[-21x^{-8}]$ (d) $\left[\frac{12}{x^3}\right]$
   (e) $\left[-\frac{20}{3x^6}\right]$ (f) $\left[\frac{-2.8}{5x^{2.4}}\right]$
5. (a) $x^{\frac{7}{2}}$ (b) $x^{\frac{3}{4}}$ (c) $x^{-\frac{3}{2}}$ (d) $\frac{1}{x^{\frac{1}{2}}}$ (e) $-\frac{1}{x^{\frac{4}{3}}}$ (f) $\frac{2}{3x^{\frac{7}{4}}}$
   (a) $[\frac{7}{2}x^{\frac{5}{2}}]$ (b) $[\frac{3}{4}x^{-\frac{1}{4}}]$ (c) $[-\frac{3}{2}x^{-\frac{5}{2}}]$ (d) $\left[\frac{-1}{2x^{\frac{3}{2}}}\right]$ (e) $\left[\frac{4}{3x^{\frac{7}{3}}}\right]$
   (f) $\left[\frac{-7}{6x^{\frac{11}{4}}}\right]$
6. (a) $\frac{\sqrt{x}}{2}$ (b) $\sqrt{x^3}$ (c) $\sqrt[3]{x^2}$ (d) $4\sqrt{x^5}$ (e) $\frac{3}{5\sqrt{x^7}}$ (f) $\frac{-1}{2\sqrt[4]{x^9}}$
   (a) $\left[\frac{1}{4\sqrt{x}}\right]$ (b) $\left[\frac{3}{2}\sqrt{x}\right]$ (c) $\left[\frac{2}{3\sqrt[3]{x}}\right]$ (d) $[10\sqrt{x^3}]$
   (e) $\left[\frac{-21}{10\sqrt{x^9}}\right]$ (f) $\left[\frac{9}{8\sqrt[4]{x^{13}}}\right]$

Differentiate the functions in Problems 7–26 with respect to the variable:

7. (a) $4u^3$ (b) $\frac{3}{2}t^4$ (a) $[12u^2]$ (b) $[6t^3]$
8. (a) $5v^2$ (b) $1.4z^5$ (a) $[10v]$ (b) $[7z^4]$
9. (a) $\dfrac{4}{a}$ (b) $\dfrac{3}{2S^2}$ (a) $\left[-\dfrac{4}{a^2}\right]$ (b) $\left[-\dfrac{3}{S^3}\right]$
10. (a) $\dfrac{7}{4y^3}$ (b) $3m^{-4}$ (a) $\left[-\dfrac{21}{4y^4}\right]$ (b) $\left[-12m^{-5}\right]$
11. (a) $\sqrt{b}$ (b) $5\sqrt{c^3}$ (a) $\left[\dfrac{1}{2\sqrt{b}}\right]$ (b) $\left[\dfrac{15}{2}\sqrt{c}\right]$
12. (a) $\dfrac{1}{\sqrt{e}}$ (b) $g^{\frac{5}{3}}$ (a) $\left[-\dfrac{1}{2\sqrt{e^3}}\right]$ (b) $[\frac{5}{3}g^{\frac{2}{3}}]$
13. (a) $4\sqrt[3]{k^2}$ (b) $\dfrac{3}{5\sqrt[4]{x^5}}$ (a) $\left[\dfrac{8}{3\sqrt[3]{k}}\right]$ (b) $\left[\dfrac{-3}{4\sqrt[4]{x^9}}\right]$
14. $5x^2 - \dfrac{1}{\sqrt{x^7}}$ $\left[10x + \dfrac{7}{2\sqrt{x^9}}\right]$
15. $3\left(2u - u^{-\frac{1}{2}} + \dfrac{4}{5u}\right)$ $\left[3\left(2 + \dfrac{u^{-\frac{3}{2}}}{2} - \dfrac{4}{5u^2}\right)\right]$
16. $\dfrac{1}{x}\left(3x^3 - \dfrac{2}{x} + \dfrac{\sqrt{x}}{5} + 1\right)$ $\left[6x + \dfrac{4}{x^3} - \dfrac{1}{10\sqrt{x^3}} - \dfrac{1}{x^2}\right]$
17. $\dfrac{3x^2 - 2\sqrt{x} - 5\sqrt[4]{x^3}}{x^2}$ $\left[\dfrac{3}{\sqrt{x^5}} + \dfrac{25}{4\sqrt[4]{x^9}}\right]$
18. $(t + 1)^2$ $[2(t + 1)]$
19. $(3\theta - 1)^2$ $[6(3\theta - 1)]$
20. $(f - 1)^4$ $[4(f^3 - 3f^2 + 3f - 1)$ or $4(f - 1)^3]$
21. (a) $5 \sin \theta$ (b) $4 \cos x$ (a) $[5 \cos \theta]$ (b) $[-4 \sin x]$
22. (a) $3(\sin t + 2 \cos t)$ (b) $7 \sin x - 2 \cos x$
(a) $[3(\cos t - 2 \sin t)]$ (b) $[7 \cos x + 2 \sin x]$
23. (a) e^{3x} (b) e^{-4y} (a) $[3e^{3x}]$ (b) $[-4e^{-4y}]$
24. (a) $6e^{2x}$ (b) $\dfrac{4}{e^{7t}}$ (a) $[12e^{2x}]$ (b) $\left[\dfrac{-28}{e^{7t}}\right]$
25. (a) $3(e^{8y} - e^{3y})$ (b) $-2(3e^{9x} - 4e^{-2x})$
(a) $[3(8e^{8y} - 3e^{3y})]$ (b) $[-2(27e^{9x} + 8e^{-2x})]$
26. (a) $\ln 5b$ (b) $4 \ln 3g$ (a) $\left[\dfrac{1}{b}\right]$ (b) $\left[\dfrac{4}{g}\right]$
27. Find the gradient of the curve $y = 4x^3 - 3x^2 + 2x - 4$ at the points (0, −4) and (1, −1). [2, 8]
28. What are the coordinates of the point on the graph of $y = 5x^2 - 2x + 1$ where the gradient is zero. $[(\frac{1}{5}, \frac{4}{5})]$
29. Find the point on the curve $f(\theta) = 4\sqrt[3]{\theta^4} + 2$ where the gradient is $10\frac{2}{3}$. [(8, 66)]
30. If $f(x) = \dfrac{5x^2}{2} - 6x + 3$ find the coordinates at the point at which the gradient is: (a) 4; and (b) −6. (a) [(2, 1)] (b) [(0, 3)]

Differentiate the products in Problems 31–45 with respect to the variable and express your answers in their simplest form:

31. $3x^3 \sin x$ $\quad [3x^2(x \cos x + 3 \sin x)]$
32. $\sqrt{t^3} \cos t$ $\quad [\sqrt{t}(\frac{3}{2} \cos t - t \sin t)]$
33. $(3x^2 - 4x + 2)(2x^3 + x - 1)$ $\quad [(30x^4 - 32x^3 + 21x^2 - 14x + 6)]$
34. $2 \sin \theta \cos \theta$ $\quad [2(\cos^2 \theta - \sin^2 \theta)]$
35. $5e^{2a} \sin a$ $\quad [5e^{2a}(\cos a + 2 \sin a)]$
36. $e^{7y} \cos y$ $\quad [e^{7y}(7 \cos y - \sin y)]$
37. $b^3 \ln 2b$ $\quad [b^2(1 + 3 \ln 2b)]$
38. $3\sqrt{x}e^{4x}$ $\quad \left[3e^{4x}\left(\frac{8x+1}{2\sqrt{x}}\right)\right]$
39. $e^t \ln t$ $\quad \left[e^t\left(\frac{1}{t} + \ln t\right)\right]$
40. $e^{2d}(4d^2 - 3d + 1)$ $\quad [e^{2d}(8d^2 + 2d - 1)]$
41. $3\sqrt{f^5} \ln 5f$ $\quad [\sqrt{f^3}(1 + \frac{5}{2} \ln 5f)]$
42. $2 \sin g \ln g$ $\quad \left[2\left(\frac{1}{g} \sin g + \ln g \cos g\right)\right]$
43. $6e^{5m} \sin m$ $\quad [6e^{5m}(\cos m + 5 \sin m)]$
44. $\sqrt{x}(1 + \sin x)$ $\quad \left[\frac{2x \cos x + \sin x + 1}{2\sqrt{x}}\right]$
45. $e^v \ln v \sin v$ $\quad \left[e^v\left\{(\sin v + \cos v) \ln v + \frac{\sin v}{v}\right\}\right]$

Differentiate the quotients in Problems 46–62 with respect to the variable and express your answers in their simplest form:

46. $\frac{4x}{x^2 - 1}$ $\quad \left[\frac{-4(x^2+1)}{(x^2-1)^2}\right]$
47. $\frac{2t - 1}{3t^2 + 5t}$ $\quad \left[\frac{5 + 6t - 6t^2}{(3t^2 + 5t)^2}\right]$
48. $\frac{2x^2 - 6x + 2}{3x^2 + 2x - 1}$ $\quad \left[\frac{2(11x^2 - 8x + 1)}{(3x^2 + 2x - 1)^2}\right]$
49. $\frac{3e^{2\theta}}{4\theta^2 - 3}$ $\quad \left[\frac{6e^{2\theta}(4\theta^2 - 4\theta - 3)}{(4\theta^2 - 3)^2}\right]$
50. $\frac{3u^4 + 2u^2 - 1}{4e^{5u}}$ $\quad \left[\frac{-15u^4 + 12u^3 - 10u^2 + 4u + 5}{4e^{5u}}\right]$
51. $\frac{4 \sin c}{5c^2 + 2c}$ $\quad \left[\frac{4(5c^2 + 2c) \cos c - 4(10c + 2) \sin c}{(5c^2 + 2c)^2}\right]$
52. $\frac{4\sqrt[3]{f^7}}{3 \sin f}$ $\quad \left[\frac{4(\sqrt[3]{f^4})(7 \sin f - 3f \cos f)}{9 \sin^2 f}\right]$
53. $\frac{6 \cos h}{h^3 + 4}$ $\quad \left[\frac{-6\{(h^3 + 4) \sin h + 3h^2 \cos h\}}{(h^3 + 4)^2}\right]$
54. $\frac{\sqrt{k^3}}{\cos k}$ $\quad \left[\frac{\sqrt{k}(\frac{3}{2} \cos k + k \sin k)}{\cos^2 k}\right]$

55. $\dfrac{4e^{6x}}{\sin x}$ $\left[\dfrac{4e^{6x}(6\sin x - \cos x)}{\sin^2 x}\right]$

56. $\dfrac{3\ln\frac{5}{2}n}{n^2 + 2n}$ $\left[\dfrac{3(n+2) - 6(n+1)\ln\dfrac{5n}{2}}{(n^2+2n)^2}\right]$

57. $\dfrac{3\sqrt{x} + x}{\frac{7}{2}\ln 4x}$ $\left[\dfrac{\left(\dfrac{3}{2\sqrt{x}} + 1\right)\ln 4x - \left(\dfrac{3}{\sqrt{x}} + 1\right)}{\frac{7}{2}(\ln 4x)^2}\right]$

58. $\dfrac{\ln 6y}{6\sin y}$ $\left[\dfrac{\dfrac{1}{y}\sin y - \ln 6y\cos y}{6\sin^2 y}\right]$

59. $\dfrac{x^2\ln 4x}{3\sin x}$ $\left[\dfrac{x\ln 4x(2\sin x - x\cos x) + x\sin x}{3\sin^2 x}\right]$

60. $\dfrac{2\sqrt{t}}{\ln 3t\cos t}$ $\left[\dfrac{(\ln 3t\cos t + 2t\ln 3t\sin t - 2\cos t)}{\sqrt{t}(\ln 3t\cos t)^2}\right]$

61. $\dfrac{x^2\sec x}{e^{2x}}$ $\left[\dfrac{x\sec x}{e^{2x}}(x\tan x + 2 - 2x)\right]$

62. $\dfrac{k}{e^k\operatorname{cosec} k}$ $\left[\dfrac{1 + k(\cot k - 1)}{e^k\operatorname{cosec} k}\right]$

63. Find the slope of the curve $y = xe^{-2x}$ at the point $\left(\dfrac{1}{2}, \dfrac{1}{2e}\right)$. [0]

64. Calculate the gradient of the curve $f(x) = \dfrac{3x^4 - 2\sqrt{x^3} + 2}{5x^2 + 1}$ at the points $(0, 2)$ and $(1, \frac{1}{2})$. $[0, \frac{2}{3}]$

Differentiation by substitution

Find the differential coefficients of the functions in Problems 65–131 with respect to the variable and express your answers in their simplest form.

65. $\sin 4x$ $[4\cos 4x]$
66. $3\tan 4x$ $[12\sec^2 4x]$
67. $\cos 3t$ $[-3\sin 3t]$
68. $5\sec 2\theta$ $[10\sec 2\theta\tan 2\theta]$
69. $4\operatorname{cosec} 5\mu$ $[-20\operatorname{cosec} 5\mu\cot 5\mu]$
70. $6\cot 3\alpha$ $[-18\operatorname{cosec}^2 3\alpha]$
71. $4\cos(2x - 5)$ $[-8\sin(2x - 5)]$
72. $2\operatorname{cosec}(5t - 1)$ $[-5\operatorname{cosec}(5t - 1)\cot(5t - 1)]$
73. $(t^3 - 2t + 3)^4$ $[4(t^3 - 2t + 3)^3(3t^2 - 2)]$
74. $\sqrt{(2v^3 - v)}$ $\left[\dfrac{6v^2 - 1}{2\sqrt{(2v^3 - v)}}\right]$
75. $\sin(3x - 2)$ $[3\cos(3x - 2)]$
76. $3\tan(5y - 1)$ $[15\sec^2(5y - 1)]$
77. $4\cos(6x + 5)$ $[-24\sin(6x + 5)]$
78. $(1 - 2u^2)^7$ $[-28u(1 - 2u^2)^6]$

79. $\dfrac{1}{2n^2 - 3n + 1}$ $\left[\dfrac{3 - 4n}{(2n^2 - 3n + 1)^2}\right]$

80. $\sin^2 t$ $[2 \sin t \cos t]$

81. $3 \cos^2 x$ $[-6 \cos x \sin x]$

82. $\dfrac{1}{(2g - 1)^6}$ $\left[\dfrac{-12}{(2g - 1)^7}\right]$

83. $3 \operatorname{cosec}^2 x$ $[-6 \operatorname{cosec}^2 x \cot x]$

84. $6 \cos^3 t$ $[-18 \cos^2 t \sin t]$

85. $\frac{3}{2} \cot (6x - 2)$ $[-9 \operatorname{cosec}^2 (6x - 2)]$

86. $\sqrt{(4x^3 + 2x^2 - 5x)}$ $\left[\dfrac{12x^2 + 4x - 5}{2\sqrt{(4x^3 + 2x^2 - 5x)}}\right]$

87. $2 \sin^4 h$ $[8 \sin^3 h \cos h]$

88. $\dfrac{3}{(x^2 + 6x - 1)^5}$ $\left[\dfrac{-30(x + 3)}{(x^2 + 6x - 1)^6}\right]$

89. $(x^2 - x + 1)^{12}$ $[12(x^2 - x + 1)^{11} (2x - 1)]$

90. $e^{4q + 3}$ $[4e^{4q + 3}]$

91. $5 e^{x - 5}$ $[5 e^{x - 5}]$

92. $\ln (3p - 1)$ $\left[\dfrac{3}{3p - 1}\right]$

93. $15 \ln \left(\dfrac{x}{3} + 5\right)$ $\left[\dfrac{15}{x + 15}\right]$

94. $3 \sec 5g$ $[15 \sec 5g \tan 5g]$

95. $4 \operatorname{cosec} (2k - 1)$ $[-8 \operatorname{cosec} (2k - 1) \cot (2k - 1)]$

96. $7\beta \tan 4\beta$ $[7(4\beta \sec^2 4\beta + \tan 4\beta)]$

97. $\sqrt{x} \sec \dfrac{x}{3}$ $\left[\dfrac{\sqrt{x}}{3} \sec \dfrac{x}{3} \tan \dfrac{x}{3} + \dfrac{1}{2\sqrt{x}} \sec \dfrac{x}{3}\right]$

98. $2 e^{5l} \operatorname{cosec} 3l$ $[2 e^{5l} \operatorname{cosec} 3l(5 - 3 \cot 3l)]$

99. $\ln 5v \cot v$ $\left[\dfrac{\cot v}{v} - \operatorname{cosec}^2 v \ln 5v\right]$

100. $(3S^4 - 2S^2 + 1) \tan \dfrac{2S}{5}$

$\left[\frac{2}{5} (3S^4 - 2S^2 + 1) \sec^2 \dfrac{2S}{5} + 4S(3S^2 - 1) \tan \dfrac{2S}{5}\right]$

101. $\dfrac{\sec 2t}{(t - 1)}$ $\left[\dfrac{\sec 2t}{(t - 1)^2}\left\{2(t - 1) \tan 2t - 1\right\}\right]$

102. $(x^2 + 1) \sin (2x^2 - 3)$ $[2x\{2(x^2 + 1) \cos (2x^2 - 3) + \sin (2x^2 - 3)\}]$

103. $(2x - 1)^9 \cos 4x$ $[-2(2x - 1)^8 \{2(2x - 1) \sin 4x - 9 \cos 4x\}]$

104. $\dfrac{2t}{\tan 3t}$ $\left[\dfrac{2(\tan 3t - 3t \sec^2 3t)}{\tan^2 3t}\right]$

105. $\dfrac{1}{\operatorname{cosec} (4v + 1)}$ $[4 \cos (4v + 1)]$

106. $\dfrac{\sin (6x - 5)}{\sqrt{(x^2 - 1)}}$ $\left[\dfrac{6(x^2 - 1) \cos (6x - 5) - x \sin (6x - 5)}{\sqrt{(x^2 - 1)^3}}\right]$

107. $\dfrac{2\sqrt{u}}{3\ \text{cosec}\ 4u}$ $\left[\dfrac{1+8u\cot 4u}{3\sqrt{u}\ \text{cosec}\ 4u}\right]$

108. $\dfrac{3\cot x}{\ln 2x}$ $\left[\dfrac{-(3\ \text{cosec}^2 x\ln 2x+\frac{3}{x}\cot x)}{(\ln 2x)^2}\right]$

109. $\dfrac{5\tan 3b}{\sqrt{b}}$ $\left[\dfrac{5}{2\sqrt{b^3}}(6b\sec^2 3b-\tan 3b)\right]$

110. $\dfrac{(3\theta^2-2)}{4\sec 2\theta}$ $\left[\dfrac{3\theta-(3\theta^2-2)\tan 2\theta}{2\sec^2 2\theta}\right]$

111. $\dfrac{3\sec^2 a}{(2a^2+3a-1)^3}$ $\left[\dfrac{3\sec^2 a[2(2a^2+3a-1)\tan a-3(4a+3)]}{(2a^2+3a-1)^4}\right]$

112. $\dfrac{3e^{7x-1}}{(x-1)^9}$ $\left[\dfrac{3e^{7x-1}(7x-16)}{(x-1)^{10}}\right]$

113. $\dfrac{\sqrt{(3c^2+4c-1)}}{2\ln 5c}$ $\left[\dfrac{2c(3c+2)\ln 5c-2(3c^2+4c-1)}{4c\sqrt{(3c^2+4c-1)}(\ln 5c)^2}\right]$

114. $3\cos(2x^2+1)$ $[-12x\sin(2x^2+1)]$

115. $5\sec(5x^3-2x^2+2)$
$[5x(15x-4)\sec(5x^3-2x^2+2)\tan(5x^3-2x^2+2)]$

116. $\sin^2(2d-1)$ $[4\sin(2d-1)\cos(2d-1)]$

117. $3\tan\sqrt{(4x-2)}$ $\left[\dfrac{6\sec^2\sqrt{(4x-2)}}{\sqrt{(4x-2)}}\right]$

118. $3\cot^3(5t^2-6)$ $[-90t\cot^2(5t^2-6)\ \text{cosec}^2(5t^2-6)]$

119. $\sin^4\sqrt{(3f^3-5)}$ $\left[\dfrac{18f^2\sin^3\sqrt{(3f^3-5)}\cos\sqrt{(3f^3-5)}}{\sqrt{(3f^3-5)}}\right]$

120. $2\ln\{\cos^3(5x^2-1)\}$ $[-60x\tan(5x^2-1)]$

121. $e^{\sec g}$ $[\sec g\tan g\, e^{\sec g}]$

122. $3e^{\text{cosec}(2x-1)}$ $[-6\ \text{cosec}(2x-1)\cot(2x-1)\, e^{\text{cosec}(2x-1)}]$

123. $6\ln\{\cot^4\sqrt{(7x^2+1)}\}$ $\left[\dfrac{-168x\ \text{cosec}^2\sqrt{(7x^2+1)}}{\sqrt{(7x^2+1)}\cot\sqrt{(7x^2+1)}}\right]$

124. $3\sqrt[3]{\{\sin(4k-2)\}^5}$ $[20\sqrt[3]{\{\sin(4k-2)\}^2}\cos(4k-2)]$

125. $(3x^3+2x)^2\sin\sqrt{(x^2-1)}$
$\left[(3x^3+2x)\left\{\dfrac{x(3x^3+2x)}{\sqrt{(x^2-1)}}\cos\sqrt{(x^2-1)}+2(9x^2+2)\sin\sqrt{(x^2-1)}\right\}\right]$

126. $(x+3)^9\cos^4(x^2+2)$
$[(x+3)^8\cos^3(x^2+2)\{9\cos(x^2+2)-8x(x+3)\sin(x^2+2)\}]$

127. $\dfrac{3\cot^2 3m}{(2m-1)^5}$ $\left[\dfrac{-6\cot 3m}{(2m-1)^6}\left\{3(2m-1)\ \text{cosec}^2 3m+5\cot 3m\right\}\right]$

128. $\dfrac{\sqrt{(3x^2-2)}}{\text{cosec}^2(3x^2-2)}$ $\dfrac{3x\{1+4(3x^2-2)\cot(3x^2-2)\}}{\sqrt{(3x^2-2)}\ \text{cosec}^2(3x^2-2)}$

129. $\ln\sqrt{(\text{cosec}\ t)}$ $[\frac{1}{2}\cot t]$

130. $\dfrac{3e^{3x^2+2x+1}}{\ln(\cos x)}$ $\left[\dfrac{3e^{3x^2+2x+1}}{[\ln(\cos x)]^2}\left\{(6x+2)\ln(\cos x)+\tan x\right\}\right]$

131. $\sqrt{\{2 \ln [\sec^5 \sqrt{(2x^2 + x - 1)}]\}}$
$$\left[\frac{5(4x+1)\tan\sqrt{(2x^2+x-1)}}{2\sqrt{\{2(2x^2+x-1)\ln[\sec^5\sqrt{(2x^2+x-1)}]\}}} \right]$$

Successive differentiation

132. If $y = 5x^3 - 6x^2 + 2x - 6$ find $\dfrac{d^2y}{dx^2}$. $[30x - 12]$

133. Find $f''(t)$ given $f(t) = \dfrac{5}{x} + \sqrt{x} - \dfrac{5}{\sqrt{x^5}} + 8$. $\left[\dfrac{10}{x^3} - \dfrac{1}{4\sqrt{x^3}} - \dfrac{175}{4\sqrt{x^9}} \right]$

134. Given $f(\theta) = 3 \sin 4\theta - 2 \cos 3\theta$ find $f'(\theta)$, $f''(\theta)$ and $f'''(\theta)$.
$[f'(\theta) = 6(2 \cos 4\theta + \sin 3\theta); f''(\theta) = 6(3 \cos 3\theta - 8 \sin 4\theta);$
$f'''(\theta) = -6(32 \cos 4\theta + 9 \sin 3\theta)]$

135. If $m = (6p + 1)\left(\dfrac{1}{p} - 3\right)$ find $\dfrac{d^2m}{dp^2}$ and $\dfrac{d^3m}{dp^3}$. $\left[\dfrac{2}{p^3}; \dfrac{-6}{p^4} \right]$

In Problems 136–147 find the second differential coefficient with respect to the variable.

136. $3 \ln 5g$ $\left[-\dfrac{3}{g^2} \right]$
137. $(h - 2)^5$ $[20(h - 2)^3]$
138. $3 \sin t - \cos 2t$ $[4 \cos 2t - 3 \sin t]$
139. $3 \tan 2y + 4 \cot 3y$ $[24(\sec^2 2y \tan 2y + 3 \operatorname{cosec}^2 3y \cot 3y)]$
140. $(3m^2 - 2)^6$ $[36(3m^2 - 2)^4 (33m^2 - 2)]$
141. $\dfrac{1}{(2r - 1)^7}$ $\left[\dfrac{224}{(2r - 1)^9} \right]$
142. $3 \cos^2 \theta$ $[6(\sin^2 \theta - \cos^2 \theta)]$
143. $\frac{1}{2} \cot (3x - 1)$ $[9 \operatorname{cosec}^2 (3x - 1) \cot (3x - 1)]$
144. $4 \sin^5 n$ $[20 \sin^3 n(4 \cos^2 n - \sin^2 n)]$
145. $4 \ln \sqrt{(\operatorname{cosec} 3c)}$ $[-18 \operatorname{cosec}^2 3c]$
146. $3x^2 \sin 2x$ $[6(1 - 2x^2) \sin 2x + 24x \cos 2x]$
147. $\dfrac{\sin t}{2t^2}$ $\left[\dfrac{1}{2t^4} \left\{ (6 - t^2) \sin t - 4t \cos t \right\} \right]$

148. $x = 3t^2 - 2\sqrt{t} + \dfrac{1}{t} - 6$. Evaluate $\dfrac{d^2x}{dt^2}$ when $t = 1$. $[8\frac{1}{2}]$

149. If $f(a) = 4 \ln \sqrt{[\cos^3 (9a^2 - 2)]}$ evaluate $f''(a)$ when $a = \dfrac{\sqrt{2}}{3}$.
$[-432]$

150. Evaluate $f''(\theta)$ when $\theta = 0$ given $f(\theta) = 5 \sec 2\theta$. $[20]$

151. If $y = \cos \alpha - \sin \alpha$ evaluate α when $\dfrac{d^2y}{d\alpha^2}$ is zero. $\left[\dfrac{\pi}{4} \right]$

152. If $y = Ae^x - Be^{-x}$ prove that: $\dfrac{e^x}{2} \left\{ \dfrac{d^2y}{dx^2} + \dfrac{dy}{dx} \right\} - e^x y = B$.

153. Show that $\dfrac{d^2b}{dS^2} + 6 \dfrac{db}{dS} + 25b = 0$ when $b = e^{-3S} \sin 4S$.

154. Show that $x = 2t\,e^{-2t}$ satisfies the equation: $\dfrac{d^2x}{dt^2} + 4\dfrac{dx}{dt} + 4x = 0$.

155. If $y = 3x^3 + 2x - 4$ prove that: $\dfrac{d^3y}{dx^3} + \dfrac{2}{9}\dfrac{d^2y}{dx^2} + x\dfrac{dy}{dx} - 3y = 30$.

156. Show that the differential equation $\dfrac{d^2y}{dx^2} - 8\dfrac{dy}{dx} + 41y = 0$ is satisfied when $y = 2e^{4x}\cos 5x$.

Chapter 5

Applications of differentiation

1. Velocity and acceleration

Let a car move a distance x metres in a time t seconds along a straight road. If the velocity v of the car is constant then

$$v = \frac{x}{t} \text{ m s}^{-1}$$

i.e. the gradient of the distance/time graph shown in Fig. 1 (a) is constant.
If, however, the velocity of the car is not constant then the distance/time graph will not be a straight line. It may be as shown in Fig. 1 (b).

The average velocity over a small time δt and distance δx is given by the gradient of the chord CD, i.e. the average velocity over time $\delta t = \dfrac{\delta x}{\delta t}$.
As $\delta t \to 0$, the chord CD becomes a tangent, such that at point C the velocity v is given by:

$$v = \frac{\mathrm{d}x}{\mathrm{d}t}$$

Hence the velocity of the car at any instant t is given by the gradient of the distance/time graph. If an expression for the distance x is known in terms of time t then the velocity is obtained by differentiating the expression.

The acceleration a of the car is defined as the rate of change of velocity.

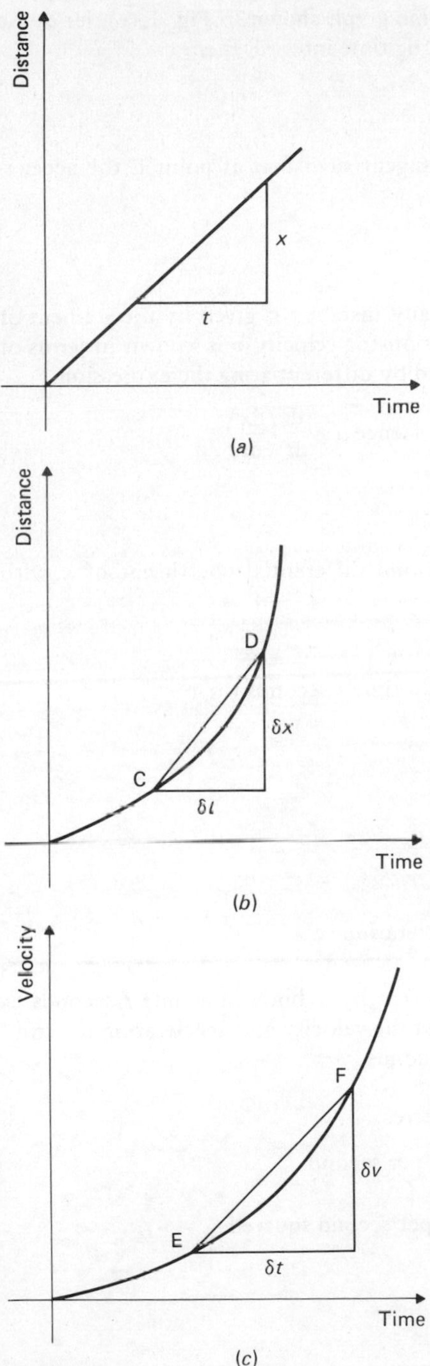
Distance
x
t
Time
(a)
Distance
D
δx
C
δt
Time
(b)
Velocity
F
δv
E
δt
Time
(c)

Figure 1

With reference to the velocity/time graph shown in Fig. 1 (c), let δv be the change in v and δt the corresponding time interval, then:

$$\text{average acceleration } a = \frac{\delta v}{\delta t}$$

As $\delta t \to 0$, the chord EF becomes a tangent such that at point E the acceleration is given by:

$$a = \frac{dv}{dt}$$

Hence the acceleration of the car at any instant t is given by the gradient of the velocity/time graph. If an expression for velocity v is known in terms of time t then the acceleration is obtained by differentiating the expression.

$$\text{Acceleration, } a = \frac{dv}{dt}\text{; but } v = \frac{dx}{dt}\text{. Hence } a = \frac{d}{dt}\left(\frac{dx}{dt}\right)$$

$$\text{which is written as: } a = \frac{d^2x}{dt^2}$$

Thus acceleration is given by the second differential coefficient of x with respect to t.

Summary

If a body moves a distance x metres in a time t seconds then:

$$\text{distance } x = f(t)$$

$$\text{velocity } v = f'(t) \text{ or } \frac{dx}{dt}$$

$$\text{and acceleration } a = f''(t) \text{ or } \frac{d^2x}{dt^2}$$

Worked problems on velocity and acceleration

Problem 1. The distance x metres moved by a body in a time t seconds is given by $x = 2t^3 + 3t^2 - 6t + 2$. Express the velocity and acceleration in terms of t and find their values when $t = 4$ seconds.

$$\text{Distance } x = 2t^3 + 3t^2 - 6t + 2 \text{ metres}$$

$$\text{Velocity } v = \frac{dx}{dt} = 6t^2 + 6t - 6 \text{ metres per second}$$

$$\text{Acceleration } a = \frac{d^2x}{dt^2} = 12t + 6 \text{ metres per second squared}$$

After 4 seconds, $v = 6(4)^2 + 6(4) - 6$

$= 96 + 24 - 6 =$ **114 m s^{-1}**

$a = 12(4) + 6 =$ **54 m s^{-2}**

Problem 2. If the distance s metres travelled by a car in time t seconds after the brakes are applied is given by $s = 15t - \frac{5}{3}t^2$: (a) what is the speed (in km h^{-1}) at the instant the brakes are applied; and (b) how far does the car travel before it stops?

(a) Distance $s = 15t - \frac{5}{3}t^2$

Velocity $v = \dfrac{ds}{dt} = 15 - \frac{10}{3}t$

At the instant the brakes are applied, $t = 0$.

Hence velocity $= 15\ \text{m s}^{-1}$

$15\ \text{m s}^{-1} = \dfrac{15}{1\,000}(60 \times 60)\ \text{km h}^{-1} = \mathbf{54\ km\ h^{-1}}$

(b) When the car finally stops, the velocity is zero,

i.e. $v = 15 - \frac{10}{3}t = 0$

i.e. $15 = \frac{10}{3}t$ or $t = 4.5$ seconds

Hence the distance travelled before the car stops is given by:

$s = 15t - \frac{5}{3}t^2$

$= 15(4.5) - \frac{5}{3}(4.5)^2$

$= \mathbf{33.75\ m}$

Problem 3. The distance x metres moved by a body in t seconds is given by:

$x = 3t^3 - \frac{11}{2}t^2 + 2t + 5$

Find:

(a) its velocity after t seconds;
(b) its velocity at the start and after 4 seconds;
(c) the value of t when the body comes to rest;
(d) its acceleration after t seconds;
(e) its acceleration after 2 seconds;
(f) the value of t when the acceleration is $16\ \text{m s}^{-2}$; and
(g) the average velocity over the third second.

(a) Distance $x = 3t^3 - \frac{11}{2}t^2 + 2t + 5$

Velocity $v = \dfrac{dx}{dt} = \mathbf{9t^2 - 11t + 2}$

(b) Velocity at the start means the velocity when $t = 0$,

i.e. $v_0 = 9(0)^2 - 11(0) + 2 = \mathbf{2\ m\ s^{-1}}$

Velocity after 4 seconds, $v_4 = 9(4)^2 - 11(4) + 2 = \mathbf{102\ m\ s^{-1}}$

(c) When the body comes to rest, $v = 0$

i.e. $9t^2 - 11t + 2 = 0$

$(9t - 2)(t - 1) = 0$

$\mathbf{t = \frac{2}{9}\ s\ or\ t = 1\ s}$

(d) **Acceleration** $a = \dfrac{d^2x}{dt^2} = \mathbf{(18t - 11)}$

(e) Acceleration after 2 seconds, $a_2 = 18(2) - 11 = \mathbf{25\ m\ s^{-2}}$

(f) When the acceleration is 16 m s^{-2} then

$18t - 11 = 16$

$18t = 16 + 11 = 27$

$t = \frac{27}{18} = 1\frac{1}{2}$ **seconds**

(g) Distance travelled in the third second = (distance travelled after 3 s) − (distance travelled after 2 s)

$= [3(3)^3 - \frac{11}{2}(3)^2 + 2(3) + 5] - [3(2)^3 - \frac{11}{2}(2)^2 + 2(2) + 5]$

$= 42\frac{1}{2} - 11$

$= 31\frac{1}{2}$ m

Average velocity over the third second $= \frac{\text{distance travelled}}{\text{time interval}}$

$= \frac{31\frac{1}{2}\text{ m}}{1\text{ s}}$

$= \mathbf{31\frac{1}{2}\ m\ s^{-1}}$

(Note that should a negative value occur for velocity it merely means that the body is moving in the direction opposite to that with which it started. Also if a negative value occurs for acceleration it indicates a deceleration (or a retardation).)

Further problems on velocity and acceleration may be found in Section 4 (Problems 1–15).

2. Maximum and minimum values

Consider the curve shown in Fig. 2.

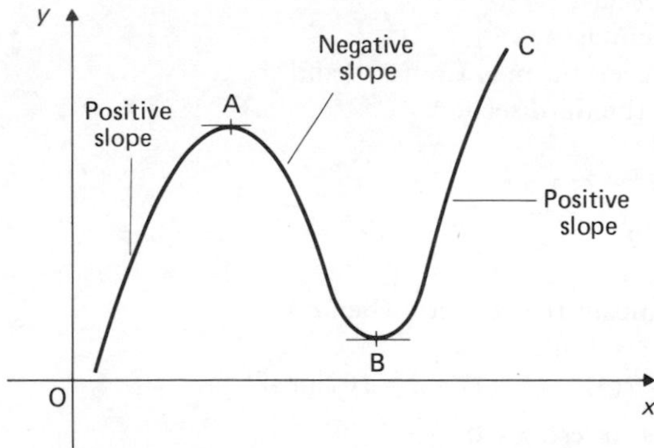

Figure 2

The slope of the curve (i.e. $\frac{dy}{dx}$) between points O and A is positive. The slope of the curve between points A and B is negative and the slope between points B and C is again positive.

At point A the slope is zero and as x increases, the slope of the curve changes from positive just before A to negative just after. Such a point is called a **maximum value.**

At point B the slope is also zero and, as x increases, the slope of the curve changes from negative just before B to positive just after. Such a point is called a **minimum value.**

Points such as A and B are given the general name of **turning-points.**

Maximum and minimum values can be confusing inasmuch as they suggest that they are the largest and smallest values of a curve. However, by their definition this is not so. A maximum value occurs at the 'crest of a wave' and the minimum value at the 'bottom of a valley'. In Fig. 2 the point C has a larger y-ordinate value than A and point O has a smaller y ordinate than B. Points A and B are turning-points and are given the special names of maximum and minimum values respectively.

Summary

1. At a maximum point the slope $\frac{dy}{dx} = 0$ and changes from positive just before the maximum point to negative just after.
2. At a minimum point the slope $\frac{dy}{dx} = 0$ and changes from negative just before the minimum point to positive just after.

Consider the function $y = x^3 - x^2 - 5x + 6$.
The turning-points (i.e. the maximum and minimum values) may be determined without going through the tedious process of drawing up a table of values and plotting the graph.

If $\quad y = x^3 - x^2 - 5x + 6$

then $\frac{dy}{dx} = 3x^2 - 2x - 5$

Now at a maximum or minimum value $\frac{dy}{dx} = 0$.

Hence $3x^2 - 2x - 5 = 0$ for a maximum or minimum value
$(3x - 5)(x + 1) = 0$

i.e. *Either* $\quad 3x - 5 = 0$ giving $x = \frac{5}{3}$

or $\quad x + 1 = 0$ giving $x = -1$

For each value of the independent variable x there is a corresponding value of the dependent variable y.

When $x = \frac{5}{3}$, $y = [\frac{5}{3}]^3 - [\frac{5}{3}]^2 - 5[\frac{5}{3}] + 6 = -\frac{13}{27}$

When $x = -1$, $y = (-1)^3 - (-1)^2 - 5(-1) + 6 = 9$

Hence turning-points occur at $(\frac{5}{3}, -\frac{13}{27})$ and $(-1, 9)$.

The next step is to determine which of the points is a maximum and which is a minimum. There are two methods whereby this may be achieved.

Consider firstly the point $(\frac{5}{3}, -\frac{13}{27})$.

$$\frac{dy}{dx} = 3x^2 - 2x - 5 = (3x - 5)(x + 1)$$

If x is slightly less than $\frac{5}{3}$ then $(3x - 5)$ becomes negative, $(x + 1)$ remains positive, making $\frac{dy}{dx} = (-) \times (+) =$ negative

If x is slightly greater than $\frac{5}{3}$ then $(3x - 5)$ becomes positive, $(x + 1)$ remains positive, making $\frac{dy}{dx} = (+) \times (+) =$ positive

Hence the slope in negative just before $(\frac{5}{3}, -\frac{13}{27})$ and positive just after. This is thus a **minimum** value.

Consider now the point $(-1, 9)$.

$$\frac{dy}{dx} = (3x - 5)(x + 1)$$

If x is slightly less than -1 (for example -1.1) then $(3x - 5)$ remains negative, $(x + 1)$ becomes negative, making $\frac{dy}{dx} = (-) \times (-) =$ positive

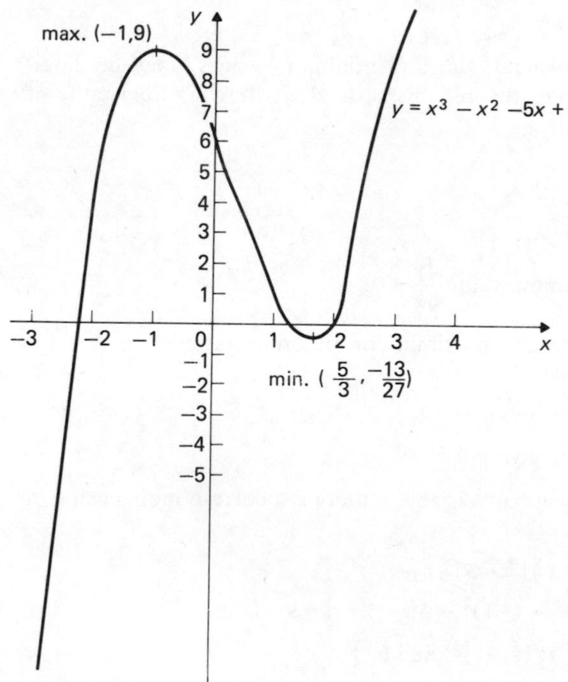

Figure 3 Graph of $y = x^3 - x^2 - 5x + 6$

If x is slightly greater than -1 (for example -0.9) then $(3x - 5)$ remains negative, $(x + 1)$ becomes positive, making $\frac{dy}{dx} = (-) \times (+) =$ negative

Hence the slope is positive just before $(-1, 9)$ and negative just after. This is thus a **maximum** value.

Figure 3 shows a graph of $y = x^3 - x^2 - 5x + 6$ with the maximum value at $(-1, 9)$ and the minimum at $(\frac{5}{3}, -\frac{13}{27})$.

Method 2

When passing through a maximum value, $\frac{dy}{dx}$ changes from positive, through zero, to negative. By convention, moving from a positive value to a negative value is moving in a negative direction. Hence the rate of change of $\frac{dy}{dx}$ is negative.

i.e. $\frac{d}{dx}\left(\frac{dy}{dx}\right) = \frac{d^2y}{dx^2}$ **is negative at a maximum value.**

Similarly, when passing through a minimum value, $\frac{dy}{dx}$ changes from negative, through zero, to positive. By convention, moving from a negative value to a positive value is moving in a positive direction. Hence the rate of change of $\frac{dy}{dx}$ is positive.

i.e. $\frac{d^2y}{dx^2}$ **is positive at a minimum value.**

Thus, in the above example, to distinguish between the points $(\frac{5}{3}, -\frac{13}{27})$ and $(-1, 9)$ the second differential is required.

Since $\frac{dy}{dx} = 3x^2 - 2x - 5$

then $\frac{d^2y}{dx^2} = 6x - 2$

When $x = \frac{5}{3}$, $\frac{d^2y}{dx^2} = 6[\frac{5}{3}] - 2 = +8$ which is **positive.**

Hence $(\frac{5}{3}, -\frac{13}{27})$ **is a minimum point.**

When $x = -1$, $\frac{d^2y}{dx^2} = 6(-1) - 2 = -8$ which is **negative.**

Hence $(-1, 9)$ **is a maximum point.**

The actual numerical value of the second differential is insignificant – the sign is the important factor. There are thus two methods of distinguishing between maximum and minimum values. Normally, the second method, that of determining the sign of the second differential, is preferred but sometimes the first method, that of examining the sign of the slope just before and just after the turning-point, is necessary either because the second differential coefficient is too difficult to obtain or because the second differential coefficient is zero.

It is possible to have a turning-point, the slope on either side of which is the same. This point is given the special name of a **point of inflexion.** At a point of inflexion $\frac{d^2y}{dx^2}$ is zero.

Maximum and minimum points and points of inflexion are given the general term of **stationary points.** Examples of each are shown in Fig. 4.

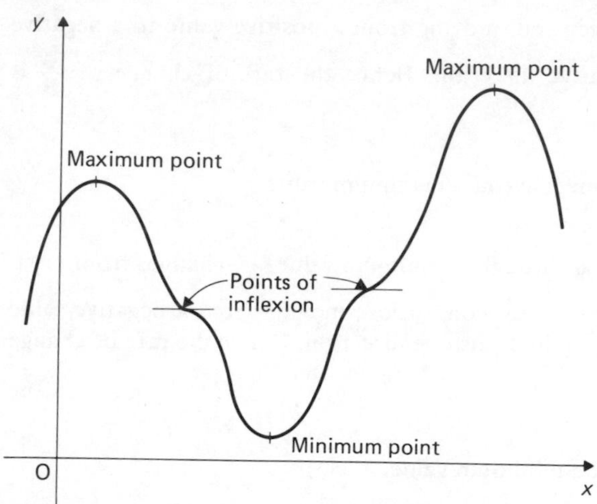

Figure 4

Procedure for finding and distinguishing between stationary points

(i) If $y = f(x)$, find $\frac{dy}{dx}$.

(ii) Let $\frac{dy}{dx} = 0$ and solve for the value(s) of x.

(iii) Substitute the value(s) of x into the original equation, $y = f(x)$, to obtain the y-ordinate value(s). Hence the coordinates of the stationary points are established.

(iv) Find $\frac{d^2y}{dx^2}$.

or

Determine the sign of the slope of the curve just before and just after the stationary point(s).

(v) Substitute values of x into $\frac{d^2y}{dx^2}$. If the result is:

(a) positive – the point is a minimum value;
(b) negative – the point is a maximum value;
(c) zero – the point is a point of inflexion.

or

If the sign change for the slope of the curve is:
(a) positive to negative – the point is a maximum value;
(b) negative to positive – the point is a minimum value;
(c) positive to positive; or
(d) negative to negative – the point is a point of inflexion.

Worked problems on maximum and minimum values

Problem 1. Find the coordinates of the maximum and minimum values of the graph $y = \frac{2x^3}{3} - 5x^2 + 12x - 7$ and distinguish between them.

From the above procedure:

(i) $y = \frac{2x^3}{3} - 5x^2 + 12x - 7$

$\frac{dy}{dx} = 2x^2 - 10x + 12$

(ii) $\frac{dy}{dx} = 0$ at a turning-point.

Therefore $2x^2 - 10x + 12 = 0$

$2(x^2 - 5x + 6) = 0$

$2(x - 2)(x - 3) = 0$

Hence $x = 2$ or $x = 3$

(iii) When $x = 2$, $y = \frac{2}{3}(2)^3 - 5(2)^2 + 12(2) - 7 = 2\frac{1}{3}$
When $x = 3$, $y = \frac{2}{3}(3)^3 - 5(3)^2 + 12(3) - 7 = 2$
The coordinates of the turning-points are thus $(2, 2\frac{1}{3})$ and $(3, 2)$.

(iv) $\frac{dy}{dx} = 2x^2 - 10x + 12$

$\frac{d^2y}{dx^2} = 4x - 10$

(v) When $x = 2$, $\frac{d^2y}{dx^2} = -2$, which is negative, giving a maximum value

When $x = 3$, $\frac{d^2y}{dx^2} = +2$, which is positive, giving a minimum value

Hence **the point $(2, 2\frac{1}{3})$ is a maximum value and the point $(3, 2)$ a minimum value.**

Note that with a quadratic equation there will be one turning-point. With a cubic equation (i.e. one containing a highest term of power 3) there may be two turning-points (i.e. one less than the highest power), and so on.

Problem 2. Locate the turning points on the following curves and distinguish between maximum and minimum values: (a) $x(5-x)$; (b) $2t-e^t$; (c) $2(\theta-\ln\theta)$.

(a) Let $y = x(5-x) = 5x - x^2$

$\frac{dy}{dx} = 5 - 2x = 0$ for a maximum or minimum value.

i.e. $x = 2\frac{1}{2}$

When $x = 2\frac{1}{2}$, $y = 2\frac{1}{2}(5 - 2\frac{1}{2}) = 6\frac{1}{4}$

Hence a turning-point occurs at $(2\frac{1}{2}, 6\frac{1}{4})$

$\frac{d^2y}{dx^2} = -2$, which is negative, giving a maximum value.

Hence **$(2\frac{1}{2}, 6\frac{1}{4})$ is a maximum point.**

(b) Let $y = 2t - e^t$

$\frac{dy}{dt} = 2 - e^t = 0$ for a maximum or minimum value.

i.e. $2 = e^t$

$\ln 2 = t$

$t = 0.693\,1$

When $t = 0.693\,1$, $y = 2(0.693\,1) - 2 = -0.613\,8$

Hence a turning-point occurs at $(0.693\,1, -0.613\,8)$

$\frac{d^2y}{dt^2} = -e^t$

When $t = 0.693\,1$, $\frac{d^2y}{dt^2} = -2$, which is negative, giving a maximum value.

Hence **$(0.693\,1, -0.613\,8)$ is a maximum point.**

(c) Let $y = 2(\theta - \ln\theta) = 2\theta - 2\ln\theta$

$\frac{dy}{d\theta} = 2 - \frac{2}{\theta} = 0$ for a maximum or minimum value.

i.e. $\theta = 1$

When $\theta = 1$, $y = 2 - 2\ln 1 = 2$

Hence a turning-point occurs at $(1, 2)$

$\frac{d^2y}{d\theta^2} = +\frac{2}{\theta^2}$

When $\theta = 1$, $\frac{d^2y}{d\theta^2} = +2$, which is positive, giving a minimum value.

Hence **$(1, 2)$ is a minimum point.**

Problem 3. Find the maximum and minimum values of the function

$$f(p) = \frac{(p-1)(p-6)}{(p-10)}$$

$$f(p) = \frac{(p-1)(p-6)}{(p-10)} = \frac{p^2 - 7p + 6}{(p-10)} \text{ (i.e. a quotient)}$$

$$f'(p) = \frac{(p-10)(2p-7)-(p^2-7p+6)(1)}{(p-10)^2}$$

$$= \frac{(2p^2-27p+70)-(p^2-7p+6)}{(p-10)^2}$$

$$= \frac{p^2-20p+64}{(p-10)^2}$$

$$= \frac{(p-4)(p-16)}{(p-10)^2} = 0 \text{ for a maximum or minimum value.}$$

Therefore $(p-4)(p-16) = 0$
i.e. $p = 4$ or $p = 16$.

When $p = 4$, $f(p) = \dfrac{(3)(-2)}{(-6)} = 1$

When $p = 16$, $f(p) = \dfrac{(15)(10)}{(6)} = 25$

Hence there are turning-points at (4, 1) and (16, 25).

To use the second-derivative approach in this case would result in a complicated and long expression. Thus the slope is investigated just before and just after the turning-point.

It will be easier to use the factorised version of $f'(p)$.

i.e. $f'(p) = \dfrac{(p-4)(p-16)}{(p-10)^2}$

Consider the point (4, 1):

When p is just less than 4, $f'(p) = \dfrac{(-)(-)}{(+)}$, i.e. positive.

When p is just greater than 4, $f'(p) = \dfrac{(+)(-)}{(+)}$, i.e. negative.

Since the slope changes from positive to negative the point (4, 1) is a maximum.

Consider the point (16, 25):

When p is just less than 16, $f'(p) = \dfrac{(+)(-)}{(+)}$, i.e. negative.

When p is just greater than 16, $f'(p) = \dfrac{(+)(+)}{(+)}$, i.e. positive.

Since the slope changes from negative to positive the point (16, 25) is a minimum.

Since, in the question, the maximum and minimum values are asked for (and not the coordinates of the turning-points) the answers are: **maximum value = 1; minimum value = 25.**

Problem 4. Find the maximum and minimum values of $y = 1.25 \cos 2\theta + \sin\theta$ for values of θ between 0 and $\dfrac{\pi}{2}$ inclusive, given $\sin 2\theta = 2\sin\theta\cos\theta$.

$y = 1.25 \cos 2\theta + \sin\theta$

$$\frac{dy}{d\theta} = -2.50 \sin 2\theta + \cos \theta = 0 \text{ for a maximum or minimum value.}$$

But $\sin 2\theta = 2 \sin \theta \cos \theta$

Therefore $-2.50(2 \sin \theta \cos \theta) + \cos \theta = 0$

$$-5.0 \sin \theta \cos \theta + \cos \theta = 0$$

$$\cos \theta(-5.0 \sin \theta + 1) = 0$$

Hence $\cos \theta = 0$, i.e. $\theta = 90°$ or $270°$

or $-5.0 \sin \theta + 1 = 0$, i.e. $\sin \theta = \frac{1}{5}$

$\theta = 11° \ 32'$ or $168° \ 28'$

Thus within the range $\theta = 0$ to $\theta = \frac{\pi}{2}$ inclusive, turning-points occur at $11° \ 32'$ and $90°$.

$$\frac{d^2y}{d\theta^2} = -5.0 \cos 2\theta - \sin \theta$$

When $\theta = 11° \ 32'$, $\frac{d^2y}{d\theta^2}$ is negative, giving a maximum value.

When $\theta = 90°$, $\frac{d^2y}{d\theta^2}$ is positive, giving a minimum value.

$y_{max} = 1.25 \cos 2(11° \ 32') + \sin (11° \ 32') = \mathbf{1.35}$

$y_{min} = 1.25 \cos 2(90°) + \sin (90°) = \mathbf{-0.25}$

Further problems on maximum and minimum values may be found in Section 4 (Problems 16–41).

3. Practical problems involving maximum and minimum values

There are many practical problems on maximum and minimum values in engineering and science which can be solved using the method(s) shown in Section 2. Often the quantity whose maximum or minimum value is required appears at first to be a function of more than one variable. It is thus necessary to eliminate all but one of the variables, and this is often the only difficult part of its solution. Once the quantity has been expressed in terms of a single variable, the procedure is identical to that used in Section 2.

Worked problems on practical problems involving maximum and minimum values

---

Problem 1. A rectangular area is formed using a piece of wire 36 cm long. Find the length and breadth of the rectangle if it is to enclose the maximum possible area.

Let the dimensions of the rectangle be x and y.

Perimeter of rectangle $= 2x + 2y = 36$

i.e. $x + y = 18$ ...(1)

Since it is the **maximum area** that is required a formula for the area A must be obtained in terms of one variable only.

Area $A = xy$

From equation (1) $y = 18 - x$

Hence $A = x(18 - x) = 18x - x^2$

Now that an expression for the area has been obtained in terms of one variable it can be differentiated with respect to that variable.

$$\frac{dA}{dx} = 18 - 2x = 0 \text{ for a maximum or minimum value}$$

i.e. $x = 9$ cm

$$\frac{d^2A}{dx^2} = -2, \text{ which is negative, giving a maximum value.}$$

$y = 18 - x = 18 - 9 = 9$ cm

Hence **the length and breadth of the rectangle of maximum area are both 9 cm,** i.e. a square gives the maximum possible area for a given perimeter length. When the perimeter of a rectangle is 36 cm the maximum area possible is 81 cm^2.

Problem 2. Find the area of the largest piece of rectangular ground that can be enclosed by 1 km of fencing if part of an existing straight wall is used as one side.

There are a large number of possible rectangular areas which can be produced from 1 000 m of fencing. Three such possibilities are shown in Fig. 5 (a) where AB represents the existing wall. All three rectangles have different areas. There must be one particular condition which gives a maximum area.

Let the dimensions of any rectangle be x and y as shown in Fig. 5 (b).

Then $2x + y = 1\,000$ m . . . (1)

Area of rectangle, $A = xy$ . . . (2)

Since it is the **maximum area** that is required, a formula for the area A must be obtained in terms of one variable only.

From equation (1) $y = 1\,000 - 2x$

Hence $A = x(1\,000 - 2x) = 1\,000x - 2x^2$

$$\frac{dA}{dx} = 1\,000 - 4x = 0 \text{ for a maximum or minimum value}$$

i.e. $x = 250$ m

$$\frac{d^2A}{dx^2} = -4, \text{ which is negative, giving a maximum value.}$$

When $x = 250$, $y = 1\,000 - 2(250) = 500$ m

Hence the maximum possible area = xy = (250)(500) = 125 000 m^2.

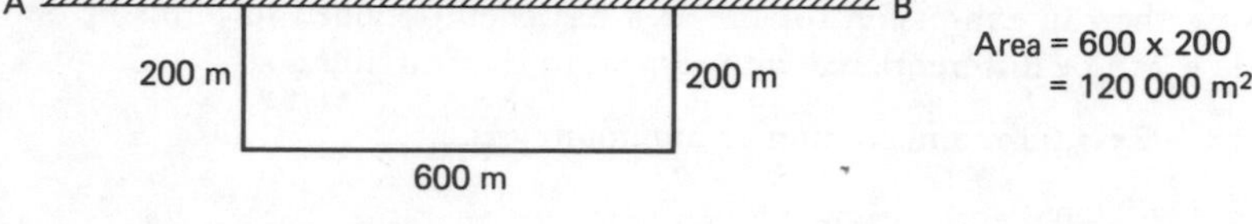

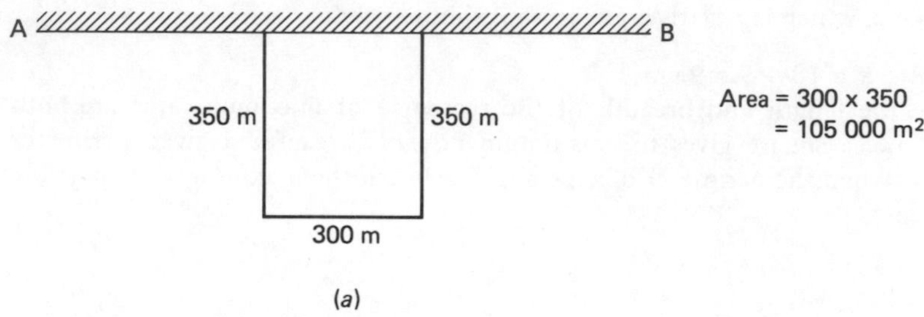

(*a*)

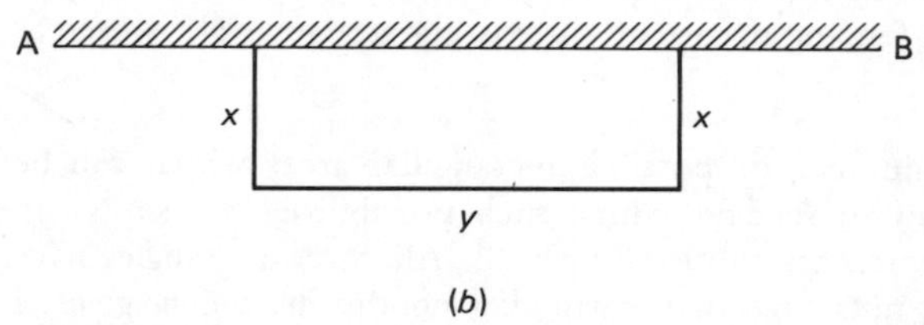

(*b*)

Figure 5

Problem 3. A lidless, rectangular box with square ends is to be made from a thin sheet of metal. What is the least area of the metal for which the volume is $4\frac{1}{2}$ m^3?

Let the dimensions of the box be x metres by x metres by y metres.

Volume of box $= x^2y = 4\frac{1}{2}$ . . . (1)

Surface area A of box consists of: two ends $= 2x^2$

two sides $= 2xy$

base $= xy$

$$A = 2x^2 + 2xy + xy = 2x^2 + 3xy \qquad \ldots (2)$$

Since it is the **least (i.e. minimum) area** that is required, a formula for the area A must be obtained in terms of one variable only.

From equation (1), $y = \dfrac{4\frac{1}{2}}{x^2} = \dfrac{9}{2x^2}$

Substituting $y = \frac{9}{2x^2}$ in equation (2) gives:

$$A = 2x^2 + 3x\left(\frac{9}{2x^2}\right) = 2x^2 + \frac{27}{2x}$$

$$\frac{dA}{dx} = 4x - \frac{27}{2x^2} = 0 \text{ for a maximum or minimum value}$$

$$4x = \frac{27}{2x^2}$$

$$x^3 = \frac{27}{8}, \text{ i.e. } x = \frac{3}{2}\text{ m}$$

$$\frac{d^2A}{dx^2} = 4 + \frac{27}{x^3}$$

When $x = \frac{3}{2}$, $\frac{d^2A}{dx^2} = 4 + \frac{27}{(\frac{3}{2})^3} = +12$, which is positive, giving a minimum (or least) value.

When $x = \frac{3}{2}$, $y = \frac{9}{2x^2} = \frac{9}{2(\frac{3}{2})^2} = 2$ m

Therefore area $A = 2x^2 + 3xy = 2[\frac{3}{2}]^2 + 3[\frac{3}{2}](2) = 13\frac{1}{2}\text{ m}^2$

Hence the least possible area of metal required to form a rectangular box with square ends of volume $4\frac{1}{2}$ m³ is $13\frac{1}{2}$ m².

Problem 4. Find the base radius and height of the cylinder of maximum volume which can be cut from a sphere of radius 10.0 cm.

A cylinder of radius r and height h is shown in Fig. 6 enclosed in a sphere of radius R = 10.0 cm.

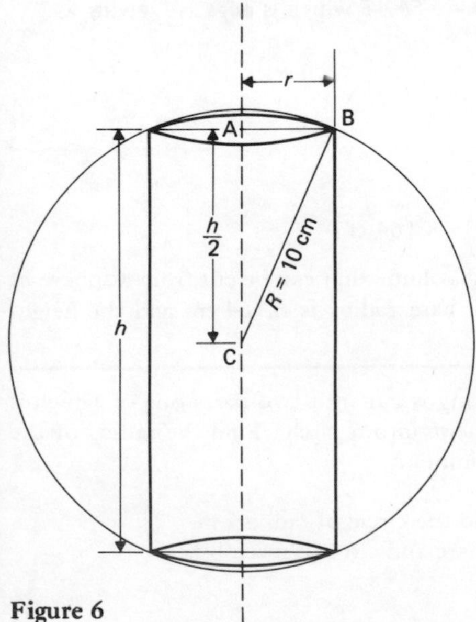

Figure 6

 Volume of cylinder, $V = \pi r^2 h$ ... (1)

Using the theorem of Pythagoras on the triangle ABC of Fig. 6 gives:

$$r^2 + \left[\frac{h}{2}\right]^2 = R^2$$

i.e. $r^2 + \frac{h^2}{4} = 100$ ... (2)

Since it is the **maximum volume** that is required, a formula for the volume V must be obtained in terms of one variable only.

From equation (2), $r^2 = 100 - \frac{h^2}{4}$

Substituting $r^2 = 100 - \frac{h^2}{4}$ in equation (1) gives:

$$V = \pi\left[100 - \frac{h^2}{4}\right]h = 100\pi\, h - \frac{\pi h^3}{4}$$

$\frac{dV}{dh} = 100\pi - \frac{3\pi}{4}h^2 = 0$ for a maximum or minimum value.

$$100\pi = \frac{3\pi}{4}h^2$$

$$h^2 = \frac{400}{3}$$

$h = 11.55$ cm ($h = -11.55$ cm is neglected for obvious reasons)

$$\frac{d^2V}{dh^2} = -\tfrac{3}{2}\pi h$$

When $h = 11.55$, $\frac{d^2V}{dh^2} = -\tfrac{3}{2}\pi(11.55) = -54.43$ which is negative, giving a maximum value.

From equation (2) $r^2 = 100 - \frac{h^2}{4}$

$$r = \sqrt{\left(100 - \frac{h^2}{4}\right)} = 8.164 \text{ cm}$$

Hence **the cylinder having the largest volume that can be cut from a sphere of radius 10.0 cm is one in which the base radius is 8.164 cm and the height 11.55 cm.**

Problem 5. A piece of wire 4.0 m long is cut into two parts one of which is bent into a square and the other bent into a circle. Find the radius of the circle if the sum of their areas is a minimum.

Let the square be of side x m and the circle of radius r m.

The sum of the perimeters of the square and circle is given by:

$$4x + 2\pi r = 4$$

or $2x + \pi r = 2$ ... (1)

Total area A of the two shapes, $A = x^2 + \pi r^2$ ... (2)

Since it is the **minimum area** that is required a formula for the area A must be obtained in terms of one variable only.

From equation (1), $x = \dfrac{2 - \pi r}{2}$

Substituting $x = \dfrac{2 - \pi r}{2}$ in equation (2) gives:

$$A = \left(\frac{2 - \pi r}{2}\right)^2 + \pi r^2 = \frac{4 - 4\pi r + \pi^2 r^2}{4} + \pi r^2$$

$$\text{i.e.}\quad A = 1 - \pi r + \frac{\pi^2 r^2}{4} + \pi r^2$$

$$\frac{dA}{dr} = -\pi + \frac{\pi^2 r}{2} + 2\pi r = 0 \text{ for a maximum or minimum value.}$$

$$\text{i.e.}\quad \pi = r\left[\frac{\pi^2}{2} + 2\pi\right]$$

$$r = \frac{\pi}{\left(\frac{\pi^2}{2} + 2\pi\right)} = \frac{1}{\left(\frac{\pi}{2} + 2\right)} = 0.280 \text{ m}$$

$$\frac{d^2A}{dr^2} = \frac{\pi^2}{2} + 2\pi = 11.22, \text{ which is positive, giving a minimum value.}$$

Hence **for the sum of the areas of the square and circle to be a minimum the radius of the circle must be 28.0 cm.**

Problem 6. Find the base radius of a cylinder of maximum volume which can be cut from a cone of height 12 cm and base radius 9 cm.

A cylinder of base radius r and height h is shown enclosed in a cone of height 12 cm and base radius 9 cm in Fig. 7.

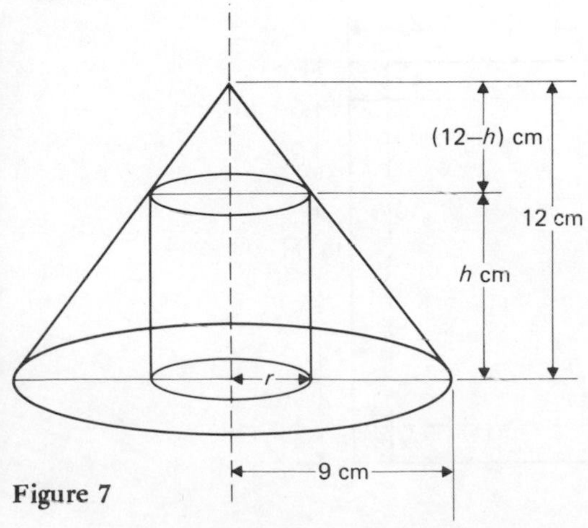

Figure 7

 Volume of the cylinder, $V = \pi r^2 h$ ... (1)

By similar triangles: $\dfrac{12 - h}{r} = \dfrac{12}{9}$ ... (2)

Since it is the **maximum volume** that is required a formula for the volume V must be obtained in terms of one variable only.

From equation (2), $9(12 - h) = 12r$

$$108 - 9h = 12r$$

$$h = \frac{108 - 12r}{9}$$

Substituting for h in equation (1) gives:

$$V = \pi r^2\left[\frac{108 - 12r}{9}\right] = 12\pi r^2 - \frac{4\pi r^3}{3}$$

$$\frac{dV}{dr} = 24\pi r - 4\pi r^2 = 0 \text{ for a maximum or minimum value.}$$

$$24\pi r = 4\pi r^2$$

i.e. $4\pi r(6 - r) = 0$

Therefore $r = 0$ or $r = 6$.

$$\frac{d^2V}{dr^2} = 24\pi - 8\pi r$$

When $r = 0$, $\dfrac{d^2V}{dr^2}$ is positive, giving a minimum value (which we would expect).

When $r = 6$, $\dfrac{d^2V}{dr^2} = 24\pi - 48\pi = -24\pi$, which is negative, giving a maximum value.

Hence **a cylinder of maximum volume having a base radius of 6 cm can be cut from a cone of height 12 cm and base radius 9 cm.**

Problem 7. A rectangular sheet of metal which measures 24.0 cm by 16.0 cm has squares removed from each of the four corners so that an open box may be formed. Find the maximum possible volume for the box.

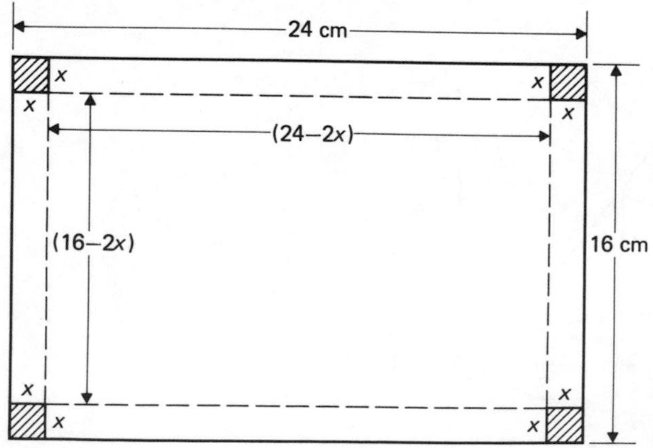

Figure 8

The squares which are to be removed are shown shaded and having side x cm in Fig. 8.

To form a box the metal has to be bent upwards along the broken lines.
The dimensions of the box will be: length $= (24.0 - 2x)$ cm
breadth $= (16.0 - 2x)$ cm
height $= x$ cm

Volume of box, $V = (24.0 - 2x)(16.0 - 2x)(x)\ \text{cm}^3$
$= 384x - 80x^2 + 4x^3$

$$\frac{dV}{dx} = 384 - 160x + 12x^2 = 0 \text{ for a maximum or minimum value.}$$

i.e. $4(3x^2 - 40x + 96) = 0$

$$x = \frac{40 \pm \sqrt{[(-4.0)^2 - 4(3)(96)]}}{6}$$

$x = 10.194$ cm or $x = 3.139$ cm

Since the breadth $= 16.0 - 2x$, $x = 10.194$ cm is an impossible solution to this problem and is thus neglected.

Hence $x = 3.139$ cm

$$\frac{d^2V}{dx^2} = -160 + 24x$$

When $x = 3.139$ cm, $\dfrac{d^2V}{dx^2} = -160 + 24(3.139) = -84.66$ which is negative, giving a maximum value.

The dimensions of the box are: length $= 24.0 - 2(3.139) = 17.72$ cm
breadth $= 16.0 - 2(3.139) = 9.722$ cm
height $= 3.139$ cm

Maximum volume $= (17.72)(9.722)(3.139) =$ **540.8 cm³**.

Further problems on practical maximum and minimum problems may be found in the following section (4) (Problems 42–69).

4. Further problems

Velocity and acceleration

1. The distance x metres moved by a body in a time t seconds is given by $x = 4t^3 - 3t^2 + 5t + 2$. Express the velocity and acceleration in terms of t and find their values when $t = 3$ s.
[$v = 12t^2 - 6t + 5$; $v_3 = 95\ \text{m s}^{-1}$; $a = 24t - 6$; $a_3 = 66\ \text{m s}^{-2}$]
2. A body obeys the equation $x = 3t - 20t^2$ where x is in metres and t in seconds. Find expressions for velocity and acceleration. Find also its velocity and acceleration when $t = 1$ s.
[$v = 3 - 40t$; $v_1 = -37\ \text{m s}^{-1}$; $a = -40$; $a_1 = -40\ \text{m s}^{-2}$]
3. If the distance x metres travelled by a vehicle in t seconds after the brakes are applied is given by: $x = 22.5t - \frac{5}{6}t^2$, then what is the speed in km h^{-1} when the brakes are applied? How far does the car travel before it stops?
[81 km h^{-1}; 151.9 m]

4. An object moves in a straight line so that after t seconds its distance x metres from a fixed point on the line is given by $x = \frac{2}{3}t^3 - 5t^2 + 8t - 6$. Obtain an expression for the velocity and acceleration of the object after t seconds and hence calculate the values of t when the object is at rest. [$v = 2t^2 - 10t + 8$; $a = 4t - 10$; $t = 1$ s or 4 s]

In Problems 5–9, x denotes the distance in metres of a body moving on a straight line, from a fixed point on the line and t denotes the time in seconds measured from a certain instant. Find the velocity and acceleration of the body when t has the given values. Find also the values of t when the body is momentarily at rest.

5. $x = \frac{4}{3}t^3 - 4t^2 + 3t - 2$; $t = 2$. [3 m s^{-1}; 8 m s^{-2}; $t = \frac{1}{2}$ or $1\frac{1}{2}$ s]
6. $x = 3\cos 2t$; $t = \frac{\pi}{4}$. [−6 m s^{-1}; 0; $t = 0, \frac{\pi}{2}, \pi, \frac{3\pi}{2}$, etc.]
7. $x = t^4 - \frac{1}{2}t^2 + 1$; $t = 1$. [3 m s^{-1}; 11 m s^{-2}; $t = 0$ or $\pm\frac{1}{2}$ s]
8. $x = 4t + 2\cos 2t$; $t = 0$. [4 m s^{-1}; −8 m s^{-2}; $t = \frac{\pi}{4}, \frac{5\pi}{4}, \frac{9\pi}{4}, \frac{13\pi}{4}$, etc.]
9. $x = \frac{t^4}{4} - \frac{5}{3}t^3 + 3t^2 + 5$; $t = 2$. [0; −2 m s^{-2}; $t = 0$, 2 or 3 s]
10. The distance s metres moved by a point in t seconds is given by $s = 5t^3 + 4t^2 - 3t + 2$. Find:
    (a) expressions for velocity and acceleration in terms of t;
    (b) the velocity and acceleration after 3 seconds; and
    (c) the average velocity over the fourth second.
    (a) [$v = (15t^2 + 8t - 3)$ m s^{-1}; $a = (30t + 8)$ m s^{-2}]
    (b) [156 m s^{-1}; 98 m s^{-2}] (c) [210 m s^{-1}]
11. The distance x metres moved by a body in t seconds is given by $x = \frac{16}{3}t^3 - 32t^2 + 39t - 16$. Find:
    (a) the velocity and acceleration at the start;
    (b) the velocity and acceleration at $t = 3$ seconds;
    (c) the values of t when the body is at rest;
    (d) the value of t when the acceleration is 16 m s^{-2}; and
    (e) the distance travelled in the second second.
    (a) [39 m s^{-1}; −64 m s^{-2}] (b) [−9 m s^{-1}; 32 m s^{-2}] (c) [$t = \frac{3}{4}$ or $3\frac{1}{4}$ s]
    (d) [$2\frac{1}{2}$ s] (e) [$-19\frac{2}{3}$ m (i.e. in the opposite direction to that in which the body initially moved)]
12. The displacement y centimetres of the slide valve of an engine is given by the expression $y = 2.6\cos 5\pi t + 3.8\sin 5\pi t$. Find an expression for the velocity v of the valve and evaluate the velocity (in metres per second) when $t = 20$ ms. [$v = 5\pi(3.8\cos 5\pi t - 2.6\sin 5\pi t)$; 0.442 m s^{-1}]
13. At any time t seconds the distance x metres of a particle moving in a straight line from a fixed point is given by: $x = 5t + \ln(1 - 2t)$. Find:
    (a) expressions for the velocity and acceleration in terms of t;
    (b) the initial velocity and acceleration;
    (c) the velocity and acceleration after 2 s; and
    (d) the time when the velocity is zero.

(a) $\left[\left(5 - \frac{2}{(1-2t)}\right) \text{m s}^{-1}; \left(\frac{-4}{(1-2t)^2}\right) \text{m s}^{-2}\right]$
(b) [3 m s^{-1}; –4 m s^{-2}] (c) [$5\frac{2}{3}$ m s^{-1}; $-\frac{4}{9}$ m s^{-2}] (d) [$\frac{3}{10}$ s]

14. If the equation $\theta = 12\pi + 27t - 3t^2$ gives the angle in radians through which a wheel turns in t seconds, find how many seconds the wheel takes to come to rest. Calculate the angle turned through in the last second of movement. [$4\frac{1}{2}$ s; 3 radians]

15. A missile fired from ground level rises s metres in t seconds, and $s = 75t - 12.5t^2$. Determine:
(a) the initial velocity of the missile;
(b) the time when the height of the missile is a maximum;
(c) the maximum height reached; and
(d) the velocity with which the missile strikes the ground.
(a) [75 m s^{-1}] (b) [3 s] (c) [112.5 m] (d) [75 ms^{-1}]

Maximum and minimum values

In Problems 16–20 find the turning-points and distinguish between them by examining the sign of the slope on either side.

16. $y = 2x^2 - 4x$ [min. (1, –2)]
17. $y = 3t^2 - 2t + 6$ [min. ($\frac{1}{3}$, $5\frac{2}{3}$)]
18. $x = \theta^3 - 3\theta + 3$ [max. (–1, 5); min. (1, 1)]
19. $y = 3x^3 + 6x^2 + 3x - 2$ [max. (–1, –2); min. ($-\frac{1}{3}$, $-2\frac{4}{9}$)]
20. $y = 7t^3 - 4t^2 - 5t + 6$ [max. ($-\frac{1}{3}$, $6\frac{26}{27}$); min. ($\frac{5}{7}$, $2\frac{46}{49}$)]

Locate the turning-points on the curves in Problems 21–39 and distinguish between them.

21. $y = x(7 - x)$ [max. ($3\frac{1}{2}$, $12\frac{1}{4}$)]
22. $y = 4x^2 - 2x + 3$ [min. ($\frac{1}{4}$, $2\frac{3}{4}$)]
23. $y = 2x^3 + 7x^2 + 4x - 3$ [max. (–2, 1); min. ($-\frac{1}{3}$, $-3\frac{17}{27}$)]
24. $2pq = 18p^2 + 8$ [max. ($-\frac{2}{3}$, –12); min. ($\frac{2}{3}$, 12)]
25. $y = 3t + e^{-t}$ [min. (–1.098 6, –2.962 5)]
26. $x = 3 \ln \theta - 4\theta$ [max. (0.75, –3.863)]
27. $S = 5t^3 - \frac{3}{2}t^2 - 12t + 6$ [max. ($-\frac{4}{5}$, $12\frac{2}{25}$); min. (1, $-2\frac{1}{2}$)]
28. $y = 4x - 2 \ln x$ [min. (0.5, 3.386)]
29. $y = 3x - e^x$ [max. (1.098 6, 0.295 8)]
30. $p = \frac{(q-1)(q-3)}{q}$ [max. (–1.732, –7.464); min. (1.732, –0.535 9)]
31. $y = \frac{(x-2)(x-5)}{(x-6)}$ [max. (4, 1); min. (8, 9)]
32. $y = 3 \sin \theta - 4 \cos \theta$ in the range θ to 2π.
[max. 5 at 143° 8′; min. –5 at 323° 8′]
33. $y = 4 \cos 2\theta + 3 \sin \theta$ in the range 0 to $\frac{\pi}{2}$ inclusive, given that $\sin 2\theta = 2 \sin \theta \cos \theta$. [max. 4.281 2 at 10° 48′; min. –1 at 90°]
34. $V = l^2(l - 1)$ [max. (0, 0); min. ($\frac{2}{3}$, $\frac{-4}{27}$)]
35. $y = 8x + \frac{1}{2x^2}$ [min. ($\frac{1}{2}$, 6)]

36. $x = t^3 + \dfrac{t^2}{2} - 7t + 4$ [max. $(-1, 10\frac{1}{2})$; min. $(\frac{2}{3}, \frac{-4}{27})$]

37. $y = \dfrac{3x}{(x-1)(x-4)}$ [max. $(2, -3)$; min. $(-2, -\frac{1}{3})$]

38. $y = (x-1)^3 + 3x(x-2)$ [max. $(-1, 1)$; min. $(1, -3)$]

39. $y = \frac{1}{2}\ln(\sin x) - \sin x$ in the range 0 to $\dfrac{\pi}{4}$ [max. $-0.846\,6$ at 30°]

40. (a) If $p + q = 7$, find the maximum value of $3pq + q^2$.
(b) If $3a - 2b = 5$, find the least value of $2a^2b$.
(a) [$55\frac{1}{8}$] (b) [$-2\frac{14}{243}$ or $-2.057\,6$]

41. The sum of a number and its reciprocal is to be a minimum. Find the number. [1]

Practical maximum and minimum problems

42. Find the maximum area of a rectangular piece of ground that can be enclosed by 200 m of fencing. [2 500 m^2]

43. A rectangular area is formed using a piece of wire of length 26 cm. Find the dimensions of the rectangle if it is to enclose the maximum possible area. [$6\frac{1}{2}$ cm by $6\frac{1}{2}$ cm]

44. A shell is projected upwards with a speed of 12 m s^{-1} and the distance vertically s metres is given by $s = 12t - 3t^2$, where t is the time in seconds. Find the maximum height reached. [12 m]

45. A length of 42 cm of thin wire is bent into a rectangular shape with one side repeated. Find the largest area that can be enclosed. [73.5 cm^2]

46. Find the area of the largest piece of rectangular ground that can be enclosed by 800 m of fencing if part of an existing wall is used on one side. [80 000 m^2]

47. The bending moment M of a beam of length l at a distance a from one end is given by $M = \dfrac{Wa}{2}(l - a)$, where W is the load per unit length. Find the maximum bending moment. $\left[\dfrac{Wl^2}{8}\right]$

48. A lidless box with square ends is to be made from a thin sheet of metal. What is the least area of the metal for which the volume of the box is 6.64 m^3? [17.50 m^2]

49. Find the height and the radius of a solid cylinder of volume 150 cm^3 which has the least surface area. [5.759 cm; 2.879 cm]

50. Find the height of a right circular cylinder of greatest volume which can be cut from a sphere of radius R. $\left[\dfrac{2R}{\sqrt{3}}\right]$

51. The power P developed in a resistor R by a battery of e.m.f. E and internal resistance r is given by $P = \dfrac{E^2r}{(R+r)^2}$.
Differentiate P with respect to r and show that the power is a maximum when $R = r$.

52. A piece of wire 5.0 m long is cut into two parts, one of which is bent

into a square and the other into a circle. Find the diameter of the circle if the sum of their areas is a minimum. [0.700 m]

53. Find the height of a cylinder of maximum volume which can be cut from a cone of height 15 cm and base radius 7.5 cm. [5 cm]
54. An alternating current is given by $i = 100 \sin(50\pi t + 0.32)$ amperes, where t is the time in seconds. Determine the maximum value of the current and the time when this maximum first occurs.
[100 amperes when $t = 7.96$ ms].
55. A frame for a box kite with a square cross-section is made of 16 pieces of wood as shown in Fig. 9. Find the maximum volume of the frame if a total length of 12 m of wood is used.
[$\frac{4}{9}$ m^3]

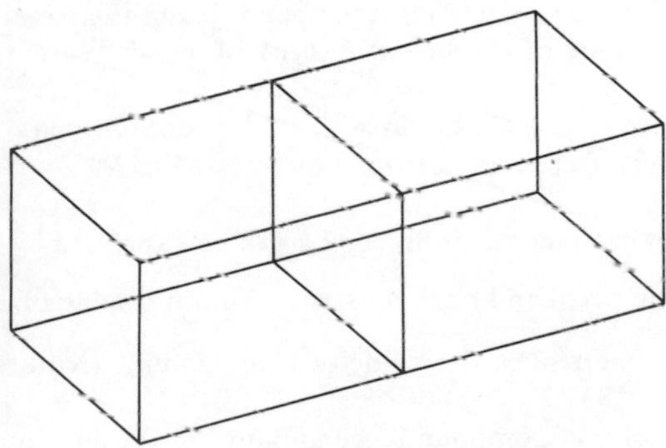

Figure 9

56. A rectangular box with a lid which covers the top and front has a volume of 150 cm^3 and the length of the base is to be $1\frac{1}{2}$ times the height. Find the dimensions of the box so that the surface area shall be a minimum.
[3.816 cm by 5.724 cm by 6.867 cm]
57. The force F required to move a body along a rough horizontal plane is given by $F = \dfrac{\mu W}{\cos\theta + \mu \sin\theta}$ where μ is the coefficient of friction and θ the angle to the direction of F. If F varies with θ show that F is a minimum when $\tan\theta = \mu$.
58. A closed cylindrical container has a surface area of 300 cm^2. Find its dimensions for maximum volume.
[radius = 3.989 cm; height = 7.981 cm]
59. A rectangular block of metal, with a square cross-section, has a total surface area of 240 cm^2. Find the maximum volume of the block of metal. [253.0 cm^3]
60. The displacement s metres in a damped harmonic oscillation is given by $s = 4e^{-2t} \sin 2t$, where t is the time in milliseconds. Find the values of t to give maximum displacements. $\left[\frac{\pi}{8}, \frac{5\pi}{8}, \frac{9\pi}{8}, \text{ and so on}\right]$
61. A square sheet of metal of side 25.0 cm has squares cut from each corner,

so that an open box may be formed. Find the surface area and the volume of the box if the volume is to be a maximum.
[555.6 cm^2; 1 157 cm^3]

62. A right circular cylinder of maximum volume is to be cut from a sphere of radius 14.0 cm. Determine the base diameter and the height of the cylinder. [22.86 cm; 16.17 cm]

63. The speed v of a signal transmitted through a cable is given by $v = kx^2 \ln \frac{1}{x}$, where x is the ratio of the inner to the outer diameters of the core and k is a constant. Find the value of x for maximum speed of the transmitted signal. [$x = e^{-\frac{1}{2}} = 0.606\ 5$]

64. An open rectangular box with square ends is fitted with an overlapping lid which covers the whole of the square ends, the open top and the front face. Find the maximum volume of the box if 8.0 m^2 of metal are used altogether. [0.871 m^3]

65. An electrical voltage E is given by: $E = 12.0 \sin 50\pi t + 36.0 \cos 50\pi t$ volts, where t is the time in seconds. Determine the maximum value of E.
[37.95 volts]

66. The velocity v of a piston of a reciprocating engine can be expressed by $v = 2\pi nr\left(\frac{\sin 2\theta}{16} + \sin\theta\right)$, where n and r are constants. Find the value of θ between 0° and 360° that makes the velocity a maximum. (Note: $\cos 2\theta = 2\cos^2\theta - 1$.) [83° 2′]

67. The periodic time T of a compound pendulum is given by $T = 2\pi\sqrt{\left[\frac{h^2 + k^2}{gh}\right]}$, where k and g are constants. Find the minimum value of T. $\left[T_{min} = 2\pi\sqrt{\left(\frac{2k}{g}\right)} \text{ when } h = k\right]$

68. The heat capacity (C) of carbon monoxide varies with absolute temperature (T) as shown: $C = 26.53 + 7.70 \times 10^{-3}T - 1.17 \times 10^{-6}T^2$. Determine the maximum value of C and the temperature at which it occurs.
[$C = 39.20$, $T = 3.291 \times 10^3$]

69. The electromotive force (E) of the Clark cell is given by
$E = 1.4 - 0.001\ 2\ (T - 288) - 0.000\ 007\ (T - 288)^2$ volts. Determine the maximum value of E. [1.451 4 volts]

Chapter 6

Methods of integration

1. Introduction to integration

The process of integration reverses the process of differentiation. In differentiation, if $f(x) = x^2$ then $f'(x) = 2x$. Since integration reverses the process of moving from $f(x)$ to $f'(x)$, it follows that the integral of $2x$ is x^2, i.e. it is the process of moving from $f'(x)$ to $f(x)$. Similarly, if $y = x^3$ then $\frac{dy}{dx} = 3x^2$. Reversing this process shows that the integral of $3x^2$ is x^3.

Integration is also a process of summation or adding parts together and an elongated 'S', shown as $\int$, is used to replace the words 'the integral of'.

Thus $\int 2x = x^2$ and $\int 3x^2 = x^3$.

In differentiation, the differential coefficient $\frac{dy}{dx}$ or $\frac{d}{dx}[f(x)]$ indicates that a function of x is being differentiated with respect to x, the dx indicating this. In integration, the variable of integration is shown by adding d (the variable) after the function to be integrated. Thus $\int 2x\, dx$ means 'the integral of $2x$ with respect to x' and $\int 3u^2\, du$ means 'the integral of $3u^2$ with respect to u'. It follows that $\int y\, dx$ means 'the integral of y with respect to x' and since only functions of x can be integrated with respect to x, y must be expressed as a function of x before the process of integration can be performed.

The arbitrary constant of integration

The differential coefficient of x^2 is $2x$, hence $\int 2x\,dx = x^2$. Also, the differential coefficient of $x^2 + 3$ is $2x$, hence $\int 2x\,dx = x^2 + 3$. Since the differential coefficient of any constant is zero, it follows that the differential coefficient of $x^2 + c$, where c is any constant, is $2x$. To allow for the possible presence of this constant, whenever the process of integration is performed the constant should be added to the result. Hence $\int 2x\,dx = x^2 + c$.

c is called the arbitrary constant of integration and it is important to include it in all work involving the process of determining integrals. Its omission will result in obtaining incorrect solutions in later work, such as in the solution of differential equations (see Chapter 9).

2. The general solution of integrals of the form x^n

$$\int x^n\,dx = \frac{x^{n+1}}{n+1} + c$$

In order to integrate x^n it is necessary to:

(a) increase the power of x by 1, i.e. the power of x^n is raised by 1 to x^{n+1};
(b) divide by the new power of x, i.e. x^{n+1} is divided by $n + 1$; and
(c) add the arbitrary constant of integration, c.

Thus to integrate x^4, the power of x is increased by 1 to 4 + 1 or 5, giving x^5, and the term is divided by $(n + 1)$ or $(4 + 1)$, i.e. 5.

So the integral of x^4 is $\dfrac{x^5}{5} + c$.

In the general solution of $\int x^n\,dx$ given above, n may be a fraction, zero or a positive or negative integer with just one exception, that being $n = -1$.

It was shown in differentiation that $\dfrac{d}{dx}(\ln x) = \dfrac{1}{x}$

Thus $\displaystyle\int \frac{1}{x}\,dx\ (= \int x^{-1}\,dx) = \ln x + c$

More generally, $$\frac{d}{dx}(\ln ax) = \frac{d}{dx}(\ln x + \ln a) = \frac{1}{x} + 0$$

Therefore, $$\int \frac{1}{x}\,dx = \ln ax + c$$

Rules of integration

Three of the basic rules of integration are:

(i) The integral of a constant k is $kx + c$. For example,

$\int 5 \, dx = \int 5x^0 \, dx$ since $x^0 = 1$.

Applying the standard integral $\int x^n \, dx = \dfrac{x^{n+1}}{n+1} + c$ gives:

$$\int 5 \, dx = \frac{5x^{0+1}}{0+1} + c = 5x + c$$

(ii) As in differentiation, constants associated with variables are carried forward, i.e. they are not involved in the integration. For example,

$$\int 3x^4 \, dx = 3 \int x^4 \, dx = 3\left(\frac{x^5}{5}\right) + c$$
$$= \tfrac{3}{5}x^5 + c$$

(iii) As in differentiation, the rules of algebra apply where functions of a variable are added or subtracted. For example,

$$\int (x^2 + x^5) \, dx = \int x^2 \, dx + \int x^5 \, dx = \frac{x^3}{3} + \frac{x^6}{6} + c$$

and $\int (2x^3 + 4) \, dx = 2 \int x^3 \, dx + \int 4 \, dx = 2\left(\dfrac{x^4}{4}\right) + 4x + c$

$$= \frac{x^4}{2} + 4x + c$$

Combining rule (ii) with the standard integral for x^n gives:

$$\int ax^n \, dx = \frac{ax^{n+1}}{n+1} + c$$

where a and n are constants and n is **not** equal to -1.

Integrals written in this form are called 'indefinite integrals', since their precise value cannot be found (i.e. c cannot be calculated) unless additional information is provided. (In differentiation there are special rules for multiplication and division of functions. However, there are no such special rules for multiplication and division in integration.)

3. Definite integrals

Limits can be applied to integrals and such integrals are then called 'definite integrals'. The increase in the value of the integral $(x^2 - 3)$ as x increases from 1 to 2 can be written as:

$$\left[\int (x^2 - 3) \, dx\right]_1^2$$

However, this is invariably abbreviated by showing the value of the upper limit at the top of the integral sign and the value of the lower limit at the bottom, i.e.

$$\left[\int (x^2 - 3) \, dx\right]_1^2 = \int_1^2 (x^2 - 3) \, dx$$

 The integral is evaluated as for an indefinite integral and then placed in the square brackets of the limit operator.

$$\text{Thus } \int_1^2 (x^2 - 3)\,dx = \left[\frac{x^3}{3} - 3x + c\right]_1^2$$

$$= \left[\frac{(2)^3}{3} - 3(2) + c\right] - \left[\frac{(1)^3}{3} - 3(1) + c\right]$$

$$= (\tfrac{8}{3} - 6 + c) - (\tfrac{1}{3} - 3 + c)$$

$$= 2\tfrac{2}{3} - 6 - \tfrac{1}{3} + 3 = -\tfrac{2}{3}$$

The arbitrary constant of integration, c, always cancels out when limits are applied to an integral and it is not usually shown when evaluating a definite integral.

4. Integrals of sin ax, cos ax, sec² ax and e^{ax}

Since integration is the reverse process of differentiation the following standard integrals may be deduced.

(a) $\dfrac{d}{dx}(\sin x) = \cos x$

Hence $\int \cos x\,dx = \sin x + c$

More generally: $\dfrac{d}{dx}(\sin ax) = a\cos ax$

Hence $\int a\cos ax\,dx = \sin ax + c$

$$\boldsymbol{\int \cos ax\,dx = \frac{1}{a}\sin ax + c}$$

(b) $\dfrac{d}{dx}(\cos x) = -\sin x$

Hence $\int -\sin x\,dx = \cos x + c$

$\int \sin x\,dx = -\cos x + c$

More generally: $\dfrac{d}{dx}(\cos ax) = -a\sin ax$

Hence $\int -a\sin ax\,dx = \cos ax + c$

$$\boldsymbol{\int \sin ax\,dx = -\frac{1}{a}\cos ax + c}$$

(c) $\dfrac{d}{dx}(\tan x) = \sec^2 x$

Hence $\int \sec^2 x\,dx = \tan x + c$

More generally: $\dfrac{d}{dx}(\tan ax) = a\sec^2 ax$

Hence $\int a \sec^2 ax \, dx = \tan ax + c$

$$\int \sec^2 ax \, dx = \frac{1}{a} \tan ax + c$$

(d) $\frac{d}{dx}(e^x) = e^x$

Hence $\int e^x \, dx = e^x + c$

More generally: $\frac{d}{dx}(e^{ax}) = ae^{ax}$

Hence $\int ae^{ax} \, dx = e^{ax} + c$

$$\int e^{ax} \, dx = \frac{1}{a} e^{ax} + c$$

Summary of standard integrals

1. $\int ax^n \, dx = \frac{ax^{n+1}}{n+1} + c$ (except when $n = -1$)

2. $\int \cos ax \, dx = \frac{1}{a} \sin ax + c$

3. $\int \sin ax \, dx = -\frac{1}{a} \cos ax + c$

4. $\int \sec^2 ax \, dx = \frac{1}{a} \tan ax + c$

5. $\int e^{ax} \, dx = \frac{1}{a} e^{ax} + c$

6. $\int \frac{1}{x} \, dx = \ln ax + c$

Worked problems on standard integrals

Problem 1. Integrate the following with respect to the variable: (a) x^7; (b) $3.0y^{1.6}$; (c) $\frac{2}{p^3}$.

(a) $\int x^7 \, dx = \frac{x^{7+1}}{7+1} + c = \frac{x^8}{8} + c$

(b) $\int 3.0y^{1.6} \, dy = \frac{3.0y^{1.6+1}}{1.6+1} + c = \frac{3.0y^{2.6}}{2.6} + c = 1.15y^{2.6} + c$

(c) $\int \frac{2}{p^3} \, dp = \int 2p^{-3} \, dp = \frac{2p^{-3+1}}{-3+1} + c$

$$= \frac{2p^{-2}}{-2} + c = \frac{-1}{p^2} + c$$

 If the final answer of an integration is differentiated then the original must result (otherwise an error has occurred). For example, in (a) above:

$$\frac{d}{dx}\left(\frac{x^8}{8}+c\right) = \frac{8x^7}{8} = x^7 \text{ (i.e. the original integral)}$$

It will be assumed that in all future integral problems such a check will be made.

Problem 2. Integrate with respect to the variable:

(a) $\left(2x^5 - 4\sqrt{x} + \frac{5}{x^4} - \frac{2}{\sqrt{x^3}} + 6\right)$;

(b) $\left(\frac{4p^5 - 3 + p}{p^3}\right)$.

(a) $\int\left(2x^5 - 4\sqrt{x} + \frac{5}{x^4} - \frac{2}{\sqrt{x^3}} + 6\right) dx = \int (2x^5 - 4x^{\frac{1}{2}} + 5x^{-4} - 2x^{-\frac{3}{2}} + 6)\, dx$

$$= \frac{2x^{5+1}}{(5+1)} - \frac{4x^{\frac{1}{2}+1}}{(\frac{1}{2}+1)} + \frac{5x^{-4+1}}{(-4+1)} - \frac{2x^{-\frac{3}{2}+1}}{(-\frac{3}{2}+1)} + 6x + c$$

$$= \frac{2x^6}{6} - \frac{4x^{\frac{3}{2}}}{\frac{3}{2}} + \frac{5x^{-3}}{-3} - \frac{2x^{-\frac{1}{2}}}{-\frac{1}{2}} + 6x + c$$

$$= \frac{x^6}{3} - \frac{8}{3}\sqrt{x^3} - \frac{5}{3x^3} + \frac{4}{\sqrt{x}} + 6x + c$$

(b) $\int\left(\frac{4p^5 - 3 + p}{p^3}\right) dp = \int\left(\frac{4p^5}{p^3} - \frac{3}{p^3} + \frac{p}{p^3}\right) dp$

$$= \int (4p^2 - 3p^{-3} + p^{-2})\, dp$$

$$= \frac{4p^3}{3} - \frac{3p^{-2}}{-2} + \frac{p^{-1}}{-1} + c$$

$$= \frac{4}{3}p^3 + \frac{3}{2p^2} - \frac{1}{p} + c$$

Problem 3. If $y = \int\left(r + \frac{1}{r}\right)^2 dr$, find the value of the arbitrary constant of integration if $y = \frac{1}{3}$ when $r = 1$.

$$y = \int\left(r + \frac{1}{r}\right)^2 dr = \int\left(r^2 + 2 + \frac{1}{r^2}\right) dr$$

$$= \frac{r^3}{3} + 2r - \frac{1}{r} + c$$

$y = \frac{1}{3}$ when $r = 1$. Hence $\frac{1}{3} = \frac{(1)^3}{3} + 2(1) - \frac{1}{(1)} + c$

$$\tfrac{1}{3} = \tfrac{1}{3} + 2 - 1 + c$$

$$c = -1$$

Hence **the arbitrary constant of integration is -1.**

Problem 4. Integrate with respect to the variable:
(a) $4\cos 3\theta$; (b) $7\sin 2x$; (c) $3\sec^2 5t$.

(a) $\int 4\cos 3\theta\, d\theta = 4(\tfrac{1}{3}\sin 3\theta) + c = \tfrac{4}{3}\sin 3\theta + c$

(b) $\int 7\sin 2x\, dx = 7(-\tfrac{1}{2}\cos 2x) + c = -\tfrac{7}{2}\cos 2x + c$

(c) $\int 3\sec^2 5t\, dt = 3(\tfrac{1}{5}\tan 5t) + c = \tfrac{3}{5}\tan 5t + c$

Problem 5. Find: (a) $\int 6e^{4x}\, dx$; (b) $\int \frac{3}{e^{2t}}\, dt$; (c) $\int \frac{3}{2u}\, du$.

(a) $\int 6e^{4x}\, dx = 6\left(\frac{e^{4x}}{4}\right) + c = \tfrac{3}{2}e^{4x} + c$

(b) $\int \frac{3}{e^{2t}}\, dt = \int 3e^{-2t}\, dt = 3\left(\frac{e^{-2t}}{-2}\right) + c = \frac{-3}{2}e^{-2t} + c = \frac{-3}{2e^{2t}} + c$

(c) $\int \frac{3}{2u}\, du = \frac{3}{2}\int \frac{1}{u}\, du = \frac{3}{2}\ln u + c$

Problem 6. Evaluate: (a) $\int_1^3 (4x-3)^2\, dx$; (b) $\int_0^4 \left(5\sqrt{b} - \frac{1}{\sqrt{b}}\right) db$.

(a) $\int_1^3 (4x-3)^2\, dx = \int_1^3 (16x^2 - 24x + 9)\, dx$

$$= \left[\frac{16x^3}{3} - 24\frac{x^2}{2} + 9x + c\right]_1^3$$

$$= \left[\frac{16}{3}(3)^3 - 12(3)^2 + 9(3) + c\right] - \left[\frac{16}{3}(1)^3 - 12(1)^2 + 9(1) + c\right]$$

$$= (144 - 108 + 27 + c) - (5\tfrac{1}{3} - 12 + 9 + c)$$

$$= (63 + c) - (2\tfrac{1}{3} + c)$$

$$= 60\tfrac{2}{3}$$

The arbitrary constant of integration, c, cancels out, thus showing it to be an unnecessary inclusion when evaluating definite integrals.

(b) $\int_0^4 \left(5\sqrt{b} - \frac{1}{\sqrt{b}}\right) db = \int_0^4 (5b^{\frac{1}{2}} - b^{-\frac{1}{2}})\, db$

$$= \left[\frac{5b^{\frac{3}{2}}}{\frac{3}{2}} - \frac{b^{\frac{1}{2}}}{\frac{1}{2}}\right]_0^4 = \left[\frac{10}{3}\sqrt{b^3} - 2\sqrt{b}\right]_0^4$$

$$= \left(\frac{10}{3}\sqrt{4^3} - 2\sqrt{4}\right) - \left(\frac{10}{3}\sqrt{0^3} - 2\sqrt{0}\right)$$

$$= \frac{10}{3}(8) - 2(2) - 0 = \frac{80}{3} - 4 = 22\tfrac{2}{3}$$

(taking positive values of square roots only).

Problem 7. Evaluate: (a) $\int_0^{\frac{\pi}{2}} 4\sin 2x\, dx$; (b) $\int_0^1 3\cos 3t\, dt$;

$$\text{(c)} \int_{\frac{\pi}{6}}^{\frac{\pi}{3}} (2 \sin \theta - 3 \cos 2\theta + 4 \sec^2 \theta)\, d\theta.$$

$$\text{(a)} \int_0^{\frac{\pi}{2}} 4 \sin 2x\, dx = \left[-\frac{4}{2} \cos 2x \right]_0^{\frac{\pi}{2}}$$

$$= \left(-2 \cos 2 \left(\frac{\pi}{2}\right)\right) - \left(-2 \cos 2(0)\right)$$

$$= (-2 \cos \pi) - (-2 \cos 0)$$

$$= (-2(-1)) - (-2(1))$$

$$= 2 + 2 = 4$$

(b) $\int_0^1 3 \cos 3t\, dt = [\frac{3}{3} \sin 3t]_0^1 = [\sin 3t]_0^1 = (\sin 3 - \sin 0)$

The limits in trigonometric functions are expressed in radians.

Thus 'sin 3' means 'the sine of 3 radians or $3\left(\frac{180}{\pi}\right)^\circ$', i.e. $171°53'$.

Hence $\quad \sin 3 - \sin 0 = \sin 171°\ 53' - \sin 0°$

$= 0.141\ 2 - 0$

Thus $\quad \int_0^1 3 \cos 3t\, dt = \mathbf{0.141\ 2}$

$$\text{(c)} \int_{\frac{\pi}{6}}^{\frac{\pi}{3}} (2 \sin \theta - 3 \cos 2\theta + 4 \sec^2 \theta)\, d\theta = \left[-2 \cos \theta - \tfrac{3}{2} \sin 2\theta + 4 \tan \theta \right]_{\frac{\pi}{6}}^{\frac{\pi}{3}}$$

$$= \left(-2 \cos \frac{\pi}{3} - \tfrac{3}{2} \sin \frac{2\pi}{3} + 4 \tan \frac{\pi}{3} \right) - \left(-2 \cos \frac{\pi}{6} - \tfrac{3}{2} \sin \frac{2\pi}{6} + 4 \tan \frac{\pi}{6} \right)$$

$$= (-2 \cos 60° - \tfrac{3}{2} \sin 120° + 4 \tan 60°) - (-2 \cos 30° - \tfrac{3}{2} \sin 60° + 4 \tan 30°)$$

$$= \left(-2(\tfrac{1}{2}) - \tfrac{3}{2}\left(\frac{\sqrt{3}}{2}\right) + 4\sqrt{3} \right) - \left(-2\left(\frac{\sqrt{3}}{2}\right) - \tfrac{3}{2}\left(\frac{\sqrt{3}}{2}\right) + 4\left(\frac{\sqrt{3}}{3}\right) \right)$$

$$= -1 - \frac{3\sqrt{3}}{4} + 4\sqrt{3} + \sqrt{3} + \frac{3\sqrt{3}}{4} - \frac{4\sqrt{3}}{3}$$

$$= -1 + \sqrt{3}(4 + 1 - \tfrac{4}{3}) = -1 + \frac{11\sqrt{3}}{3}$$

$$= \mathbf{5.351}$$

Problem 8. Evaluate: (a) $\int_1^2 3e^{4x}\, dx$; (b) $\int_3^4 \frac{5}{x}\, dx$.

(a) $\int_1^2 3e^{4x}\, dx = [\frac{3}{4} e^{4x}]_1^2 = \frac{3}{4} e^8 - \frac{3}{4} e^4 = \frac{3}{4} e^4 (e^4 - 1) = \mathbf{2\ 195}$

$$\text{(b)} \int_3^4 \frac{5}{x}\, dx = 5[\ln x]_3^4 = 5[\ln 4 - \ln 3]$$

$$= 5 \ln \tfrac{4}{3} = 5(0.287\ 4) = \mathbf{1.437\ 0}$$

Further problems on standard integrals may be found in Section 6 (Problems 1–65).

5. Integration by substitution

Functions which require integrating are not usually in the standard integral form previously met. However, by using suitable substitutions some functions can be changed into a form which can be readily integrated. The substitution usually made is to let u be equal to $f(x)$, such that $f(u)\,du$ is a standard integral.

A most important point in the use of substitution is that once a substitution has been made the original variable must be removed completely, because a variable can only be integrated with respect to itself, i.e. we cannot integrate, for example, a function of t with respect to x.

A concept that $\frac{du}{dx}$ is a single entity (indicating the change of u with respect to x) has been established in the work done on differentiation. Frequently, in work on integration and differential equations, $\frac{du}{dx}$ is split. Provided that when this is done, the original differential coefficient can be re-formed by applying the rules of algebra, then it is in order to do it. For example, if $\frac{dy}{dx} = x$ then it is in order to write $dy = x\,dx$ since dividing both sides by dx re-forms the original differential coefficient. This principle is shown in the following worked problems.

Worked problems on integration by substitution

Problem 1. Find: $\int \cos(5x + 2)\,dx$.

Let $u = 5x + 2$

then $\frac{du}{dx} = 5$, i.e. $dx = \frac{du}{5}$

$$\int \cos(5x+2)\,dx = \int \cos u \frac{du}{5} = \tfrac{1}{5}\int \cos u\,du$$

$$= \tfrac{1}{5}(\sin u) + c$$

Since the original integral is given in terms of x, the result should be stated in terms of x.

$u = 5x + 2$

Hence $\boldsymbol{\int \cos(5x+2)\,dx = \tfrac{1}{5}\sin(5x+2) + c}$

Problem 2. Find: $\int (4t - 3)^7\,dt$.

Let $u = 4t - 3$

then $\frac{du}{dt} = 4$, i.e. $dt = \frac{du}{4}$

$$\int (4t-3)^7\,dt = \int u^7 \frac{du}{4} = \tfrac{1}{4}\int u^7\,du$$

$$= \tfrac{1}{4}\left(\frac{u^8}{8}\right) + c$$

$$= \frac{u^8}{32} + c$$

Since $u = (4t - 3)$,

$\int (4t - 3)^7 \, dt = \frac{1}{32} (4t - 3)^8 + c$

Problem 3. Integrate with respect to x: $\dfrac{1}{7x + 2}$.

Let $u = 7x + 2$

then $\dfrac{du}{dx} = 7$, i.e. $dx = \dfrac{du}{7}$

$$\int \frac{1}{7x + 2} \, dx = \int \frac{1}{u} \frac{du}{7}$$
$$= \tfrac{1}{7} \ln u + c$$

Since $u = (7x + 2)$,

$$\int \frac{1}{7x + 2} \, dx = \tfrac{1}{7} \ln (7x + 2) + c$$

From Problems 1–3 above it may be seen that:

If 'x' in a standard integral is replaced by $(ax + b)$ where a and b are constants, then $(ax + b)$ is written for x in the result and the result is multiplied by $\dfrac{1}{a}$.

For example, $\int (ax + b) \, dx = \dfrac{1}{2a} (ax + b)^2 + c$ and, more generally,

$\int (ax + b)^n \, dx = \dfrac{1}{a(n + 1)} (ax + b)^{n+1} + c$ (except when $n = -1$, see Problem 3).

Problem 4. Integrate the following with respect to x, using the general rule (i.e. without making a substitution): (a) $3 \sin (2x - 1)$; (b) $2e^{8x+3}$; (c) $\dfrac{5}{9x - 2}$

(a) $\int 3 \sin (2x - 1) \, dx = 3(- \cos (2x - 1)) \tfrac{1}{2} + c$
$= -\tfrac{3}{2} \cos (2x - 1) + c$

(b) $\int 2e^{8x+3} \, dx = 2(e^{8x+3})(\tfrac{1}{8}) + c = \tfrac{1}{4} e^{8x+3} + c$

(c) $\displaystyle\int \frac{5}{9x - 2} \, dx = 5 [\ln (9x - 2)] \tfrac{1}{9} + c = \tfrac{5}{9} \ln (9x - 2) + c$

Problem 5. Find: $\int 3x(x^2 + 2)^6 \, dx$.

Let $u = x^2 + 2$

then $\dfrac{du}{dx} = 2x$, i.e. $dx = \dfrac{du}{2x}$

Hence $\int 3x(x^2 + 2)^6 \, dx = \displaystyle\int 3x(u)^6 \frac{du}{2x} = \tfrac{3}{2} \int u^6 \, du$

The original variable, x, has been removed completely and the integral is now only in terms of u.

$\frac{3}{2}\int u^6\, du = \frac{3}{2}\left(\frac{u^7}{7}\right) + c$

Since $u = x^2 + 2$,

$\boldsymbol{\int 3x(x^2+2)^6\, dx = \frac{3}{14}(x^2+2)^7 + c}$

Problem 6. Find: $\int \sin\theta\cos\theta\, d\theta$.

Let $u = \sin\theta$

then $\frac{du}{d\theta} = \cos\theta$, i.e. $d\theta = \frac{du}{\cos\theta}$

Hence $\int \sin\theta\cos\theta\, d\theta = \int u\cos\theta\,\frac{du}{\cos\theta} = \int u\, du = \frac{u^2}{2} + c$

Since $u = \sin\theta$,

$\boldsymbol{\int \sin\theta\cos\theta\, d\theta = \frac{1}{2}\sin^2\theta + c}$

Another solution to this integral is possible.

Let $u = \cos\theta$

then $\frac{du}{d\theta} = -\sin\theta$, i.e. $d\theta = \frac{-du}{\sin\theta}$

Hence $\int \sin\theta\cos\theta\, d\theta = \int \sin\theta\,(u)\left(\frac{-du}{\sin\theta}\right) = -\int u\, du = -\frac{u^2}{2} + c$

Since $u = \cos\theta$,

$\boldsymbol{\int \sin\theta\cos\theta\, d\theta = -\frac{1}{2}\cos^2\theta + c}$

From Problems 5 and 6 above it may be seen that:
Integrals of the form $k\int [f(x)]^n f'(x)\, dx$ (where k is a constant) can be integrated by substituting u for $f(x)$.

Problem 7. Find: $\int \frac{(2x+3)}{\sqrt{(2x^2+6x-1)}}\, dx$

Let $u = 2x^2 + 6x - 1$

then $\frac{du}{dx} = 4x + 6$, i.e. $dx = \frac{du}{4x+6} = \frac{du}{2(2x+3)}$

Hence $\int \frac{(2x+3)}{\sqrt{(2x^2+6x-1)}}\, dx = \int \frac{(2x+3)}{\sqrt{u}}\,\frac{du}{2(2x+3)} = \frac{1}{2}\int \frac{du}{\sqrt{u}}$

$= \frac{1}{2}\int u^{-\frac{1}{2}}\, du = \frac{1}{2}\left(\frac{u^{\frac{1}{2}}}{\frac{1}{2}}\right) + c = u^{\frac{1}{2}} + c$

Since $u = 2x^2 + 6x - 1$,

$\boldsymbol{\int \frac{(2x+3)}{\sqrt{(2x^2+6x-1)}}\, dx = \sqrt{(2x^2+6x-1)} + c}$

 Problem 8. Find: $\int \tan\theta \, d\theta$.

$$\int \tan\theta\, d\theta = \int \frac{\sin\theta}{\cos\theta}\, d\theta$$

Let $u = \cos\theta$

then $\dfrac{du}{d\theta} = -\sin\theta$, i.e. $d\theta = \dfrac{-du}{\sin\theta}$

Hence $\displaystyle\int \frac{\sin\theta}{\cos\theta}\, d\theta = \int \frac{\sin\theta}{u}\left(\frac{-du}{\sin\theta}\right) = -\int \frac{1}{u}\, du = -\ln u + c$

Since $u = \cos\theta$,

$$\begin{aligned}\int \tan\theta\, d\theta &= -\ln(\cos\theta) + c\\ &= \ln(\cos\theta)^{-1} + c\\ &= \mathbf{\ln(\sec\theta) + c}\end{aligned}$$

From Problems 7 and 8 above it may be seen that:

Integrals of the form $k\displaystyle\int \frac{f'(x)}{[f(x)]^n}\, dx$ (where k and n are constants) can be integrated by substituting u for $f(x)$.

Problem 9. Evaluate the following:

(a) $\int_0^1 3\sec^2(4\theta - 1)\, d\theta$

(b) $\int_0^4 5x\sqrt{(2x^2+4)}\, dx$ taking positive values of roots only

(c) $\displaystyle\int_1^3 \frac{e^t}{3+e^t}\, dt$

(a) $$\begin{aligned}\int_0^1 3\sec^2(4\theta-1)\, d\theta &= \left[\tfrac{3}{4}\tan(4\theta-1)\right]_0^1 = \tfrac{3}{4}[\tan 3 - \tan(-1)]\\ &= \tfrac{3}{4}[\tan 171^\circ 53' - \tan(-57^\circ 18')]\\ &= \tfrac{3}{4}[(-0.142\,5) - (-1.557\,4)]\\ &= \tfrac{3}{4}(1.414\,9) = \mathbf{1.061\,2}\end{aligned}$$

(b) $\int_0^4 5x\sqrt{(2x^2+4)}\, dx = \int_0^4 5x(2x^2+4)^{\frac{1}{2}}\, dx$

Let $u = 2x^2 + 4$

then $\dfrac{du}{dx} = 4x$, i.e. $dx = \dfrac{du}{4x}$

$$\begin{aligned}\int 5x(2x^2+4)^{\frac{1}{2}}\, dx &= \int 5x(u^{\frac{1}{2}})\frac{du}{4x} = \tfrac{5}{4}\int u^{\frac{1}{2}}\, du\\ &= \tfrac{5}{4}\left(\frac{u^{\frac{3}{2}}}{\frac{3}{2}}\right) + c = \tfrac{5}{6}(\sqrt{u^3}) + c\end{aligned}$$

Since $u = 2x^2 + 4$,

$$\begin{aligned}\int_0^4 5x\sqrt{(2x^2+4)}\, dx &= \left[\tfrac{5}{6}\sqrt{(2x^2+4)^3}\right]_0^4\\ &= \tfrac{5}{6}\left\{\sqrt{[(2(4)^2+4)]^3} - \sqrt{(4)^3}\right\}\\ &= \tfrac{5}{6}(216 - 8), \text{ taking positive values of roots only}\\ &= \mathbf{173\tfrac{1}{3}}\end{aligned}$$

(c) $\int_1^3 \frac{e^t}{3 + e^t}\, dt$

Let $u = 3 + e^t$

then $\frac{du}{dt} = e^t$, i.e. $dt = \frac{du}{e^t}$

Hence $\int \frac{e^t}{3 + e^t}\, dt = \int \frac{e^t}{u}\frac{du}{e^t} = \int \frac{du}{u} = \ln u + c$

Since $u = 3 + e^t$,

$$\int_1^3 \frac{e^t}{3 + e^t}\, dt = [\ln (3 + e^t)]_1^3$$

$$= [\ln (3 + e^3) - \ln (3 + e^1)]$$

$$= \ln\left[\frac{3 + e^3}{3 + e^1}\right] = \ln\left[\frac{23.085}{5.718\ 3}\right]$$

$$= \mathbf{1.395\ 5}$$

Further problems on integration by substitution may be found in the following section (6) (Problems 66–125).

6. Further problems

Standard integrals

In Problems 1–35 integrate with respect to the variable.

1. x^5 $\left[\frac{x^6}{6} + c\right]$
2. $2p^3$ $\left[\frac{p^4}{2} + c\right]$
3. $3k^6$ $[\frac{3}{7}k^7 + c]$
4. $4u^{2.3}$ $\left[\frac{4}{3.3}u^{3.3} + c\right]$
5. $x^{-2.1}$ $\left[\frac{-x^{-1.1}}{1.1} + c\right]$
6. $\frac{2}{x^2}$ $\left[\frac{-2}{x} + c\right]$
7. $\frac{3}{p}$ $[3 \ln p + c]$
8. $\sqrt{y}$ $[\frac{2}{3}y^{\frac{3}{2}} + c]$
9. $2\sqrt{S^3}$ $[\frac{4}{5}\sqrt{S^5} + c]$
10. $\frac{1}{3\sqrt{t}}$ $[\frac{2}{3}\sqrt{t} + c]$
11. $\frac{4}{\sqrt[3]{k^2}}$ $[12\sqrt[3]{k} + c]$
12. $3a^3 - \frac{2}{3}\sqrt{a}$ $\left[\frac{3a^4}{4} - \frac{4}{9}\sqrt{a^3} + c\right]$

13. $\dfrac{-4}{v^{1.4}}$ $\left[\dfrac{10}{v^{0.4}}+c\right]$

14. $\dfrac{x}{3}(2x+\sqrt{x})$ $\left[\dfrac{2x^3}{9}+\dfrac{2}{15}\sqrt{x^5}+c\right]$

15. $\dfrac{r^3+2r-1}{r^2}$ $\left[\dfrac{r^2}{2}+2\ln r+\dfrac{1}{r}+c\right]$

16. $(x+2)^2$ $\left[\dfrac{x^3}{3}+2x^2+4x+c\right]$

17. $(1+\sqrt{w})^2$ $\left[w+\frac{4}{3}\sqrt{w^3}+\dfrac{w^2}{2}+c\right]$

18. $\sin 2\theta$ $[-\frac{1}{2}\cos 2\theta+c]$

19. $\cos 4\alpha$ $[\frac{1}{4}\sin 4\alpha+c]$

20. $2\sin 3t$ $[-\frac{2}{3}\cos 3t+c]$

21. $-4\cos 5x$ $[-\frac{4}{5}\sin 5x+c]$

22. $\sec^2 6\beta$ $[\frac{1}{6}\tan 6\beta+c]$

23. $-3\sec^2 t$ $[-3\tan t+c]$

24. $4(\cos 2\theta-3\sin\theta)$ $[2(\sin 2\theta+6\cos\theta)+c]$

25. e^{3x} $\left[\dfrac{e^{3x}}{3}+c\right]$

26. $2e^{-4t}$ $[-\frac{1}{2}e^{-4t}+c]$

27. $\dfrac{6}{e^t}$ $\left[-\dfrac{6}{e^t}+c\right]$

28. $3(e^x-e^{-x})$ $[3(e^x+e^{-x})+c]$

29. $3(e^t-1)^2$ $\left[3\left(\dfrac{e^{2t}}{2}-2e^t+t\right)+c\right]$

30. $\dfrac{4}{e^{2x}}+e^x$ $\left[\dfrac{-2}{e^{2x}}+e^x+c\right]$

31. $\dfrac{1}{4t}$ $[\frac{1}{4}\ln t+c]$

32. $\dfrac{3}{5t}+\sqrt{t^5}$ $[\frac{3}{5}\ln t+\frac{2}{7}\sqrt{t^7}+c]$

33. $\left(\dfrac{1}{x}+x\right)^2$ $\left[-\dfrac{1}{x}+2x+\dfrac{x^3}{3}+c\right]$

34. $3\sin 50\pi t+4\cos 50\pi t$ $\left[\dfrac{1}{50\pi}(4\sin 50\pi t-3\cos 50\pi t)+c\right]$

35. $(e^{2x}-1)(e^{-2x}+1)$ $[\frac{1}{2}(e^{2x}+e^{-2x})+c]$

In Problems 36–65 evaluate the definite integrals. (Where roots are involved in the solution, take positive values only when evaluating.)

36. $\int_1^3 2\,dt$ $[4]$

37. $\int_3^5 4x\,dx$ $[32]$

38. $\int_{-4}^2 -3u^2\,du$ $[-72]$

39. $\int_{-1}^1 \frac{3}{4}f^2\,df$ $[\frac{1}{2}]$

40. $\int_1^4 x^{-1.5}\,dx$ $[1]$

41. $\int_1^9 \frac{dx}{\sqrt{x}}$ [4]

42. $\int_2^5 \frac{4}{x}\,dx$ [3.665]

43. $\int_0^2 (x^2 + 2x - 1)\,dx$ [$4\frac{2}{3}$]

44. $\int_1^4 \left(\sqrt{r} - \frac{1}{\sqrt{r}}\right) dr$ [$2\frac{2}{3}$]

45. $\int_1^4 (3x^3 - 4x^2 + x - 2)\,dx$ [$108\frac{3}{4}$]

46. $\int_1^3 (m - 2)(m - 1)\,dm$ [$\frac{2}{3}$]

47. $\int_1^2 \left(\frac{1}{x^2} + \frac{1}{x} + \frac{1}{2}\right) dx$ [1.693]

48. $\int_{-2}^2 (3x - 1)\,dx$ [−4]

49. $\int_1^3 \left(\frac{2}{t^2} - 3t^2 + 4\right) dt$ [$-16\frac{2}{3}$]

50. $\int_0^{\frac{\pi}{2}} \sin\theta\,d\theta$ [1]

51. $\int_0^{\frac{\pi}{3}} 3\sin 2x\,dx$ [$2\frac{1}{4}$]

52. $\int_0^{\frac{\pi}{6}} 4\sin 3\theta\,d\theta$ [$1\frac{1}{3}$]

53. $\int_{\frac{\pi}{6}}^{\frac{\pi}{3}} 2\cos t\,dt$ [0.732 1]

54. $\int_0^1 5\sin 2\theta\,d\theta$ [3.540 4]

55. $\frac{1}{2}\int_1^2 \cos 3\alpha\,d\alpha$ [−0.070 1]

56. $\int_{0.1}^{0.6} (\frac{1}{4}\sin 3\beta + \frac{1}{2}\cos 2\beta)\,d\beta$ [0.281 9]

57. $\int_{-\frac{\pi}{2}}^{\frac{\pi}{2}} 3\cos\theta\,d\theta$ [6]

58. $\int_0^{\frac{\pi}{4}} 3\sec^2\theta\,d\theta$ [3]

59. $\int_{-1}^1 3\sec^2 2t\,dt$ [−6.555]

60. $\int_1^2 \frac{e^{3x}}{5}\,dx$ [25.56]

61. $\int_{0.4}^{0.7} 3e^{2t}\,dt$ [2.744]

62. $\int_0^1 \frac{2}{e^{3t}}\,dt$ [0.633 5]

63. $\int_1^4 \left(\frac{t+2}{\sqrt{t}}\right) dt$ [$8\frac{2}{3}$]

64. $\int_1^3 \frac{(3x + 2)(x - 4)}{x}\,dx$ [−16.789]

65. $\int_0^1 2\sqrt{x}(x+2)^2\,dx$ [9.105]

Integration by substitution

In Problems 65–105 integrate with respect to the variable.

66. $\sin(3x+2)$ $[-\frac{1}{3}\cos(3x+2)+c]$
67. $2\cos(4t+1)$ $[\frac{1}{2}\sin(4t+1)+c]$
68. $3\sec^2(t+5)$ $[3\tan(t+5)+c]$
69. $4\sin(6\theta-3)$ $[-\frac{2}{3}\cos(6\theta-3)+c]$
70. $(2x+1)^5$ $[\frac{1}{12}(2x+1)^6+c]$
71. $3(4S-7)^4$ $[\frac{3}{20}(4S-7)^5+c]$
72. $\frac{1}{12}(9x+5)^8$ $[\frac{1}{108}(9x+5)^9+c]$
73. $\dfrac{1}{3a+1}$ $[\frac{1}{3}\ln(3a+1)+c]$
74. $\dfrac{5}{5f-2}$ $[\ln(5f-2)+c]$
75. $\dfrac{7}{2x+1}$ $[\frac{7}{2}\ln(2x+1)+c]$
76. $\dfrac{-1}{6x+5}$ $[-\frac{1}{6}\ln(6x+5)+c]$
77. $\dfrac{3}{15y-2}$ $[\frac{1}{5}\ln(15y-2)+c]$
78. e^{3x+2} $[\frac{1}{3}e^{3x+2}+c]$
79. $4e^{7t-1}$ $[\frac{4}{7}e^{7t-1}+c]$
80. $2e^{2-3x}$ $[-\frac{2}{3}e^{2-3x}+c]$
81. $4x(2x^2+3)^5$ $[\frac{1}{6}(2x^2+3)^6+c]$
82. $5t(t^2-1)^7$ $[\frac{5}{16}(t^2-1)^8+c]$
83. $(3x^2+4)^8x$ $[\frac{1}{54}(3x^2+4)^9+c]$
84. $\sin^2\theta\cos\theta$ $[\frac{1}{3}\sin^3\theta+c]$
85. $\sin^3 t\cos t$ $[\frac{1}{4}\sin^4 t+c]$
86. $2\cos^2\beta\sin\beta$ $[-\frac{2}{3}\cos^3\beta+c]$
87. $\sec^2\theta\tan\theta$ $[\frac{1}{2}\tan^2\theta+c]$
88. $3\tan 2x\sec^2 2x$ $[\frac{3}{4}\tan^2 2x+c]$
89. $\frac{6}{5}\sin^5\theta\cos\theta$ $[\frac{1}{5}\sin^6\theta+c]$
90. $6x\sqrt{(3x^2+2)}$ $[\frac{2}{3}\sqrt{(3x^2+2)^3}+c]$
91. $(4x^2-1)\sqrt{(4x^3-3x+1)}$ $[\frac{2}{9}\sqrt{(4x^3-3x+1)^3}+c]$
92. $\dfrac{3\ln t}{t}$ $[\frac{3}{2}(\ln t)^2+c]$
93. $\dfrac{6x+2}{(3x^2+2x-1)^5}$ $\left[\dfrac{-1}{4(3x^2+2x-1)^4}+c\right]$
94. $\dfrac{4y-1}{(4y^2-2y+5)^7}$ $\left[\dfrac{-1}{12(4y^2-2y+5)^6}+c\right]$
95. $\dfrac{2x}{\sqrt{(x^2+1)}}$ $[2\sqrt{(x^2+1)}+c]$
96. $\dfrac{3a}{\sqrt{(3a^2+5)}}$ $[\sqrt{(3a^2+5)}+c]$

97. $\dfrac{12x^2+1}{\sqrt{(4x^3+x-1)}}$ $[2\sqrt{(4x^3+x-1)}+c]$

98. $\dfrac{r^2-1}{\sqrt{(r^3-3r+2)}}$ $[\frac{2}{3}\sqrt{(r^3-3r+2)}+c]$

99. $\dfrac{3e^t}{\sqrt{(1+e^t)}}$ $[6\sqrt{(1+e^t)}+c]$

100. $2x\sin(x^2+1)$ $[-\cos(x^2+1)+c]$

101. $(4\theta+1)\sec^2(4\theta^2+2\theta)$ $[\frac{1}{2}\tan(4\theta^2+2\theta)+c]$

102. $\frac{1}{3}(4x+1)\cos(2x^2+x-1)$ $[\frac{1}{3}\sin(2x^2+x-1)+c]$

103. $4te^{2t^2-3}$ $[e^{2t^2-3}+c]$

104. $3\tan\beta$ $[3\ln(\sec\beta)+c]$

105. $(5x-2)e^{5x^2-4x+1}$ $[\frac{1}{2}e^{5x^2-4x+1}+c]$

In Problems 106–125 evaluate the definite integrals.

106. $\int_0^1(3x-1)^4\,dx$ $[2\frac{1}{5}]$

107. $\int_0^2(8x-3)(4x^2-3x)^3\,dx$ $[2\,500]$

108. $\int_1^3 x\sqrt{(x^2+1)}\,dx$ $[9.598]$

109. $\displaystyle\int_0^{\frac{\pi}{4}}\sin\left(4\theta+\frac{\pi}{3}\right)d\theta$ $[\frac{1}{4}]$

110. $\displaystyle\int_{\frac{1}{3}}^{1}\sec^2(3x-1)\,dx$ $[-0.728\,3]$

111. $\int_1^2 3\cos(5t-2)\,dt$ $[0.508\,9]$

112. $\displaystyle\int_{\frac{1}{2}}^{2}\frac{1}{(4s-1)}\,ds$ $[0.486\,5]$

113. $\int_0^2(9x^2-4)\sqrt{(3x^3-4x)}\,dx$ $[42\frac{2}{3}]$

114. $\displaystyle\int_1^3\frac{4\ln x}{x}\,dx$ $[2.413\,9]$

115. $\displaystyle\int_0^2\frac{t}{\sqrt{(2t^2+1)}}\,dt$ $[1]$

116. $\displaystyle\int_1^2\frac{4x-3}{(2x^2-3x-1)^4}\,dx$ $[-\frac{3}{8}]$

117. $\int_1^2 3\theta\sin(2\theta^2+1)\,d\theta$ $[-0.059\,1]$

118. $\int_0^1 2te^{3t^2-1}\,dt$ $[2.340\,4]$

119. $\displaystyle\int_0^{\frac{\pi}{2}}3\sin^4\theta\cos\theta\,d\theta$ $[\frac{3}{5}]$

120. $\displaystyle\int_1^2\frac{dx}{(2x-1)^3}$ $[\frac{2}{9}]$

121. $\displaystyle\int_1^2\frac{2e^{3\theta}}{e^{3\theta}-5}\,d\theta$ $[2.182\,5]$

122. $\int_0^1 2t\sec^2(3t^2)\,dt$ $[-0.047\,5]$

123. $\int_1^2 x\sin(2x^2-1)\,dx$ $[-0.053\,4]$

124. $\int_{\frac{\pi}{6}}^{\frac{\pi}{3}} \frac{2}{3} \sin t \cos^3 t \, dt$ [0.083 3]

125. $\int_1^2 \frac{e^{3\theta} - e^{-3\theta}}{2} \, d\theta$ [63.88]

Chapter 7

Application of integration to areas and volumes

1. The area between a curve, the x-axis and given ordinates

There are several instances in branches of engineering and science where the area under a curve is required to be accurately determined. For example, the areas, between given limits, of:

(a) velocity/time graphs give distances travelled;
(b) force/distance graphs give work done;
(c) acceleration/time graphs give velocity;
(d) voltage/current graphs give power;
(e) pressure/volume graphs give work done;
(f) normal distribution curves give frequency.

Provided there is a known relationship between the variables forming the axes of the above graphs then the areas may be calculated exactly using integral calculus. If a relationship between variables is not known then areas have to be approximately determined using such techniques as the trapezoidal rule, the mid-ordinate rule or Simpson's rule. Let A be the area enclosed between the curve $y = f(x)$, the x-axis and the ordinates $x = a$ and $x = b$. Also let A be subdivided into a number of elemental strips each of width δx as shown in Fig. 1.

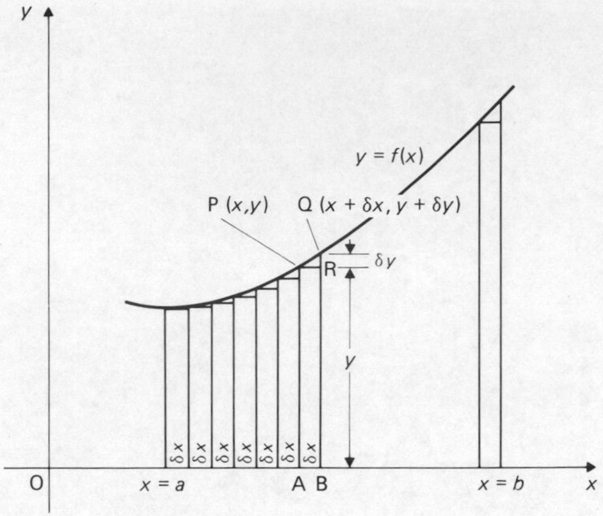

Figure 1

One such strip is shown as PQRBA, with point P having coordinates (x, y) and point Q having coordinates $(x + \delta x, y + \delta y)$. Let the area PQRBA be δA, which can be seen from Fig. 1 to consist of a rectangle PRBA, of area of $y\delta x$, and PQR, which approximates to a triangle of area $\frac{1}{2}\delta x \delta y$, i.e.

$$\delta A \simeq y\delta x + \tfrac{1}{2}\delta x \delta y$$

Dividing both sides by δx gives:

$$\frac{\delta A}{\delta x} \simeq y + \tfrac{1}{2}\delta y$$

As δx is made smaller and smaller, the number of rectangles increases and all such areas as PQR becomes smaller and smaller. Also δy becomes smaller and in the limit as δx approaches zero, $\dfrac{\delta A}{\delta x}$ becomes the differential coefficient $\dfrac{\mathrm{d}A}{\mathrm{d}x}$, and δy becomes zero.

i.e. $$\lim_{\delta x \to 0} \left(\frac{\delta A}{\delta x} \right) = \frac{\mathrm{d}A}{\mathrm{d}x} = y + \tfrac{1}{2}(0) = y$$

Hence $$\frac{\mathrm{d}A}{\mathrm{d}x} = y \qquad \ldots (1)$$

This shows that when a limiting value is taken, all such areas as PQR become zero. Hence the area beneath the curve is given by the sum of all such rectangles as PRBA.

i.e. Area $= \Sigma y \delta x$.
Between the limits $x = a$ and $x = b$.

$$\text{Area, } A = \lim_{\delta x \to 0} \sum_{x=a}^{x=b} y\,\delta x \qquad \ldots (2)$$

From equation (1), $\dfrac{dA}{dx} = y$ and by integration:

$$\int \frac{dA}{dx}\, dx = \int y\, dx$$

Hence $A = \int y\, dx$

The ordinates $x = a$ and $x = b$ limit the area and such ordinate values are shown as limits.

$$\text{Thus } A = \int_a^b y\, dx \qquad \ldots (3)$$

Equations (2) and (3) show that:

$$\textbf{Area, } A = \lim_{\delta x \to 0} \sum_{x=a}^{x=b} y\,\delta x = \int_a^b y\, dx$$

This statement, that the limiting value of a sum is equal to the integral between the same limits, forms a fundamental theorem of integration. This can be illustrated by considering simple shapes of known areas. For example, Fig. 2 (a) shows a rectangle bounded by the line $y = h$, ordinates $x = a$ and $x = b$ and the x-axis.

Let the rectangle be divided into n equal vertical strips of width δx. The area of strip PQAB is $h\delta x$ and since there are n strips making up the total area the total area $= nh\delta x$. The base length of the rectangle, i.e. $(b - a)$, is made up of n strips, each δx in width, hence $n\delta x = (b - a)$. Therefore the total area $= h(b - a)$.

The total area is also obtained by adding the areas of all such strips at PQAB and is independent of the value of n, that is, n can be infinitely large.

$$\text{Hence total area} = \lim_{\delta x \to 0} \sum_{x=a}^{x=b} h\,\delta x = h(b - a) \qquad \ldots (4)$$

Also the total area is given by $\displaystyle\int_a^b y\, dx = \int_a^b h\, dx$

$$= [hx]_a^b = h(b - a) \qquad \ldots (5)$$

But this is the area obtained from equation (4).

$$\text{Hence } \lim_{\delta x \to 0} \sum_{x=a}^{x=b} h\,\delta x = \int_a^b h\, dx$$

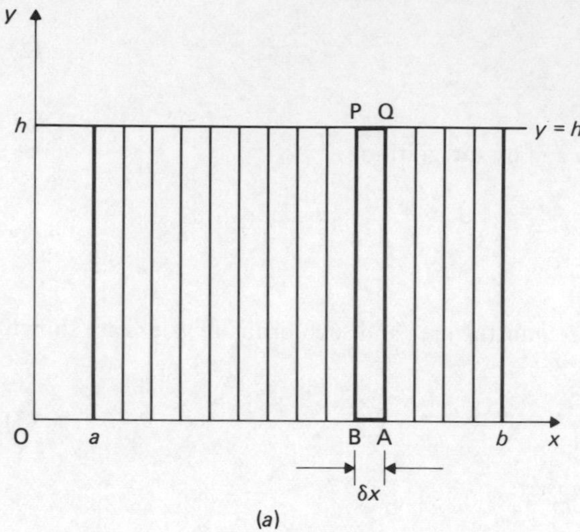

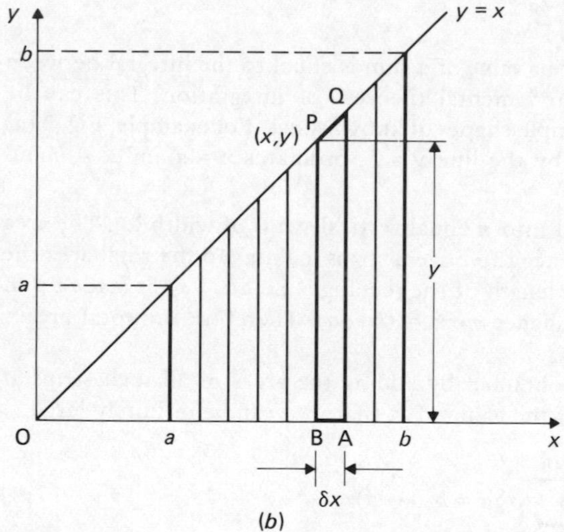

Figure 2

Similarly for, say, a trapezium bounded by the line $y = x$, the ordinates $x = a$ and $x = b$ and the x-axis (as shown in Fig. 2 (b)), the total area is given by:

(half the sum of the parallel sides)(perpendicular distance between these sides)

i.e. $\frac{1}{2}(a + b)(b - a)$ or $\frac{1}{2}(b^2 - a^2)$ . . . (6)

Also, the total area will be given by the sum of all areas such as PQAB which each have an area of $y\delta x$ provided δx is infinitely small.

i.e. total area $= \lim_{\delta x \to 0} \sum_{x=a}^{x=b} y\delta x = \frac{1}{2}(b^2 - a^2)$ from above ... (7)

Also, the total area $= \int_a^b y \, dx = \int_a^b x \, dx = \left[\frac{x^2}{2}\right]_a^b = \frac{1}{2}(b^2 - a^2)$ ... (8)

Hence from equation (7) and (8), it has been shown that

$$\lim_{\delta x \to 0} \sum_{x=a}^{x=b} y\delta x = \int_a^b y \, dx \qquad \text{... (9)}$$

The two simple illustrations used above show that equation (9) is valid in these two cases and we will assume that it is generally true, although a more rigorous proof is beyond the scope of this book.

If the area between a curve $x = f(y)$, the y-axis and ordinates $y = m$ and $y = n$ is required, then by similar reasoning to the above:

$$\text{Area} = \int_m^n x \, dy.$$

Thus finding the area beneath a curve is the same as determining the value of a definite integral previously discussed in Chapter 6.

A part of the curve $y = 2x^2 + 3$ is shown in Fig. 3, which is produced from the table of values shown below.

x	-2	-1	0	1	2	3
$y = 2x^2 + 3$	11	5	3	5	11	21

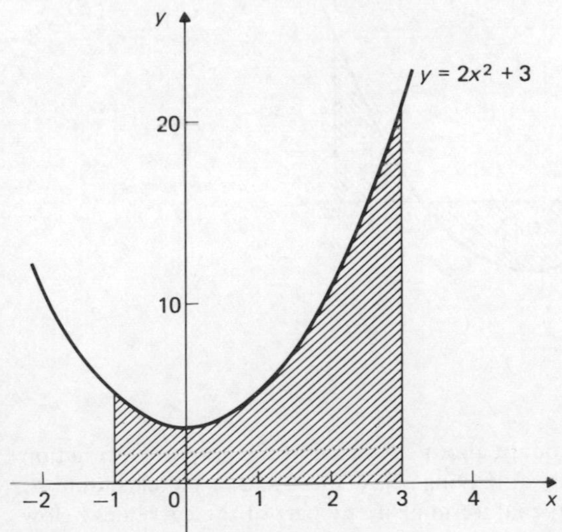

Figure 3 Graph of $y = 2x^2 + 3$

196 The area between the curve, the x-axis and the ordinates $x = -1$ and $x = 3$ is shown shaded. This area is given by:

$$\text{Area} = \int_{-1}^{3} y\,dx = \int_{-1}^{3} (2x^2 + 3)\,dx$$

$$= \left[\frac{2x^3}{3} + 3x\right]_{-1}^{3}$$

$$= \left[\frac{2(3)^3}{3} + 3(3)\right] - \left[\frac{2(-1)^3}{3} + 3(-1)\right]$$

$$= \mathbf{30\tfrac{2}{3}\ square\ units}$$

With the curve $y = 2x^2 + 3$ shown in Fig. 3 all values of y are positive. Hence all the terms in $\Sigma y\delta x$ are positive and $\int_a^b y\,dx$ is positive. However, if a curve should drop below the x-axis, then y becomes negative, all terms in $\Sigma y\delta x$ become negative and $\int_a^b y\,dx$ is negative.

In Fig. 4 the total area between the curve $y = f(x)$, the x-axis and the ordinates $x = a$ and $x = b$ is given by: area P + area Q + area R.

i.e. $\int_a^c f(x)\,dx - \int_c^d f(x)\,dx + \int_d^b f(x)\,dx$

This is **not** the same as the area given by $\int_a^b f(x)\,dx$.

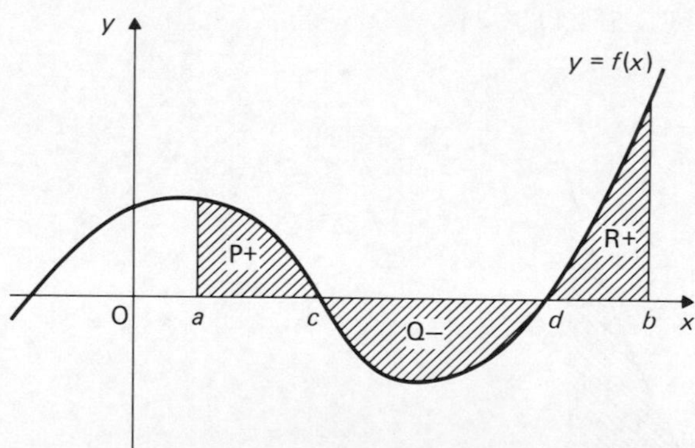

Figure 4

For this reason, if there is any doubt about the shape of the graph of a function or any possibility of all or part of it lying below the x-axis, a sketch should be made over the required limits to determine if any part of the curve lies below the x-axis.

2. The area between two curves

Let the graphs of the functions $y = f_1(x)$ and $y = f_2(x)$ intersect at points A and B as shown in Fig. 5.

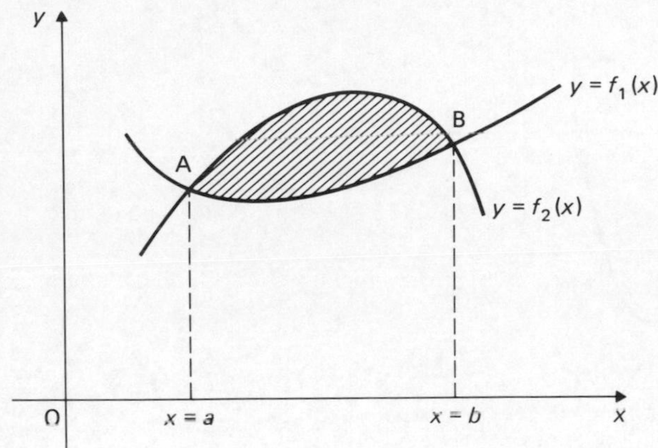

Figure 5

At the points of intersection $f_1(x) = f_2(x)$.

The area enclosed between the curve $y = f_2(x)$, the x-axis and the ordinates $x = a$ and $x = b$ is given by $\int_a^b f_2(x)\,\mathrm{d}x$.

The area enclosed between the curve $y = f_1(x)$, the x-axis and the ordinates $x = a$ and $x = b$ is given by $\int_a^b f_1(x)\,\mathrm{d}x$.

It follows that the area enclosed between the two curves (shown shaded in Fig. 5) is given by:

$$\textbf{Shaded area} = \int_a^b f_2(x)\,\mathrm{d}x - \int_a^b f_1(x)\,\mathrm{d}x$$

$$= \int_a^b [f_2(x) - f_1(x)]\,\mathrm{d}x$$

Worked problems on finding areas under and between curves

Problem 1. Find the area enclosed by the given curves, the x-axis and the given ordinates: (a) $y = \sin 2x$, $x = 0$, $x = \frac{\pi}{2}$; (b) $y = 3\cos\frac{1}{2}x$, $x = 0$, $x = \frac{2\pi}{3}$. Sketch the curves in the given ranges.

(a) A sketch of $y = \sin 2x$ in the range $x = 0$ to $x = \pi$ is shown in Fig. 6 (a).

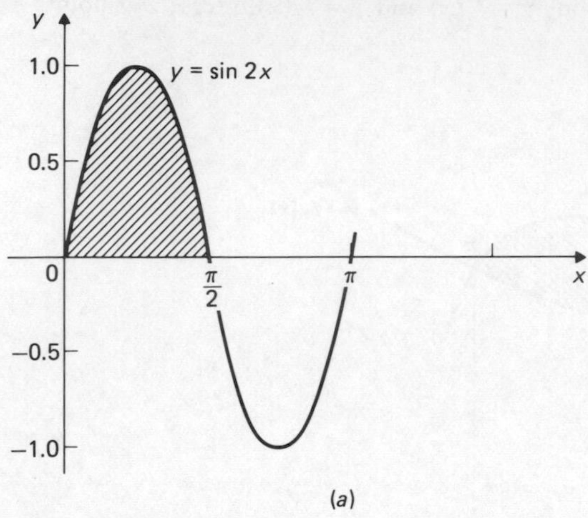

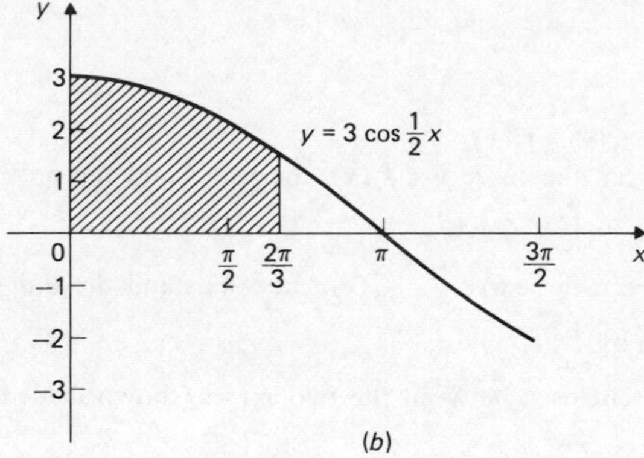

Figure 6 Graphs of $y = \sin 2x$ and $y = 3 \cos \frac{1}{2}x$

The area shown shaded is given by:

$$\textbf{Area} = \int_0^{\frac{\pi}{2}} \sin 2x \, dx$$

$$= \left[-\frac{\cos 2x}{2} \right]_0^{\frac{\pi}{2}} = \left(\frac{-\cos 2(\pi/2)}{2} \right) - \left(\frac{-\cos 0}{2} \right)$$

$$= \left(\frac{-\cos \pi}{2} \right) - \left(\frac{-\cos 0}{2} \right) = (- -\tfrac{1}{2}) - (-\tfrac{1}{2})$$

$= \textbf{1 square unit}$

(b) A sketch of $y = 3 \cos \frac{1}{2}x$ in the range $x = 0$ to $x = \dfrac{2\pi}{3}$ is shown in Fig. 6 (b).
The area shown shaded is given by:

$$\text{Area} = \int_0^{2\pi/3} 3 \cos \tfrac{1}{2}x \, dx$$

$$= \left[6 \sin \tfrac{1}{2}x \right]_0^{2\pi/3} = \left(6 \sin \frac{\pi}{3} \right) - (6 \sin 0)$$

$$= 6 \sin 60° = \mathbf{3\sqrt{3} \text{ or } 5.196 \text{ square units}}$$

Problem 2. Find the area enclosed by the curve $y = 2x^2 - x + 3$, the x-axis and the ordinates $x = -1$ and $x = 2$.

A table of values is produced as shown below.

x	-1	0	1	2
y	6	3	4	9

The area between the curve, the x-axis and the ordinates $x = -1$ and $x = 2$ is wholly above the x-axis, since all values of y in the table are positive. Thus the area is positive. In such cases as this it is unnecessary to actually draw the graph.

$$\text{Area} = \int_{-1}^{2} (2x^2 - x + 3) \, dx$$

$$= \left[\frac{2x^3}{3} - \frac{x^2}{2} + 3x \right]_{-1}^{2}$$

$$= (\tfrac{16}{3} - 2 + 6) - (-\tfrac{2}{3} - \tfrac{1}{2} - 3)$$

$$= (9\tfrac{1}{3}) - (-4\tfrac{1}{6}) = \mathbf{13\tfrac{1}{2} \text{ square units}}$$

Problem 3. Calculate the area of the figure bounded by the curve $y = 2e^{t/2}$, the t-axis and ordinates $t = -1$ and $t = 3$.

A table of values is produced as shown below.

t	-1	0	1	2	3
$y = 2e^{t/2}$	1.213	2.000	3.297	5.437	8.963

Since all the values of y are positive, the area required is wholly above the t-axis. Hence the area enclosed by the curve, the t-axis and the ordinates $t = -1$ and $t = 3$ is given by:

$$\text{Area} = \int_{-1}^{3} 2e^{t/2}\, dt$$

$$= \left[4e^{t/2} \right]_{-1}^{3} = 4[e^{\frac{3}{2}} - e^{-\frac{1}{2}}]$$

$$= 4[4.481\,7 - 0.606\,5]$$

$$= \mathbf{15.50\ square\ units}$$

Problem 4. Find the area enclosed by the curve $y = x^2 + 3$, the x-axis and the ordinates $x = 0$ and $x = 3$. Sketch the curve within these limits. Find also the area enclosed by the curve and the y-axis, between the same limits.

A table of values is produced as shown below.

x	0	1	2	3
y	3	4	7	12

(a) Part of the curve $y = x^2 + 3$ is shown in Fig. 7.

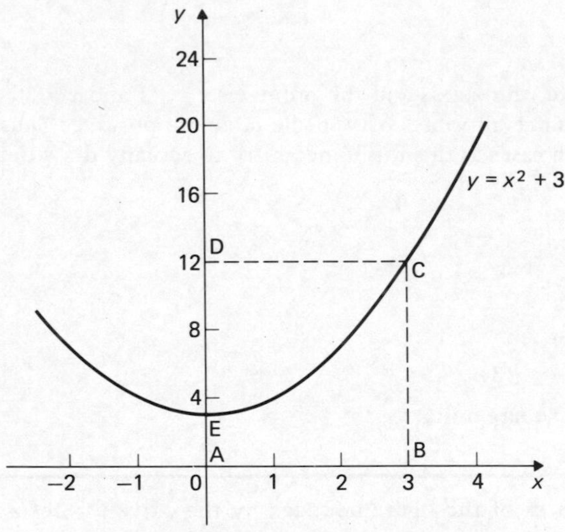

Figure 7 Graph of $y = x^2 + 3$

The area enclosed by the curve, the x-axis and ordinates $x = 0$ and $x = 3$ (i.e. area ABCE of Fig. 7) is given by:

$$\text{Area} = \int_{0}^{3} (x^2 + 3)\, dx = \left[\frac{x^3}{3} + 3x \right]_{0}^{3}$$

$$= \mathbf{18\ square\ units}$$

(b) When $x = 3$, $y = 3^2 + 3 = 12$
when $x = 0$, $y = 3$

If $y = x^2 + 3$ then $x^2 = y - 3$ and $x = \sqrt{(y - 3)}$

Hence the area enclosed by the curve $y = x^2 + 3$ (i.e. the curve $x = \sqrt{(y - 3)}$), the y-axis and the ordinates $y = 3$ and $y = 12$ (i.e. area CDE of Fig. 7) is given by:

$$\text{Area} = \int_{y=3}^{y=12} x\, dy = \int_3^{12} \sqrt{(y-3)}\, dy$$

Let $u = y - 3$

then $\dfrac{du}{dy} = 1$, i.e. $dy = du$

Hence $\int (y-3)^{\frac{1}{2}}\, dy = \int u^{\frac{1}{2}}\, du = \dfrac{2u^{\frac{3}{2}}}{3}$

Since $u = y - 3$ then

$$\text{Area} = \int_3^{12} \sqrt{(y-3)}\, dy = \left[\tfrac{2}{3}(y-3)^{\frac{3}{2}}\right]_3^{12}$$

$= \frac{2}{3}[\sqrt{9^3} - 0]$

$=$ **18 square units**

The sum of the areas in parts (a) and (b) is 36 square units, which is equal to the area of the rectangle ABCD.

Problem 5. Calculate the area between the curve $y = x^3 - x^2 - 6x$ and the x-axis.

$$y = x^3 - x^2 - 6x = x(x^2 - x - 6)$$
$$= x(x-3)(x+2)$$

Thus when $y = 0$, $x = 0$ or $x - 3 = 0$ or $x + 2 = 0$

i.e. $x = 0$, $x = 3$ or $x = -2$.

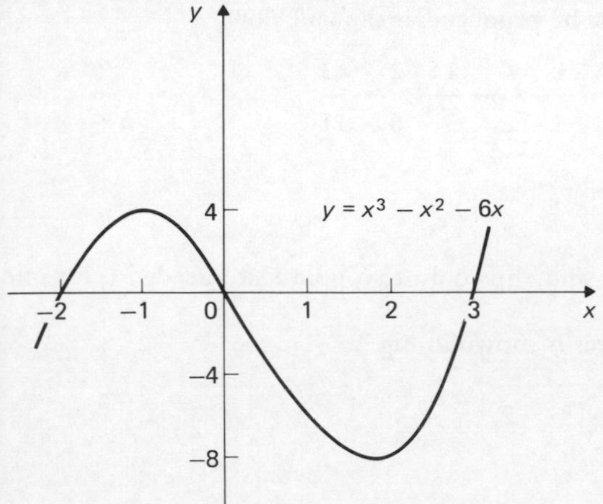

Figure 8 Graph of $y = x^3 - x^2 - 6x$

Hence the curve cuts the x-axis at $x = 0$, 3 and -2. Since the curve is a continuous function, only one other value need be calculated before a sketch of the curve can be produced. For example, when $x = 1$, $y = -6$, which shows that the portion of the curve between ordinates $x = 0$ and $x = 3$ is negative. Hence the portion of the curve between ordinates $x = 0$ and $x = -2$ must be positive.

A sketch of part of the curve $y = x^3 - x^2 - 6x$ is shown in Fig. 8.

If $y = f(x)$ had not factorised as above, then a table of values could have been produced and the graph sketched in the usual manner.

The sketch shows that the area needs to be calculated in two parts, one part being positive and the other negative.

The area between the curve and the x-axis is given by:

$$\textbf{Area} = \int_{-2}^{0} (x^3 - x^2 - 6x)\,dx - \int_{0}^{3} (x^3 - x^2 - 6x)\,dx$$

$$= \left[\frac{x^4}{4} - \frac{x^3}{3} - 3x^2\right]_{-2}^{0} - \left[\frac{x^4}{4} - \frac{x^3}{3} - 3x^2\right]_{0}^{3}$$

$$= (5\tfrac{1}{3}) - (-15\tfrac{3}{4})$$

$$= \mathbf{21\tfrac{1}{12}\ square\ units}$$

Problem 6. Find the area enclosed between the curves $y = x^2 + 2$ and $y + x = 14$.

The first step is to find the points of intersection of the two curves. This will enable us to limit the range of values when drawing up a table of values in order to sketch the curves. At the points of intersection the curves are equal (i.e. their coordinates are the same). Since $y = x^2 + 2$ and $y + x = 14$ (i.e. $y = 14 - x$), then $x^2 + 2 = 14 - x$ at the points of intersection.

i.e. $x^2 + x - 12 = 0$

$(x - 3)(x + 4) = 0$

Hence $x = 3$ or $x = -4$ at the points of intersection.

Tables of values may now be produced as shown below.

x	-4	-3	-2	-1	0	1	2	3
$y = x^2 + 2$	18	11	6	3	2	3	6	11

x	-4	0	3
$y = 14 - x$	18	14	11

$y = 14 - x$ is a straight line and thus only two points are needed (plus one more to check).

A sketch of the two curves is shown in Fig. 9.

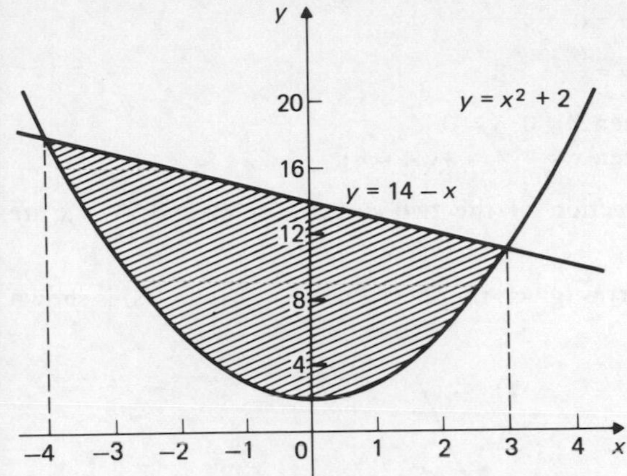

Figure 9 Graphs of $y = x^2 + 2$ and $y = 14 - x$

The area between the two curves (shown shaded) is given by:

$$\textbf{Shaded area} = \int_{-4}^{3} (14 - x)\,dx - \int_{-4}^{3} (x^2 + 2)\,dx$$

$$= \left[14x - \frac{x^2}{2}\right]_{-4}^{3} - \left[\frac{x^3}{3} + 2x\right]_{-4}^{3}$$

$$= 101\tfrac{1}{2} - 44\tfrac{1}{3}$$

$$= 57\tfrac{1}{6} \textbf{ square units}$$

Problem 7. Find the points of intersection of the two curves $x^2 = 2y$ and $\dfrac{y^2}{16} = x$. Sketch the two curves and calculate the area enclosed by them.

$x^2 = 2y$, i.e. $y = \dfrac{x^2}{2}$

$\dfrac{y^2}{16} = x$, i.e. $y = \sqrt{(16x)} = \pm 4\sqrt{x}$

At the points of intersection, $\dfrac{x^2}{2} = \pm 4\sqrt{x}$

Squaring both sides gives: $\left(\dfrac{x^2}{2}\right)^2 = (\pm 4\sqrt{x})^2$

i.e. $\dfrac{x^4}{4} = 16x$

$x^4 = 64x$

Hence $x^4 - 64x = 0$

$x(x^3 - 64) = 0$

i.e. $x = 0$ or $x^3 - 64 = 0$

Hence at the points of intersection $x = 0$ and $x = 4$.

Using $y = \frac{x^2}{2}$: When $x = 0, y = 0$

When $x = 4, y = \frac{(4)^2}{2} = 8$

[Check, using $y = 4\sqrt{x}$: When $x = 0, y = 0$

When $x = 4, y = 4\sqrt{4} = 8$]

Hence the points of intersection of the two curves $x^2 = 2y$ and $\frac{y^2}{16} = x$ are (0, 0) and (4, 8).

A sketch of the two curves (given the special name of parabolas) is shown in Fig. 10.

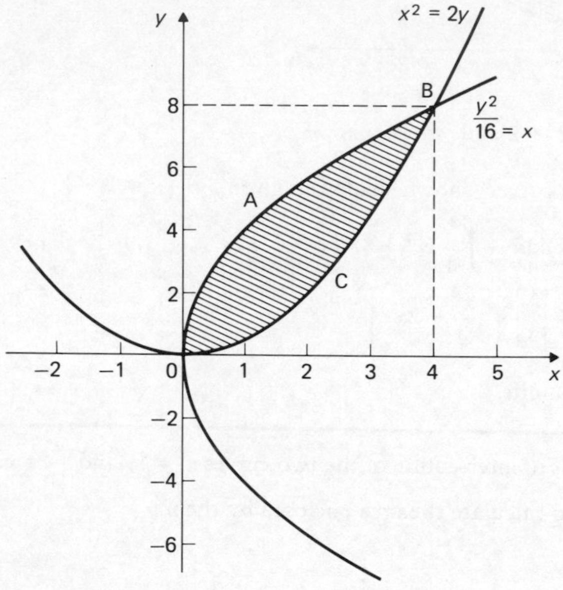

Figure 10 Graphs of $x^2 = 2y$ and $\frac{y^2}{16} = x$

The area enclosed by the two curves, i.e. OABC (shown shaded), is given by:

$$\text{Area} = \int_0^4 4\sqrt{x}\,\mathrm{d}x - \int_0^4 \frac{x^2}{2}\,\mathrm{d}x$$

(Note that for one curve $y = \pm 4\sqrt{x}$. The $-4\sqrt{x}$ is neglected since the shaded area required is above the x-axis, and hence positive.)

$$\textbf{Area} = 4\left[\tfrac{2}{3}x^{\frac{3}{2}}\right]_0^4 - \tfrac{1}{2}\left[\frac{x^3}{3}\right]_0^4$$

$$= 21\tfrac{1}{3} - 10\tfrac{2}{3}$$

$$= \mathbf{10\tfrac{2}{3}}\ \textbf{square units}$$

Further problems on areas under and between curves may be found in Section 4 (Problems 1–46).

3. Volumes of solids of revolution

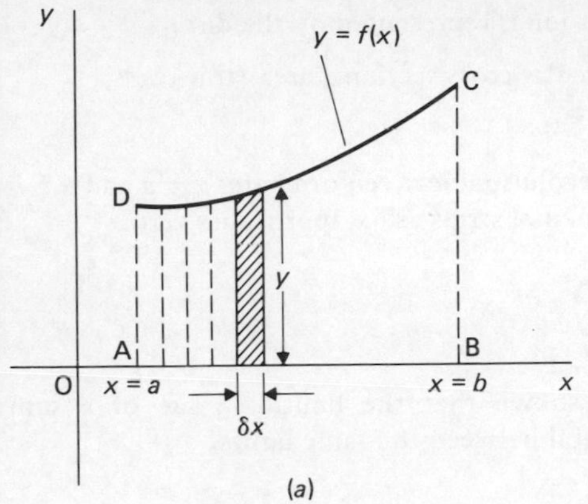

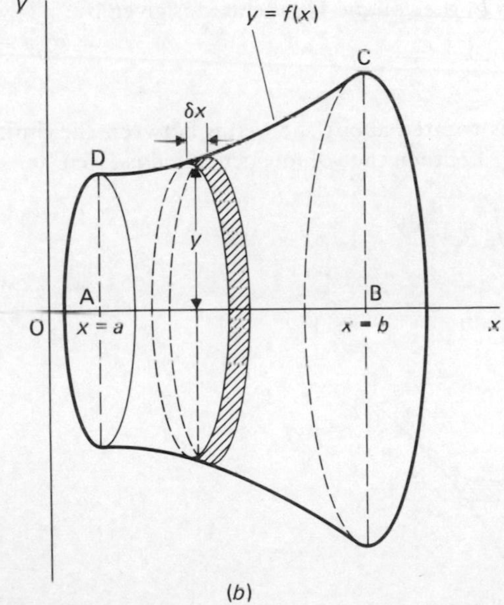

Figure 11

Figure 11 (a) shows a plane area ABCD bounded by the curve $y = f(x)$, the x-axis and the ordinates $x = a$ and $x = b$. If this area is rotated 360° about the x-axis, then a volume known as a **solid of revolution** is produced, as shown in Figure 11 (b).

Let the area ABCD be divided into a large number of strips, each of width δx. A typical strip is shown shaded. When the area ABCD is rotated 360°

 about the x-axis, each strip produces a solid of revolution which approximates to a circular disc of radius y and thickness δx. The smaller δx becomes, the more accurately the solid of revolution is represented by the disc.

The volume of one such disc = (circular cross-sectional area)(thickness)

$$= (\pi y^2)(\delta x)$$

The total volume of the solid of revolution between ordinates $x = a$ and $x = b$ is given by the sum of all such elemental strips as δx approaches zero.

$$\text{i.e. Total volume} = \lim_{\delta x \to 0} \sum_{x=a}^{x=b} \pi y^2 \delta x$$

When dealing with areas, it was shown that the limiting value of a sum between limits is equal to the integral between the same limits.

$$\text{i.e.} \lim_{\delta x \to 0} \sum_{x=a}^{x=b} \pi y^2 \delta x = \int_a^b \pi y^2 \, dx$$

Hence, when the curve $y = f(x)$ is rotated one revolution about the x-axis between the limits $x = a$ and $x = b$, the volume V generated is given by:

Volume, $\boldsymbol{V = \int_a^b \pi y^2 \, dx}$

Similarly, if a curve $x = f(y)$ is rotated about the y-axis between the limits $y = c$ and $y = d$, as shown in Fig. 12, then the volume generated is given by:

$$\textbf{volume} = \lim_{\delta y \to 0} \sum_{y=c}^{y=d} \pi x^2 \delta y = \int_c^d \pi x^2 \, dy$$

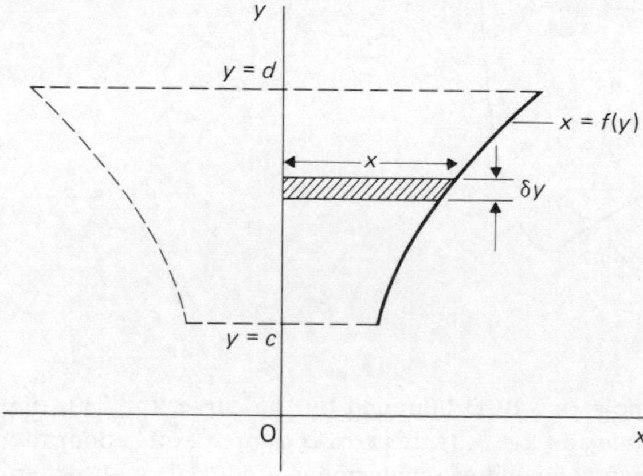

Figure 12

Problem 1. Find the volume of the solid of revolution between the limits $x = 0$ and $x = 4$ when the following curves are rotated about the x-axis: (a) $y = 3$; (b) $y = x$.

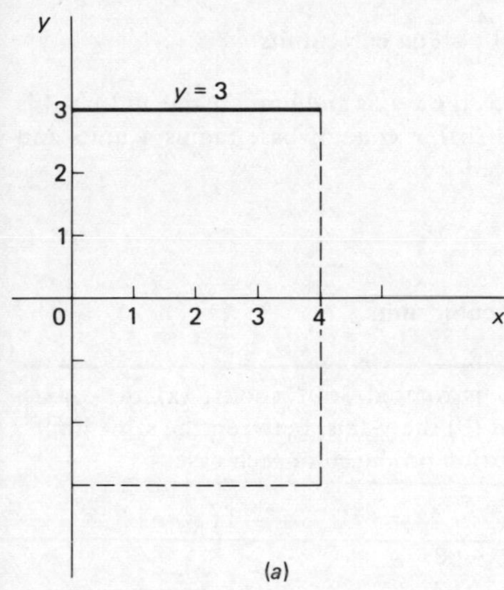

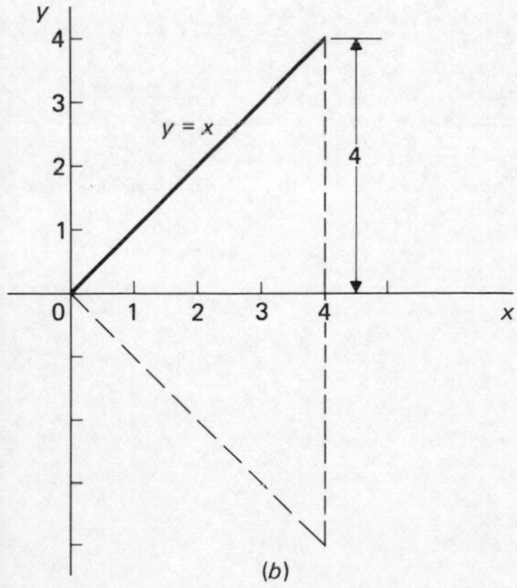

Figure 13

(a) If the area bounded by $y = 3$, the x-axis and the limits $x = 0$ and $x = 4$ is rotated about the x-axis (see Fig. 13 (a)), a cylinder of radius 3 units and height 4 units is produced.

$$\text{Volume generated} = \int_0^4 \pi y^2\, dx = \int_0^4 \pi(3)^2\, dx$$

$$= 9\pi \int_0^4 dx = 9\pi [x]_0^4 = \mathbf{36\pi \text{ cubic units}}$$

(b) If the area bounded by $y = x$, the x-axis and limits $x = 0$ and $x = 4$ is rotated about the x-axis (see Fig. 13 (b)), a cone of base radius 4 units and perpendicular height 4 units is produced.

$$\text{Volume generated} = \int_0^4 \pi y^2\, dx = \int_0^4 \pi x^2\, dx$$

$$= \pi \left[\frac{x^3}{3} \right]_0^4 = \mathbf{\frac{64\pi}{3} \text{ cubic units}}$$

Problem 2. The curve $y = 2x^2 + 3$ is rotated 360° about: (a) the x-axis, between the limits $x = 1$ and $x = 3$; and (b) the y-axis, between the same limits. Find the volume of the solid of revolution produced in each case.

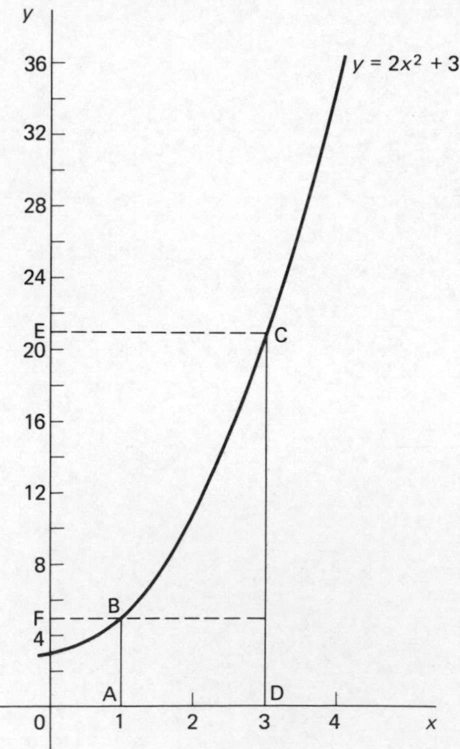

Figure 14

The relevant portion of the curve $y = 2x^2 + 3$ is shown in Fig. 14.

(a) The volume produced when the curve $y = 2x^2 + 3$ is rotated about the x-axis between $x = 1$ and $x = 3$ (i.e. rotating area ABCD) is given by:

$$\text{volume} = \int_1^3 \pi y^2 \, dx$$

$$= \int_1^3 \pi(2x^2 + 3)^2 \, dx = \pi \int_1^3 (4x^4 + 12x^2 + 9) \, dx$$

$$= \pi \left[\frac{4x^5}{5} + \frac{12x^3}{3} + 9x \right]_1^3$$

$$= \pi[(194.4 + 108 + 27) - (0.8 + 4 + 9)]$$

$$= \mathbf{315.6\pi \text{ cubic units}}$$

(b) When $x = 1$, $y = 2(1)^2 + 3 = 5$
When $x = 3$, $y = 2(3)^2 + 3 = 21$

Since $y = 2x^2 + 3$, then $x^2 = \dfrac{y - 3}{2}$

The volume produced when the curve $y = 2x^2 + 3$ is rotated about the y-axis between $y = 5$ and $y = 21$ (i.e. rotating area BCEF is given by:

$$\text{volume} = \int_5^{21} \pi x^2 \, dy = \int_5^{21} \pi \frac{(y - 3)}{2} \, dy$$

$$= \frac{\pi}{2} \left[\frac{y^2}{2} - 3y \right]_5^{21}$$

$$= \frac{\pi}{2} \left[\left\{ \frac{(21)^2}{2} - 3(21) \right\} - \left\{ \frac{(5)^2}{2} - 3(5) \right\} \right]$$

$$= \mathbf{80.0\pi \text{ cubic units}}$$

Problem 3. Find the volume generated when the area above the x-axis bounded by the curve $x^2 + y^2 = 4$ and the ordinates $x = 2$ and $x = -2$ is rotated about the x-axis.

Figure 15 shows the part of the curve $x^2 + y^2 = 4$ lying above the x-axis.

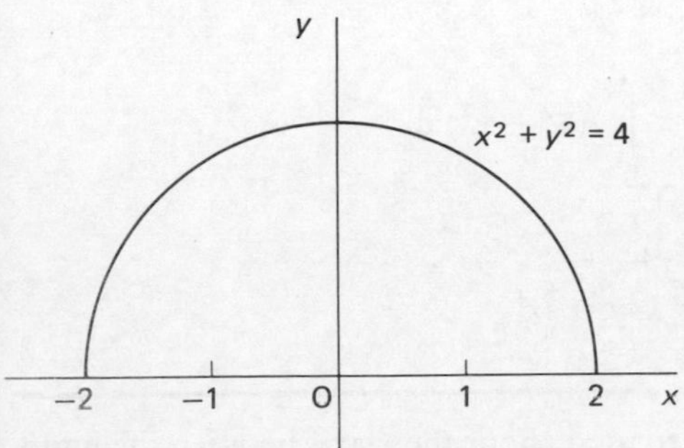

Figure 15

 The curve $x^2 + y^2 = 4$ is a circle, centre O and radius 2. (In general $x^2 + y^2 = r^2$ represents a circle, centre O and radius r.)

If the semicircle shown in Fig. 15 is rotated about the x-axis then the volume generated is given by:

$$\text{volume} = \int_{-2}^{2} \pi y^2 \, dx = \int_{-2}^{2} \pi(4 - x^2) \, dx$$

$$= \pi \left[4x - \frac{x^3}{3} \right]_{-2}^{2} = \frac{32\pi}{3} \text{ cubic units}$$

Problem 4. Calculate the volume of the frustum of a sphere of radius 5 cm which lies between two parallel planes at 1 cm and 3 cm from the centre and on the same side of it.

The volume of the frustum of the sphere may be found by rotating the curve $x^2 + y^2 = 5^2$ (i.e. a circle, centre O, radius 5) about the x-axis between the limits $x = 1$ and $x = 3$ (i.e. rotating the area shown shaded in Fig. 16).

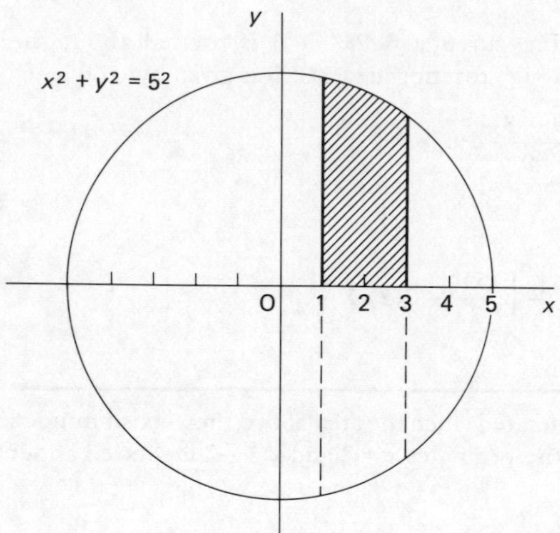

Figure 16

$$\text{Volume of frustum} = \int_{1}^{3} \pi y^2 \, dx = \pi \int_{1}^{3} (5^2 - x^2) \, dx$$

$$= \pi \left[25x - \frac{x^3}{3} \right]_{1}^{3}$$

$$= \pi[(75 - 9) - (25 - \tfrac{1}{3})]$$

$$= 41\tfrac{1}{3} \text{ cm}^3$$

Problem 5. The curve $y = 3 \sec \dfrac{x}{2}$ is rotated about the x-axis between the limits

$x = 0$ and $x = \frac{\pi}{3}$. Find the volume of the solid formed.

Using natural secant tables the following table of values is produced.

x	0	$\frac{\pi}{6}$	$\frac{\pi}{3}$
$y = 3 \sec \frac{x}{2}$	3.000	3.106	3.464

A part of the curve $y = 3 \sec \frac{x}{2}$ is shown in Fig. 17. If the area shown shaded

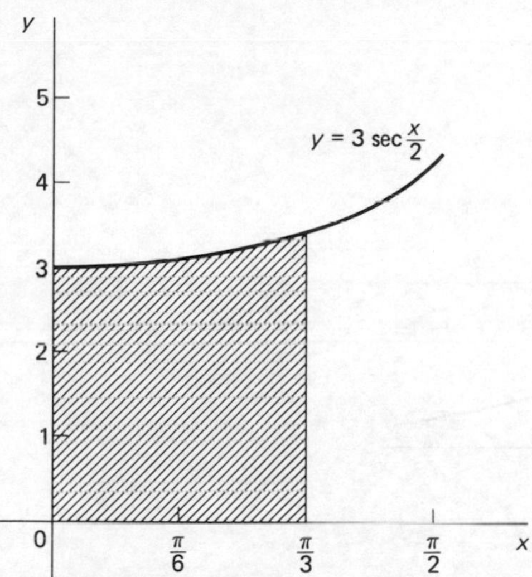

Figure 17

(i.e. between limits $x = 0$ and $x = \frac{\pi}{3}$ is rotated about the x-axis then the volume formed is given by:

$$\text{volume} = \int_0^{\frac{\pi}{3}} \pi y^2 \, dx = \int_0^{\frac{\pi}{3}} \pi \left(3 \sec \frac{x}{2} \right)^2 dx$$

$$= 9\pi \int_0^{\frac{\pi}{3}} \sec^2 \frac{x}{2} \, dx$$

$$= 9\pi \left[2 \tan \frac{x}{2} \right]_0^{\frac{\pi}{3}}$$

$$= 18\pi \left[\tan \frac{\pi}{6} - \tan 0 \right] = \frac{18\pi}{\sqrt{3}}$$

$$= \mathbf{6\sqrt{3}\pi \text{ cubic units}}$$

 Problem 6. Find the volume of the solid formed by revolving the area enclosed between the curve $y = \frac{1}{x^2}$ and the lines $y = 2$ and $y = 4$ about the y axis.

$y = \frac{1}{x^2}$. As $x \to 0$, $y \to \infty$. As $x \to \infty$, $y \to 0$.

When $x = \frac{1}{2}$, $y = 4$ and when $x = 1$, $y = 1$. Using these values a sketch of $y = \frac{1}{x^2}$ may be drawn as shown in Fig. 18.

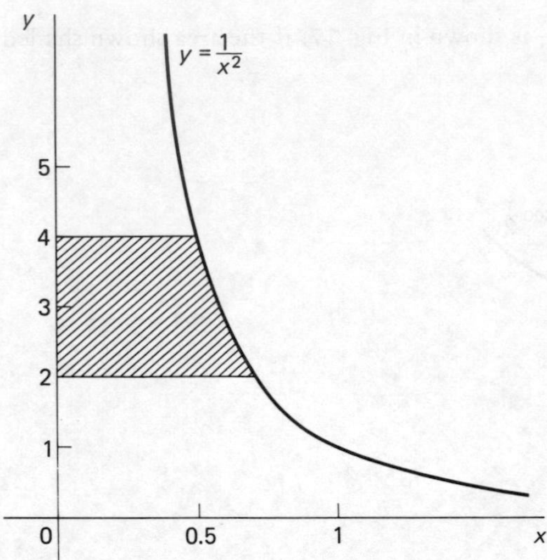

Figure 18

If the area bounded by the curve $y = \frac{1}{x^2}$ and the lines $y = 2$ and $y = 4$ (shown shaded in Fig. 18) is rotated about the y-axis the volume is given by:

$$\text{volume} = \int_2^4 \pi x^2 \, dy$$

Since $y = \frac{1}{x^2}$, then $x^2 = \frac{1}{y}$

$$\text{Hence volume} = \int_2^4 \pi \frac{1}{y} \, dy$$

$$= \pi [\ln y]_2^4$$

$$= \pi [\ln 4 - \ln 2] = \pi \ln \tfrac{4}{2}$$

$$= \boldsymbol{\pi \ln 2 \text{ or } 0.693\pi \text{ cubic units}}$$

Problem 7. The area enclosed by the curve $y = 2e^{x/2}$, the x-axis and ordinates $x = -2$ and $x = 3$ is rotated about the x-axis. Calculate the volume generated.

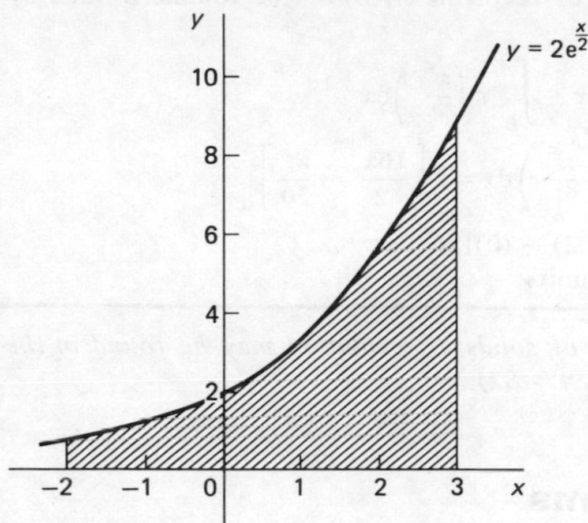

Figure 19

If the area shown shaded in Fig. 19 is rotated about the x-axis, the volume generated is given by:

$$\text{volume} = \int_{-2}^{3} \pi y^2 \, dx = \int_{-2}^{3} \pi(2e^{\frac{x}{2}})^2 \, dx$$

$$= 4\pi \int_{-2}^{3} e^x \, dx = 4\pi [e^x]_{-2}^{3}$$

$$= 4\pi [e^3 - e^{-2}] = \mathbf{79.8\pi \text{ cubic units}}$$

Problem 8. The area enclosed between the two parabolas $x^2 = 2y$ and $y^2 = 16x$ is rotated about the x-axis. Find the volume of the solid produced.

At the points of intersection the curves are equal.

Curve 1: $x^2 = 2y$ or $y = \dfrac{x^2}{2}$, i.e. $y^2 = \dfrac{x^4}{4}$

Curve 2: $y^2 = 16x$

At the points of intersection $\dfrac{x^4}{4} = 16x$

$$\frac{x^4}{4} - 16x = 0$$

$$x\left(\frac{x^3}{4} - 16\right) = 0$$

Hence $x = 0$ or $\dfrac{x^3}{4} - 16 = 0$, i.e. $x = 4$.

The area enclosed by the two curves is shown as OABC in Fig. 10.

The volume produced by revolving the shaded area about the x-axis is given by: (the volume formed by revolving OAB) – (the volume formed by revolving OCB).

$$\text{i.e. Volume} = \int_0^4 \pi(16x)\,dx - \int_0^4 \pi\left(\frac{x^4}{4}\right)dx$$
$$= \pi\int_0^4 \left(16x - \frac{x^4}{4}\right)dx = \pi\left[\frac{16x^2}{2} - \frac{x^5}{20}\right]_0^4$$
$$= \pi[(128 - 51.2) - (0)]$$
$$= \mathbf{76.8\pi \text{ cubic units}}$$

Further problems on volumes of solids of revolution may be found in the following section (4) (Problems 47–82).

4. Further problems

Areas under and between curves

(All answers are in square units.)

In Problems 1–22 find the area enclosed between the given curve, the horizontal axis and the given ordinates. Sketch the curve in the given range for each.

1. $y = 2x; x = 0, x = 5$ [25]
2. $y = x^2 - x + 2; x = -1, x = 2$ $[7\frac{1}{2}]$
3. $y = \frac{x}{2}; x = 3, x = 7$ [10]
4. $y = p - 1; p = 1, p = 5$ [8]
5. $F = 8S - 2S^2; S = 0, S = 2$ $[10\frac{2}{3}]$
6. $y = (x - 1)(x - 2); x = 0, x = 3$ $[1\frac{5}{6}]$
7. $y = 8 + 2x - x^2; x = -2, x = 4$ [36]
8. $u = 2(4 - t^2); t = -2, t = 2$ $[21\frac{1}{3}]$
9. $y = x(x - 1)(x + 3); x = -2, x = 1$ $[7\frac{11}{12}]$
10. $x = 4a^3; a = -2, a = 2$ [32]
11. $y = x(x - 1)(x - 3); x = 0, x = 3$ $[3\frac{1}{12}]$
12. $a = t^3 + t^2 - 4t - 4; t = -3, t = 3$ $[24\frac{1}{3}]$
13. $y = \sin\theta; \theta = 0, \theta = \frac{\pi}{2}$ [1]
14. $y = \cos x; x = \frac{\pi}{4}, x = \frac{\pi}{2}$ [0.292 9]
15. $y = 3\sin 2\beta; \beta = 0, \beta = \frac{\pi}{4}$ $[1\frac{1}{2}]$
16. $y = 5\cos 3\alpha; \alpha = 0, \alpha = \frac{\pi}{6}$ $[1\frac{2}{3}]$
17. $y = \sin x - \cos x; x = 0, x = \frac{\pi}{4}$ [0.414 2]

18. $2y^2 = x$; $x = 0$, $x = 2$ [$2\frac{2}{3}$]
19. $5 = xy$; $x = 2$, $x = 5$ [4.581]
20. $y = 2e^{2t}$; $t = 0$, $t = 2$ [53.60]
21. $ye^{4x} = 3$; $x = 1$, $x = 3$ [0.013 7]
22. $y = 2x + e^x$; $x = 0$, $x = 3$ [28.09]
23. Find the area between the curve $y = 3x - x^2$ and the x-axis. [$4\frac{1}{2}$]
24. Calculate the area enclosed between the curve $y = 12 - x - x^2$ and the x-axis. [$57\frac{1}{6}$]
25. Sketch the curve $y = \sec^2 2x$ from $x = 0$ to $x = \dfrac{\pi}{4}$ and calculate the area enclosed between the curve, the x-axis and the ordinates $x = 0$ and $x = \dfrac{\pi}{6}$. [0.866]
26. Find the area of the template enclosed between the curve $y = \dfrac{1}{x-2}$, the x-axis and the ordinates $x = 3$ cm and $x = 5$ cm. [1.098 6 cm^2]
27. Sketch the curves $y = x^2 + 4$ and $y + x = 10$ and find the area enclosed by them. [$20\frac{5}{6}$]
28. Calculate the area enclosed between the curves $y = \sin\theta$ and $y = \cos\theta$ and the θ-axis between the limits $\theta = 0$ and $\theta = \dfrac{\pi}{4}$. [0.414 2]
29. Find the area between the two parabolas $9y^2 = 16x$ and $x^2 = 6y$. [$3\frac{5}{9}$]
30. Calculate the area of the metal plate enclosed between $y = x(x - 4)$ and the x-axis, where x is in metres. [$10\frac{2}{3}$ m^2]
31. Sketch the curve $x^2 - y = 3x + 10$ and find the area enclosed between it and the x-axis. [$57\frac{1}{6}$]
32. Find the area enclosed by the curve $y = 4(x^2 - 1)$, the x-axis and the ordinates $x = 0$ and $x = 2$. Find also the area enclosed by the curve and the y-axis between the same limits. [8, $21\frac{1}{3}$]
33. Calculate the area between the curve $y = x(x^2 - 2x - 3)$ and the x-axis. [$11\frac{5}{6}$]
34. Find the area of the figure bounded by the curve $y = 3e^{2x}$, the x-axis and the ordinates $x = -2$ and $x = 2$. [81.87]
35. Find the area enclosed between the curves $y = x^2 - 3x + 5$ and $y - 1 = 2x$. [$4\frac{1}{2}$]
36. Find the points of intersection of the two curves $\dfrac{x^2}{2} = (\sqrt{2})\,y$ and $y^2 = 8x$ and calculate the area enclosed by them. [(0, 0); (4, 5.657); 7.542]
37. Calculate the area bounded by the curve $y = x^2 + x + 4$ and the line $y = 2(x + 5)$. [$20\frac{5}{6}$]
38. Find the area enclosed between the curves $y = x^2$ and $y = 8 - x^2$. [$21\frac{1}{3}$]
39. Calculate the area between the curve $y = 3x^3$ and the line $\dfrac{y}{12} = x$ in the first quadrant. [24]

40. Find the area bounded by the three straight lines $y = 4(2 - x)$, $y = 4x$ and $3y = 4x$. [2]

41. A vehicle has an acceleration a of $(30 + 2t)$ metres per second squared after t seconds. If the vehicle starts from rest find its velocity after 10 seconds. $\left(\text{Velocity} = \int_{t_1}^{t_2} a\, dt.\right)$ [400 m s^{-1}]

42. A car has a velocity v of $(3 + 4t)$ metres per second after t seconds. How far does it move in the first four seconds? Find the distance travelled in the fifth second. $\left(\text{Distance travelled} = \int_{t_1}^{t_2} v\, dt.\right)$ [44 m; 21 m]

43. A gas expands according to the law pv = constant. When the volume is 2 m^3 the pressure is 200 kPa. Find the work done as the gas expands from 2 m^3 to a volume of 5 m^3. $\left(\text{Work done} = \int_{v_1}^{v_2} p\, dv.\right)$ [367 kJ]

44. The brakes are applied to a train and the velocity v at any time t seconds after applying the brakes is given by $(16 - 2.5t)$ m s^{-1}. Calculate the distance travelled in 8 seconds. $\left(\text{Distance travelled} = \int_{t_1}^{t_2} v\, dt.\right)$ [48 m]

45. The force F newtons acting on a body at a distance x metres from a fixed point is given by $F = 3x + \frac{1}{x^2}$. Find the work done when the body moves from the position where $x = 1$ m to that where $x = 3$ m. $\left(\text{Work done} = \int_{x_1}^{x_2} F\, dx.\right)$ [$12\frac{2}{3}$ Nm]

46. The velocity of a body t seconds after a certain instant is $(4t^2 + 3)$ m s^{-1}. Find how far it moves in the interval from $t = 2$ s to $t = 6$ s. $\left(\text{Distance travelled} = \int_{t_1}^{t_2} v\, dt.\right)$ [$289\frac{1}{3}$ m]

Volumes of solids of revolution

(All answers are in cubic units and are left in terms of π.)

In Problems 47–61 find the volume of the solids of revolution formed by revolving the areas enclosed by the given curves, the x-axis and the given ordinates through 360° about the x-axis.

47. $y = 2; x = 1, x = 5$ [16π]
48. $y = 2x; x = 0, x = 6$ [288π]
49. $y = x^2; x = -1, x = 2$ [$6\frac{3}{5}\pi$]
50. $y = x + 1; x = 1, x = 2$ [$6\frac{1}{3}\pi$]
51. $y + 1 = x^2; x = -1, x = 1$ [$1\frac{1}{15}\pi$]
52. $y = 3x^2 + 4; x = 0, x = 2$ [$153\frac{3}{5}\pi$]
53. $\frac{y^2}{3} = x; x = 1, x = 6$ [$52\frac{1}{2}\pi$]
54. $y = \frac{2}{x}; x = 2, x = 4$ [π]

55. $3xy = 5; x = 1, x = 3$ $[1\frac{23}{27}\pi]$
56. $x^2 + y^2 = 16; x = -4, x = 4$ $[85\frac{1}{3}\pi]$
57. $x = \sqrt{(9 - y^2)}; x = 0, x = 3$ $[18\pi]$
58. $y = 3e^x; x = 0, x = 2$ $\left[\frac{9e^4\pi}{2} \text{ or } 245.7\pi\right]$
59. $y = 2 \sec x; x = 0, x = \frac{\pi}{4}$ $[4\pi]$
60. $y = 3 \operatorname{cosec} x; x = \frac{\pi}{6}, x = \frac{\pi}{3}$ $[10.39\pi]$
61. $\frac{y}{6} = \sqrt{x^3}; x = 2, x = 4$ $[2\,160\pi]$

In Problems 62–70 find the volume of the solids of revolution formed by revolving the areas enclosed by the given curves, the y-axis and the given ordinates through 360° about the y-axis.

62. $y = x^2; y = 1, y = 4$ $[7\frac{1}{2}\pi]$
63. $y = 2x^2 - 3; y = 0, y = 2$ $[4\pi]$
64. $2y = x^4; y = 1, y = 4$ $[6.60\pi]$
65. $y = \frac{3}{x}; y = 2, y = 3$ $[1.5\pi]$
66. $x^2 + y^2 = r^2; y = 0, y = r$ $[\frac{2}{3}\pi r^3]$
67. $y = \sqrt{(25 - x^2)}; y = -5, y = 5$ $[166\frac{2}{3}\pi]$
68. $y = \sqrt{x^3}; y = 0, y = 1$ $\left[\frac{3\pi}{7}\right]$
69. $x\sqrt{y} = 1; y = 2, y = 3$ $[\pi \ln \frac{3}{2} \text{ or } 0.405\pi]$
70. $\sqrt{x} = (\sqrt{3})y; y = 0, y = 2$ $[57.6\pi]$
71. The curve $y = 3x^2 - 4$ is rotated about: (a) the x-axis between the limits $x = 0$ and $x = 2$; and (b) the y-axis between the same limits. Find the volume generated in each case. (a) $[25.6\pi]$ (b) $[24\pi]$
72. Find the volume of a pressure vessel generated when the area above the x-axis bounded by the curve $x^2 + y^2 = 36$ and the ordinates $x = 6$ m and $x = -6$ m is rotated about the x-axis. $[288\pi \text{ m}^3]$
73. Calculate the volume of the plug formed by the frustum of a sphere of radius 7 cm which lies between two parallel planes at 2 cm and 5 cm from the centre and on the same side of it. $[108\pi \text{ cm}^3]$
74. The area enclosed between the two curves $x^2 = 4y$ and $y^2 = 4x$ is rotated about the x-axis. Find the volume of the solid formed. $[19\frac{1}{5}\pi]$
75. Calculate the volume of the solid gun-mounting formed by revolving the area between the curve $y = \frac{3}{2x^2}$ and the lines $y = 1$ m and $y = 4$ m about the y-axis. $[2.079\pi \text{ m}^3]$
76. The curve $y = x^2 + 2x + 1$ is rotated about the x-axis. Find the volume generated between the limits $x = 0$ and $x = 2$. $[48\frac{2}{5}\pi]$
77. Find the volume of the solid obtained by rotating about the x-axis the part of the curve $y = 4x - x^2$ lying above the x-axis. $[34\frac{2}{15}\pi]$
78. The portion of the curve $y = x^2 + \frac{2}{x}$ lying between $x = 1$ cm and $x = 2$ cm

is revolved about the x-axis. Calculate the volume of the solid for the component formed. [14.2π cm^3]

79. The curve $y = \dfrac{5}{x+1}$ is rotated about the x-axis between the limits $x = 0$ and $x = 4$ m. Find the volume of the podium generated. [20π m^3]

80. Calculate the volume of a frustum of a sphere of radius 8 cm which lies between two parallel planes at 3 cm and 4 cm from the centre and on opposite sides of it. [$417\frac{2}{3}\pi$ cm^3]

81. The area enclosed between the curves $x^2 = (2\sqrt{2})y$ and $\dfrac{y^2}{8} = x$ is rotated 360° about the x-axis. Find the volume produced. [38.4π]

82. The area between $\dfrac{y}{x^2} = 1$ and $y + x^2 = 8$ is rotated through 4 right angles about the x-axis. Calculate the volume generated. [$170\frac{2}{3}\pi$]

Chapter 8

Application of integration to centroids and second moments of area

1. Centroids of plane areas

Centroid

A **lamina** is a thin sheet of uniform thickness. If, when supported at a particular point, a lamina balances perfectly, then this point is called the **centre of gravity**. It is through the centre of gravity that the mass of the lamina is considered to act. If a lamina of negligible thickness, and hence negligible mass, is considered, then the term centre of gravity is inappropriate. As we are now dealing only with a shape or area, the term **centre of area** or simply **centroid** is used for the point where the centre of gravity of a lamina of that shape would lie.

The first moment of area

A 'moment' (in mechanics) is the measure of the power of a force in causing rotation (i.e. the 'moment of a force' is the product of the force and the perpendicular distance from a fixed point).

i.e. moment = force × distance
But force = mass × acceleration,

and in the case of a thin uniform lamina,

mass $\propto$ area

 Hence force $\propto$ area $\times$ acceleration

Since acceleration due to gravity can be taken to be a constant, then:

force $\propto$ area

The first moment of area is defined as the product of the area and the perpendicular distance of its centroid from a given axis in the plane of the area.

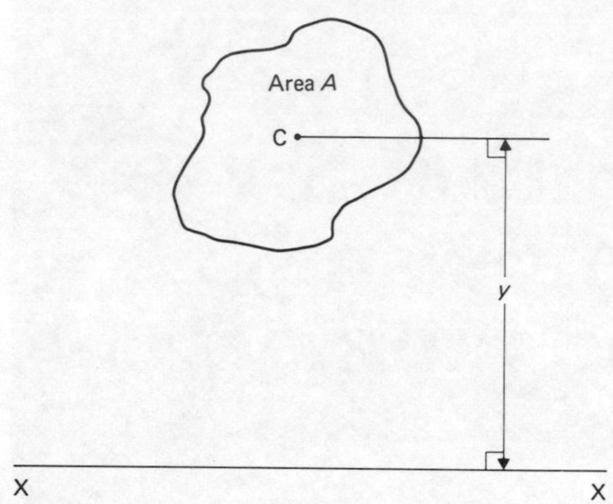

Figure 1

In Fig. 1 any area A is shown with its centroid at point C. XX is any axis in the same plane as A.

The first moment of area about the axis XX (which is at a perpendicular distance y from the centroid C) is given by (area A)(distance y).

i.e. **first moment of area = Ay cubic units.**

When dealing with centroids, the expression 'taking moments about XX' is often used for 'first moment of area about axis XX'.

The centroids of rectangles, triangles, circles and semicircles, together with composite areas of these shapes were dealt with in *Technician Mathematics, Level 2*, Chapter 6.

The centroid of: (a) a rectangle lies on the intersection of the diagonals; (b) a triangle, of perpendicular height h, lies at a point $\frac{h}{3}$ from the base; (c) a circle lies at its centre; (d) a semicircle, of radius r, lies on the centre line at a distance $\frac{4r}{3\pi}$ from the diameter.

If the centroid of an area between a curve and given limits is required then integration may be used.

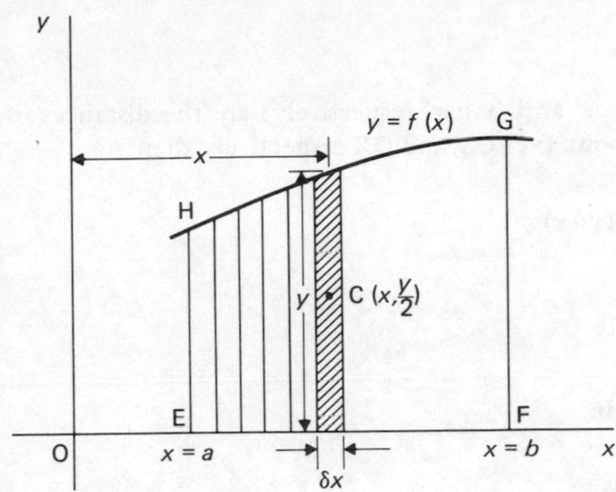

Figure 2

Figure 2 shows a plane area EFGH bounded by the curve $y = f(x)$, the x-axis and ordinates $x = a$ and $x = b$. Let this area be divided into a large number of strips each of width δx. A typical strip is shown shaded. The centroid of the shaded strip is at its centre, i.e. at coordinates $\left(x, \frac{y}{2}\right)$. The area of the strip is given by $y\delta x$.

Therefore the first moment of area of the strip about axis OY

= (area of strip)(perpendicular distance between the centroid and axis OY)

$= (y\delta x)x = xy\delta x$

The total first moment of area EFGH about axis OY is given by the sum of the first moments of area of all such strips, i.e. in the limit $\sum_{x=a}^{x=b} xy\delta x = \int_a^b xy\,dx$

The first moment of area of the strip about axis OX

= (area of strip)(perpendicular distance between the centroid and axis OX)

$= (y\delta x)\frac{y}{2} = \frac{1}{2}y^2\delta x$

The total first moment of area EFGH about axis OX is given by the sum of the first moments of area of all such strips, i.e. in the limit $\sum_{x=a}^{x=b} \frac{1}{2}y^2\delta x = \frac{1}{2}\int_a^b y^2\,dx$

The area A bounded by the curve $y = f(x)$, the x-axis and the limits $x = a$ and $x = b$, i.e. the area EFGH, is given by:

$$A = \int_a^b y \, dx$$

If $\bar{x}$ and $\bar{y}$ (pronounced 'x bar' and 'y bar' respectively) are the distances of the centroid of the area A about axes OY and OX respectively then:

$$(\bar{x})(\text{total area } A) = \sum_{x=a}^{x=b} x(y\delta x)$$

$$\text{i.e. in the limit } A\bar{x} = \int_a^b xy\,dx$$

$$\bar{x} = \frac{\int_a^b xy \, dx}{\int_a^b y \, dx}$$

$$\text{Similarly, } (\bar{y})(\text{total area } A) = \sum_{x=a}^{x=b} \frac{y}{2}(y\delta x)$$

$$\text{i.e. in the limit} \quad A\bar{y} = \tfrac{1}{2}\int_a^b y^2 \, dx$$

$$\bar{y} = \frac{\tfrac{1}{2}\int_a^b y^2 \, dx}{\int_a^b y \, dx}$$

Centroid of area between a curve and the y-axis

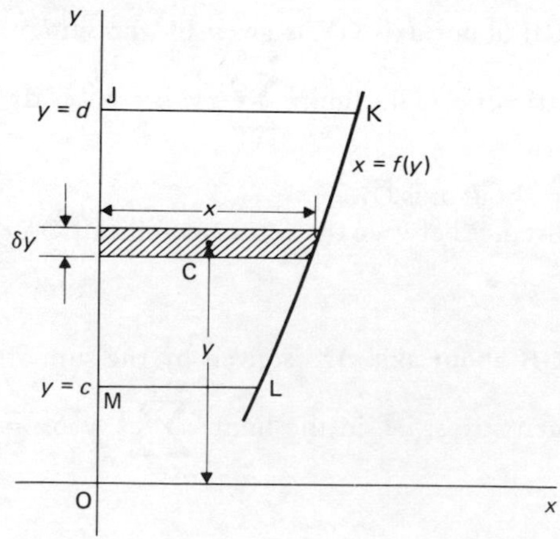

Figure 3

If the position of the centroid of the area between a curve ($x = f(y)$) and the y-axis is required, say area JKLM of Fig. 3, then with similar reasoning to above:

$$(\bar{x})(\text{total area}) = \sum_{y=c}^{y=d} \frac{x}{2}(x\delta y)$$

i.e.
$$\bar{x} = \frac{\frac{1}{2}\int_c^d x^2\,dy}{\int_c^d x\,dy}$$

Similarly,
$$(\bar{y})(\text{total area}) = \sum_{y=c}^{y=d} y(x\delta y)$$

i.e.
$$\bar{y} = \frac{\int_c^d xy\,dy}{\int_c^d x\,dy}$$

The expressions for $\bar{x}$ and $\bar{y}$ are the same as those used for finding the position of the centroid of the area between a curve and the x-axis except that the x's and y's have been interchanged. (See Problem 4 below.)

Theorem of Pappus

In *Technician Mathematics, Level 2*, Chapter 9, a theorem of Pappus was discussed. The theorem essentially enables volumes of solids to be calculated. However, if the volume is known then the centroid of an area may be calculated (see Problems 8 and 9 below). The theorem states:

If a plane area is rotated about an axis in its own plane but not intersecting it, the volume of the solid formed is given by the product of the area and the distance moved through by the centroid of the area.

i.e. volume generated = area × distance moved through by the centroid.

Let A be any area whose centroid is at C (see Fig. 4).

If the distance of the centroid C from the axis OX is $\bar{y}$ then the distance moved through by the centroid when the area A is rotated 360° about the x-axis is $2\pi\bar{y}$ (i.e. the circumference of a circle of radius $\bar{y}$) and the volume V of the solid produced is given by:

volume generated = area × $2\pi\bar{y}$

i.e.
$$\mathbf{V = 2\pi A\bar{y}\ \text{cubic units}}$$

or
$$\bar{y} = \frac{V}{2\pi A}\ \text{units}$$

Similarly, if $\bar{x}$ is the perpendicular distance of the centroid C from the y-axis then rotation of area A about the y-axis will generate a volume V given by $V = A\ (2\pi\bar{x})$, i.e. $\bar{x} = \dfrac{V}{2\pi A}$ **units.**

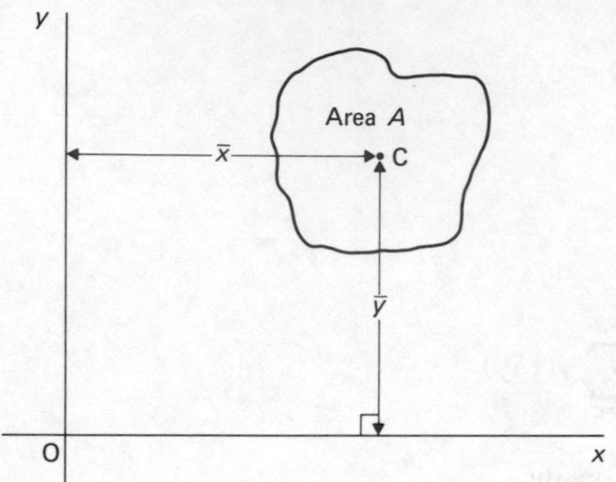

Figure 4

Worked problems on centroids

Problem 1. Find the first moment of area about axis XX for each of the shapes shown in Fig. 5.

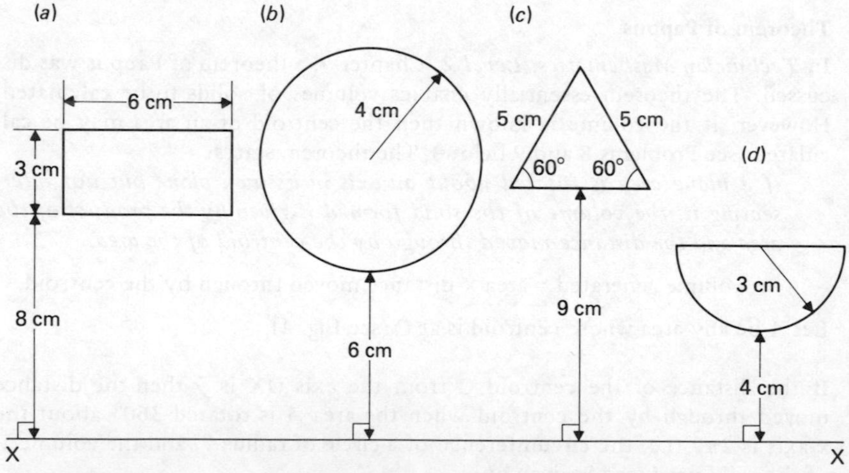

Figure 5

The first moment of area about axis XX = (area of shape) × (perpendicular distance between the centroid of the shape and the axis XX).

(a) The centroid of a rectangle lies at the intersection of the diagonals.

First moment of area of rectangle $= (3 \times 6)(8 + 1\tfrac{1}{2})$

$= (18)(9\tfrac{1}{2}) = \mathbf{171\ cm^3}$

(b) The centroid of a circle lies at its centre.
First moment of area of circle = $[\pi(4)^2]\,[6 + 4]$
$= (16\pi)(10) = \mathbf{160\pi\ cm^3}$

(c) The centroid of a triangle, of perpendicular height h, lies at a point $\frac{h}{3}$ from the base. The perpendicular height of the triangle is 5 sin 60° or 4.330 cm. Since there are 180° in any triangle, the third angle of the triangle is 60°. The triangle is thus equilateral, each side being 5 cm in length.
First moment of area of triangle = $[\frac{1}{2}(5)(4.330)]\,[9 + \frac{1}{3}(4.330)]$
$= \frac{1}{2}(5)(4.330)(10.443) = \mathbf{113.0\ cm^3}$

(d) The centroid of a semicircle, of radius r, lies on the centre line at a distance $\frac{4r}{3\pi}$ from the diameter.

First moment of area of semicircle = $[\frac{1}{2}\pi(3)^2]\left[4 + \quad 3 - \left(\frac{4(3)}{3\pi}\right)\right]$
$= \frac{1}{2}(9)\pi[4 + (3 - 1.273)]$
$= \mathbf{80.96\ cm^3}$

Problem 2. Find the position of the centroid of the area bounded by the curve $y = 2x^2$, the x-axis and the ordinates $x = 0$ and $x = 3$.

$$\bar{x} = \frac{\int_0^3 xy\,dx}{\int_0^3 y\,dx} = \frac{\int_0^3 x(2x^2)\,dx}{\int_0^3 2x^2\,dx} = \frac{\int_0^3 2x^3\,dx}{\int_0^3 2x^2\,dx}$$

$$= \frac{\left[\frac{2x^4}{4}\right]_0^3}{\left[\frac{2x^3}{3}\right]_0^3} = \frac{\frac{81}{2}}{18} = 2\tfrac{1}{4}$$

$$\bar{y} = \frac{\frac{1}{2}\int_0^3 y^2\,dx}{\int_0^3 y\,dx} = \frac{\frac{1}{2}\int_0^3 (2x^2)^2\,dx}{18} = \frac{1}{36}\int_0^3 4x^4\,dx$$

$$= \frac{1}{9}\left[\frac{x^5}{5}\right]_0^3 = \frac{1}{9}\left(\frac{243}{5}\right) = 5\tfrac{2}{5}$$

Hence **the centroid of the area bounded by the curve $y = 2x^2$, the x-axis and the ordinates $x = 0$ and $x = 3$ is at $(2\frac{1}{4}, 5\frac{2}{5})$.**

[Note that functions within an integral in a numerator must not be 'cancelled' with a function within an integral in the denominator. For example, from above,

$\bar{x} = \dfrac{\int_0^3 2x^3\,dx}{\int_0^3 2x^2\,dx}$. Whereas constants (which do not affect integration) may be cancelled, the x^2 in the denominator **must not** be cancelled with the x^3 in the numerator to produce $\bar{x} = \int_0^3 x\, dx$. This is incorrect; each integration must be evaluated separately.]

Problem 3. Calculate the coordinates of the centroid of the area lying between the curve $y = 4x - x^2$ and the x-axis.

It is necessary to obtain the limits of integration.
$y = 4x - x^2 = x(4 - x)$
When $y = 0$ (i.e. the x-axis), $x = 0$ or $(4 - x) = 0$, i.e. $x = 4$. Hence the curve cuts the x-axis at $x = 0$ and $x = 4$ as shown in Fig. 6.

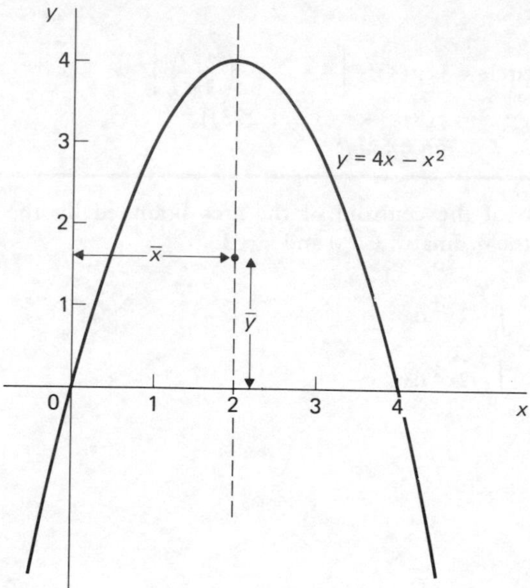

Figure 6

$$\bar{x} = \frac{\int_0^4 xy\, dx}{\int_0^4 y\, dx} = \frac{\int_0^4 x(4x - x^2)\, dx}{\int_0^4 (4x - x^2)\, dx} = \frac{\int_0^4 (4x^2 - x^3)\, dx}{\int_0^4 (4x - x^2)\, dx}$$

$$= \frac{\left[\frac{4x^3}{3} - \frac{x^4}{4}\right]_0^4}{\left[\frac{4x^2}{2} - \frac{x^3}{3}\right]_0^4} = \frac{(\frac{256}{3} - 64)}{(32 - \frac{64}{3})} = \frac{21\frac{1}{3}}{10\frac{2}{3}} = 2$$

$$\bar{y} = \frac{\frac{1}{2}\int_0^4 y^2\,dx}{\int_0^4 y\,dx} = \frac{\frac{1}{2}\int_0^4 (4x - x^2)^2\,dx}{10\frac{2}{3}} = \frac{3}{64}\int_0^4 (16x^2 - 8x^3 + x^4)\,dx$$

$$= \frac{3}{64}\left[\frac{16x^3}{3} - \frac{8x^4}{4} + \frac{x^5}{5}\right]_0^4 = \frac{3}{64}\left[\frac{16(64)}{3} - 512 + \frac{16(64)}{5}\right]$$

$$= 1.6$$

Hence **the centroid of the area lying between $y = 4x - x^2$ and the x-axis is at (2, 1.6).**

From the sketch shown in Fig. 6 it can be seen that the curve is symmetrical about $x = 2$. Thus the calculation for $\bar{x}$ need not have been made. However, if there is any doubt as to whether an axis of symmetry exists, the above calculations should be made.

Problem 4. Determine the position of the centroid of the area enclosed by the curve $y = 4x^2$, the y-axis and ordinates $y = 1$ and $y = 9$, correct to 3 decimal places.

$$\bar{x} = \frac{\frac{1}{2}\int_1^9 x^2\,dy}{\int_1^9 x\,dy} = \frac{\frac{1}{2}\int_1^9 \frac{y}{4}\,dy}{\int_1^9 \sqrt{\frac{y}{4}}\,dy} = \frac{\frac{1}{2}\left[\frac{y^2}{8}\right]_1^9}{\frac{1}{2}\left[\frac{2y^{\frac{3}{2}}}{3}\right]_1^9}$$

$$= \frac{\frac{1}{2}\left[\frac{81}{8} - \frac{1}{8}\right]}{\frac{1}{2}\left[\frac{2(27)}{3} - \frac{2}{3}\right]} = \frac{5}{8\frac{2}{3}} = 0.577$$

$$\bar{y} = \frac{\int_1^9 xy\,dy}{\int_1^9 x\,dy} = \frac{\int_1^9 \left(\sqrt{\frac{y}{4}}\right) y\,dy}{8\frac{2}{3}} = \frac{\frac{1}{2}\int_1^9 y^{\frac{3}{2}}\,dy}{8\frac{2}{3}}$$

$$= \frac{\frac{1}{2}\left[\frac{2y^{\frac{5}{2}}}{5}\right]_1^9}{8\frac{2}{3}} = \frac{\frac{1}{5}[243 - 1]}{8\frac{2}{3}} = 5.585$$

Hence **the position of the centroid of the area is at (0.577, 5.585).**

Problem 5. Find the position of the centroid of the area enclosed between the curve $y = 9 - x^2$ and the x-axis.

$y = 9 - x^2$. When $y = 0$ (i.e. the x-axis) then $9 - x^2 = 0$ or $9 = x^2$

Hence $x = \pm 3$.

Thus the curve $y = 9 - x^2$ cuts the x-axis at $x = -3$ and $x = +3$. Also, when $x = 0$, $y = 9$. A sketch of a part of $y = 9 - x^2$ is shown in Fig. 7.

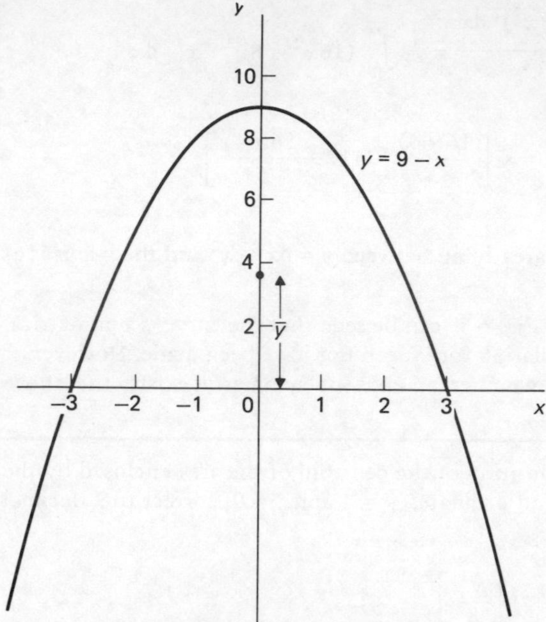

Figure 7

It may be seen from Fig. 7 that the area is symmetrical about the y-axis. Hence $\bar{x} = 0$.

$$\bar{y} = \frac{\frac{1}{2}\int_{-3}^{3} y^2 \, dx}{\int_{-3}^{3} y \, dx} = \frac{\frac{1}{2}\int_{-3}^{3} (9 - x^2)^2 \, dx}{\int_{-3}^{3} (9 - x^2) \, dx} = \frac{\frac{1}{2}\int_{-3}^{3} (81 - 18x^2 + x^4) \, dx}{\int_{-3}^{3} (9 - x^2) \, dx}$$

$$= \frac{\frac{1}{2}\left[81x - \frac{18x^3}{3} + \frac{x^5}{5}\right]_{-3}^{3}}{\left[9x - \frac{x^3}{3}\right]_{-3}^{3}}$$

$$= \frac{\frac{1}{2}[(243 - 162 + \frac{243}{5}) - (-243 + 162 - \frac{243}{5})]}{[(27 - 9) - (-27 + 9)]}$$

$$= \frac{\frac{1}{2}[(129.6) - (-129.6)]}{[(18) - (-18)]}$$

$$= \frac{129.6}{36} = 3.6$$

Hence the centroid of the area enclosed by $y = 9 - x^2$ and the x-axis lies at (0, 3.6).

Problem 6. Prove by integration that the centroid of a triangle of perpendicular height h and base length b lies at a point $\frac{h}{3}$ from the base.

The equation $y = mx + c$ represents a straight line of slope m and y-axis intercept c. Hence the equation $y = \frac{b}{h}x$ represents a straight line of slope $\frac{b}{h}$ and intercept O.

The area bounded by $y = \frac{b}{h}x$, the x-axis and ordinates $x = 0$ and $x = h$ forms a right-angled triangle OAB as shown in Fig. 8.

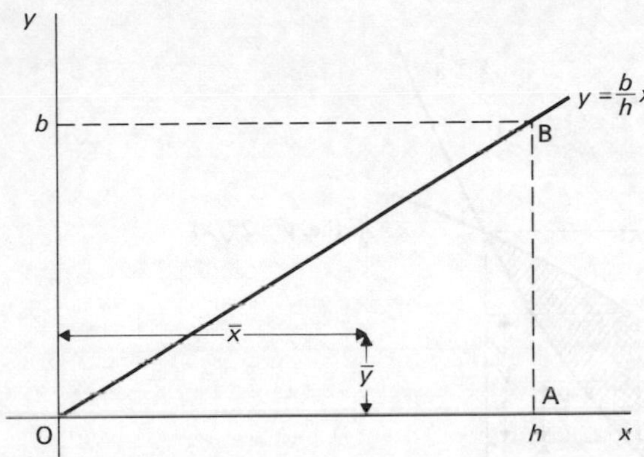

Figure 8

$$\bar{x} = \frac{\int_0^h xy\,dx}{\int_0^h y\,dx} = \frac{\int_0^h x\left(\frac{b}{h}x\right)\,dx}{\text{area of triangle OAB}} = \frac{\frac{b}{h}\left[\frac{x^3}{3}\right]_0^h}{\frac{1}{2}bh}$$

$$= \frac{\frac{b}{h}\left(\frac{h^3}{3}\right)}{\frac{bh}{2}} = \frac{\frac{bh^2}{3}}{\frac{bh}{2}} = \frac{2h}{3}$$

$\bar{x}$ is the perpendicular distance of the centroid from the vertex of triangle OAB. Hence **the distance of the centroid from the base AB is $\frac{1}{3}h$.**

$$[\text{Also } \bar{y} = \frac{\frac{1}{2}\int_0^h y^2\,dx}{\int_0^h y\,dx} = \frac{\frac{1}{2}\left(\frac{b}{h}\right)^2\int_0^h x^2\,dx}{\frac{bh}{2}}$$

$$= \frac{\frac{1}{2}\left(\frac{b}{h}\right)^2\left[\frac{x^3}{3}\right]_0^h}{\frac{bh}{2}} = \frac{\frac{1}{2}\left(\frac{b}{h}\right)^2\left(\frac{h^3}{3}\right)}{\frac{bh}{2}} = \frac{b}{3}$$

Hence if OA is considered as the base of triangle OAB then the centroid is at a distance of $\frac{b}{3}$ from it.]

Problem 7. Locate the position of the centroid of the area enclosed by the curves $x = \frac{y^2}{4}$ and $y = \frac{x^2}{4}$.

Figure 9 shows the two curves intersecting at (0, 0) and (4, 4) and enclosing a shaded area, the centroid of which is required.

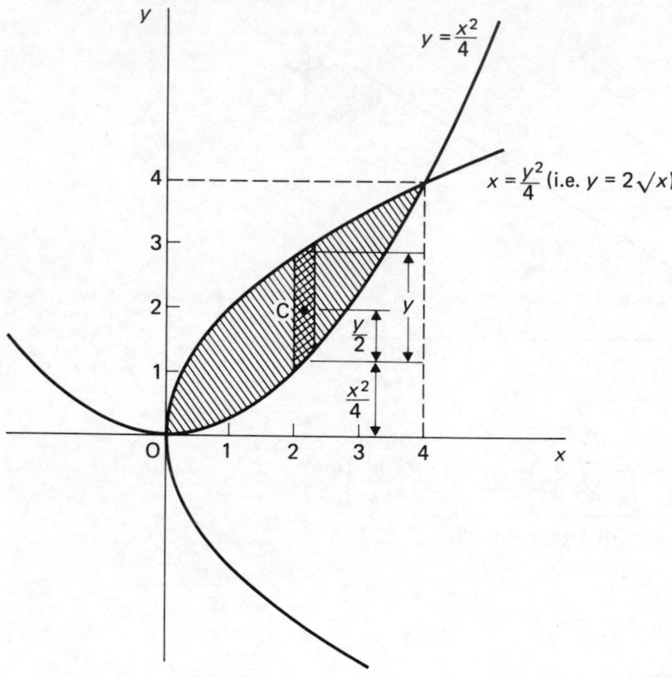

Figure 9

$$\bar{x} = \frac{\int_0^4 xy\, dx}{\int_0^4 y\, dx}$$

The value of y in this integral is given by the height of the typical strip shown i.e. $\left(2\sqrt{x} - \frac{x^2}{4}\right)$.

Hence $\bar{x} = \dfrac{\int_0^4 x\left(2\sqrt{x} - \frac{x^2}{4}\right)\mathrm{d}x}{\int_0^4 \left(2\sqrt{x} - \frac{x^2}{4}\right)\mathrm{d}x} = \dfrac{\int_0^4 \left(2x^{\frac{3}{2}} - \frac{x^3}{4}\right)\mathrm{d}x}{\int_0^4 \left(2x^{\frac{1}{2}} - \frac{x^2}{4}\right)\mathrm{d}x}$

$$= \frac{\left[\frac{4x^{\frac{5}{2}}}{5} - \frac{x^4}{16}\right]_0^4}{\left[\frac{4x^{\frac{3}{2}}}{3} - \frac{x^3}{12}\right]_0^4} = \frac{[\frac{4}{5}(32) - 16]}{[\frac{4}{3}(8) - \frac{64}{12}]}$$

$$= \frac{9\frac{3}{5}}{5\frac{1}{3}} = 1.8$$

Care must be taken when finding $\bar{y}$ since the centroid of the typical strip is not now at a distance $\frac{y}{2}$ from the x-axis.

Taking moments about OX gives:

$$(\text{Total area})(\bar{y}) = \sum_{x=0}^{x=4} (\text{area of strip})(\text{perpendicular distance of the centroid of the strip to axis OX})$$

$y = 2\sqrt{x} - \frac{x^2}{4}$. Therefore $\frac{y}{2} = \frac{1}{2}\left(2\sqrt{x} - \frac{x^2}{4}\right)$

The perpendicular distance from the centroid C of the strip to the axis OX is given by $\frac{1}{2}\left(2\sqrt{x} - \frac{x^2}{4}\right) + \frac{x^2}{4}$

$$\therefore (\text{area})(\bar{y}) = \int_0^4 \left(2\sqrt{x} - \frac{x^2}{4}\right)\left[\frac{1}{2}\left(2\sqrt{x} - \frac{x^2}{4}\right) + \frac{x^2}{4}\right]\mathrm{d}x$$

$$5\tfrac{1}{3}(\bar{y}) = \int_0^4 \left(2\sqrt{x} - \frac{x^2}{4}\right)\left(\sqrt{x} + \frac{x^2}{8}\right)\mathrm{d}x$$

$$\bar{y} = \frac{\int_0^4 \left(2x - \frac{x^4}{32}\right)\mathrm{d}x}{5\frac{1}{3}}$$

$$= \frac{\left[x^2 - \frac{x^5}{5(32)}\right]_0^4}{5\frac{1}{3}} = \frac{16 - 6\frac{2}{5}}{5\frac{1}{3}}$$

$$= 1.8$$

Hence the position of the centroid of the area enclosed by the curves $x = \frac{y^2}{4}$ and $y = \frac{x^2}{4}$ is at (1.8, 1.8).

 Thus when finding centroids of areas enclosed by two curves, say $y_1 = f(x)$ and $y_2 = g(x)$, then the formula derived for $\bar{x}$ $\left(\text{i.e. } \dfrac{\int xy\,dx}{\int y\,dx}\right)$ may be used, where y represents the difference between y_1 and y_2. However, the formula derived for $\bar{y}$ $\left(\text{i.e. } \dfrac{\frac{1}{2}\int y^2\,dx}{\int y\,dx}\right)$ must not be used. $\bar{y}$ is obtained from first principles as shown above.

Problem 8. Calculate the position of the centroid of a semicircle of radius r: (a) by using the theorem of Pappus; and (b) by using integration (given that the equation of a circle, centre O and radius r is $x^2 + y^2 = r^2$.

Figure 10 shows a semicircle with its diameter lying on the x-axis, with its centre at the origin.

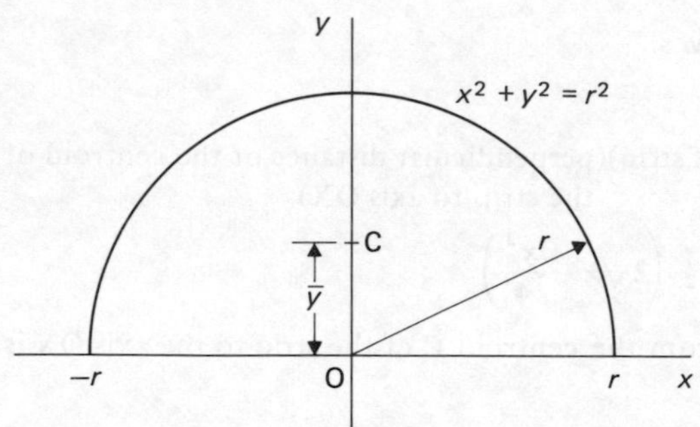

Figure 10

(a) Using the theorem of Pappus

If the semicircular area is rotated about the x-axis then the volume of the solid generated is that of a sphere, i.e. $\frac{4}{3}\pi r^3$. The area of the semicircle is $\dfrac{\pi r^2}{2}$. The centroid lies on the axis of symmetry OY. Pappus's theorem states:

volume generated = area × distance moved through by centroid

$$\text{i.e. } \tfrac{4}{3}\pi r^3 = \left(\frac{\pi r^2}{2}\right)(2\pi\bar{y})$$

$$\bar{y} = \frac{\frac{4}{3}\pi r^3}{\left(\frac{\pi r^2}{2}\right)(2\pi)} = \frac{4r}{3\pi}$$

(b) Using integration

The curve $x^2 + y^2 = r^2$ is symmetrical about $x = 0$. Hence the centroid lies on the y-axis, i.e. $\bar{x} = 0$.

$$\bar{y} = \frac{\frac{1}{2}\int_{-r}^{r} y^2\,dx}{\text{area}} = \frac{\frac{1}{2}\int_{-r}^{r} (r^2 - x^2)\,dx}{\frac{\pi r^2}{2}} = \frac{\frac{1}{2}\left[r^2x - \frac{x^3}{3}\right]_{-r}^{r}}{\frac{\pi r^2}{2}}$$

$$= \frac{1}{\pi r^2}\left[\left(r^3 - \frac{r^3}{3}\right) - \left(-r^3 + \frac{r^3}{3}\right)\right] = \frac{1}{\pi r^2}\left[2\left(r^3 - \frac{r^3}{3}\right)\right]$$

$$= \frac{2}{\pi r^2}\left(\tfrac{2}{3}r^3\right) = \frac{4r}{3\pi}$$

Hence **the centroid lies on the axis of symmetry at a distance of $\frac{4r}{3\pi}$ from the diameter.**

Problem 9. (a) Find the area bounded by the curve $y = 3x^2$, the x-axis and ordinates $x = 0$ and $x = 2$.
(b) If this area is revolved: (i) about the x-axis; (ii) about the y-axis, find the volume of the solid produced in each case.
(c) Find the position of the centroid of the area: (i) by using integration; (ii) by using the theorem of Pappus.

(a) The relevant area is shown shaded in Fig. 11.

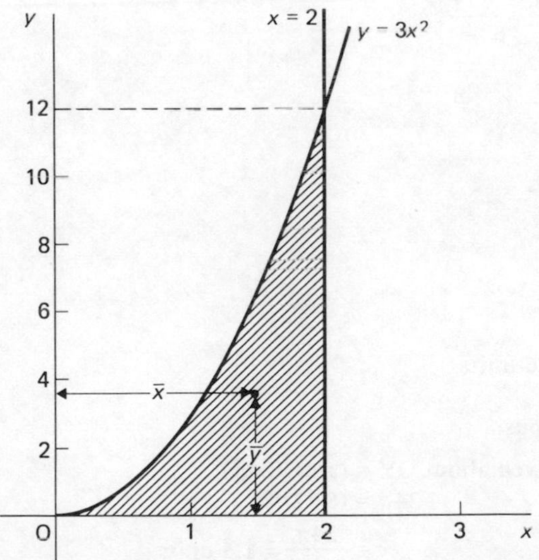

Figure 11

$$\text{Area} = \int_0^2 y\,dx = \int_0^2 3x^2\,dx = [x^3]_0^2 = \mathbf{8 \text{ square units}}$$

(b) (i) When revolved about the x-axis, volume of shaded area = $\int_0^2 \pi y^2\,dx$

i.e. **volume $_{Ox}$** $= \pi \int_0^2 9x^4\,dx = 9\pi \left[\frac{x^5}{5}\right]_0^2$ **= 57.6π cubic units**

(ii) When revolved about the y-axis the limits are $y = 0$ and $y = 12$.

Since $y = 3x^2$ then $x^2 = \frac{y}{3}$.

The volume generated by the shaded area when revolved about the y-axis is given by: (volume generated by $x = 2$) – (volume generated by $y = 3x^2$).

When a curve is revolved about the y-axis to produce a solid of revolution, the volume produced is given by: $\int_{y_1}^{y_2} \pi x^2\,dy$.

Thus, **volume$_{Oy}$** $= \int_0^{12} \pi(2)^2\,dy - \int_0^{12} \pi\left(\frac{y}{3}\right)dy$

$$= \pi \int_0^{12} \left[(2)^2 - \frac{y}{3}\right] dy = \pi \left[4y - \frac{y^2}{6}\right]_0^{12}$$

$= \pi[48 - \frac{144}{6}] = \pi[48 - 24] =$ **24π cubic units**

(c) (i) Using integration:

$$\bar{x} = \frac{\int_0^2 xy\,dx}{\int_0^2 y\,dx} = \frac{\int_0^2 3x^3\,dx}{8} = \frac{\left[\frac{3x^4}{4}\right]_0^2}{8}$$

$= \frac{(3)\frac{16}{4}}{8} =$ **1.5 units**

$$\bar{y} = \frac{\frac{1}{2}\int_0^2 y^2\,dx}{\int_0^2 y\,dx} = \frac{\frac{1}{2}\int_0^2 9x^4\,dx}{8}$$

$= \frac{\frac{9}{2}\left[\frac{x^5}{5}\right]_0^2}{8} = \frac{9}{16}\left(\frac{32}{5}\right) =$ **3.6 units**

(ii) Using the theorem of Pappus:

Volume generated when revolved about OY = (area)$(2\pi\bar{x})$

$24\pi = (8)(2\pi\bar{x})$

$\bar{x} = \frac{24\pi}{8(2\pi)} =$ **1.5 units**

Volume generated when revolved about OX = (area)$(2\pi\bar{y})$

$57.6\pi = (8)(2\pi\bar{y})$

$\bar{y} = \frac{57.6\pi}{8(2\pi)} =$ **3.6 units**

Hence the centroid of the shaded area is at (1.5, 3.6).

Further problems on centroids may be found in Section 5 (Problems 1–24).

2. Second moment of area

Consider three small bodies rotating at an angular velocity ω rad s^{-1} around an axis DD at distances r_1, r_2 and r_3 as shown in Fig. 12 (a). Let the bodies have cross-sectional areas a_1, a_2 and a_3, and corresponding masses of m_1, m_2 and m_3, moving with tangential velocities v_1, v_2 and v_3 respectively as shown.

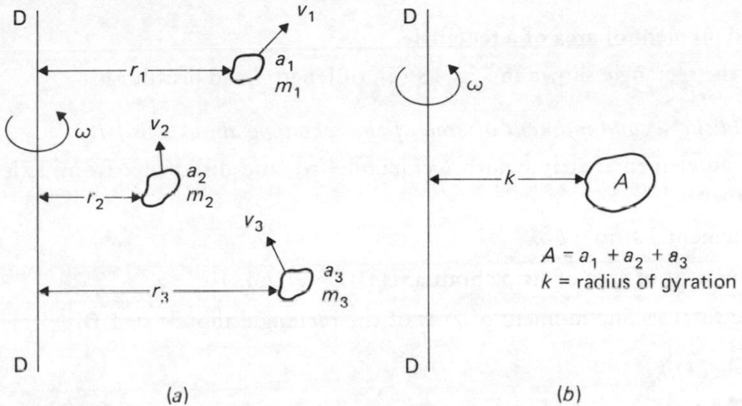

Figure 12

The kinetic energy of the system $= \frac{1}{2} m_1 v_1{}^2 + \frac{1}{2} m_2 v_2{}^2 + \frac{1}{2} m_3 v_3{}^2$

Since $v = \omega r$, then $v_1 = \omega r_1$, $v_2 = \omega r_2$ and $v_3 = \omega r_3$

Hence the kinetic energy $= \frac{1}{2} m_1 (\omega r_1)^2 + \frac{1}{2} m_2 (\omega r_2)^2 + \frac{1}{2} m_3 (\omega r_3)^2$

$= \frac{1}{2} \omega^2 (m_1 r_1{}^2 + m_2 r_2{}^2 + m_3 r_3{}^2)$

If such a system contained many bodies then, in general, **kinetic energy** $= \frac{1}{2} \omega^2 \Sigma m r^2$.

The expression $\Sigma m r^2$ is known as the **moment of inertia** and is defined as the second moment of the total mass about axis DD of the system. It is a measure of the amount of work done to give the system an angular velocity of ω rad s^{-1}, or the amount of work which can be done by a system turning at ω rad s^{-1}. When mass is proportional to area then the moment of inertia (i.e. $\Sigma m r^2$) becomes $\Sigma a r^2$. The term $\Sigma a r^2$ is called the **second moment of area** and it is a quantity much used, particularly in mechanical engineering, for example, in the theory of bending of beams (where the beam equation is $\frac{M}{I} = \frac{\sigma}{y} = \frac{E}{R}$, and I denotes the second moment of area) and in torsion of shafts, and also in naval architecture, for example, in calculations involving water planes and centres of pressure. Whereas the first moment of area is given by Σar, the second moment of area is given by Σar^2 (i.e. the distance is *'squared'* when finding the *'second'* moment).

If the areas a_1, a_2 and a_3 of Fig. 12 (a) are replaced by a single area A, such that $A = a_1 + a_2 + a_3$, at a distance k from axis DD such that $Ak^2 = \Sigma ar^2$, then k is called the **radius of gyration** of area A about axis DD (see Fig. 12 (b)).

In order to find the second moment of area of regular sections about a given axis, the second moment of area of a typical element is first determined and then the sum of all such second moment of areas found by integrating between appropriate limits. This is shown in the following derivations for rectangles, triangles, circles and semicircles.

(i) Second moment of area of a rectangle

Consider the rectangle shown in Fig. 13 (a), of length l and breadth b.

(a) To find the second moment of area of the rectangle about axis DD:

Consider an elemental strip, width δx, parallel to, and distance x from, axis DD as shown.

Area of elemental strip = $b\delta x$

Second moment of area of strip about axis DD = $x^2 (b\delta x)$

Hence the total second moment of area of the rectangle about axis DD

$$= \sum_{0=x}^{x=l} x^2 b\delta x$$

$$= \int_0^l x^2 b \, dx = b\left[\frac{x^3}{3}\right]_0^l = \frac{bl^3}{3}$$

But the total area A of the rectangle = lb

Hence **the second moment of area about axis DD** = $lb\left(\frac{l^2}{3}\right)$

$$= A\frac{l^2}{3}$$

Since the second moment of area = $Ak_{DD}{}^2$

$$k_{DD}{}^2 = \frac{l^2}{3}$$

i.e. **the radius of gyration about axis DD,** $k_{DD} = \frac{l}{\sqrt{3}}$

(b) To find the second moment of area of the rectangle about axis EE:

Consider an elemental strip, width δx, parallel to, and distance x from, axis EE as shown in Fig. 13 (b).

Area of elemental strip = $l\delta x$

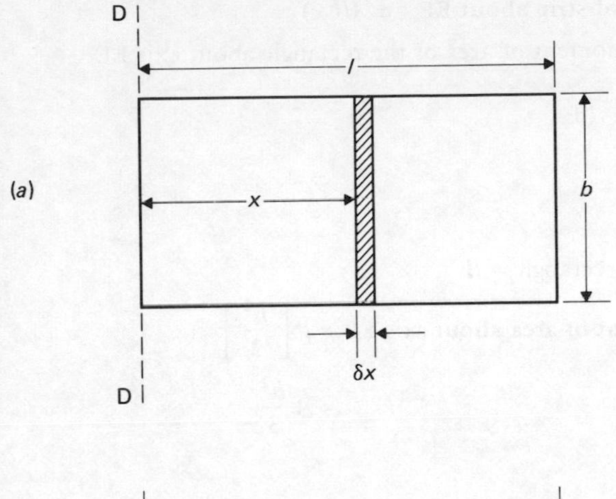

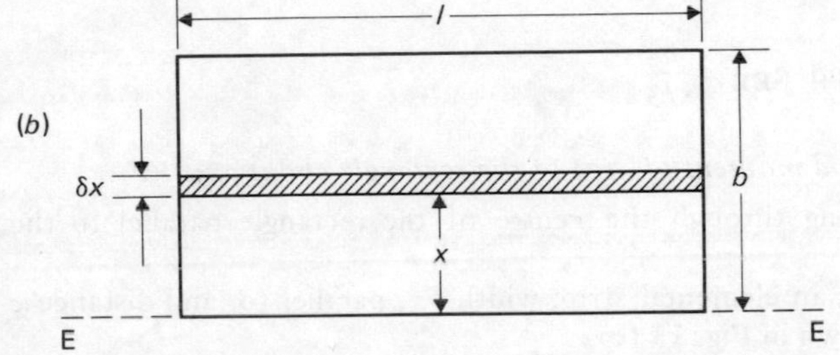

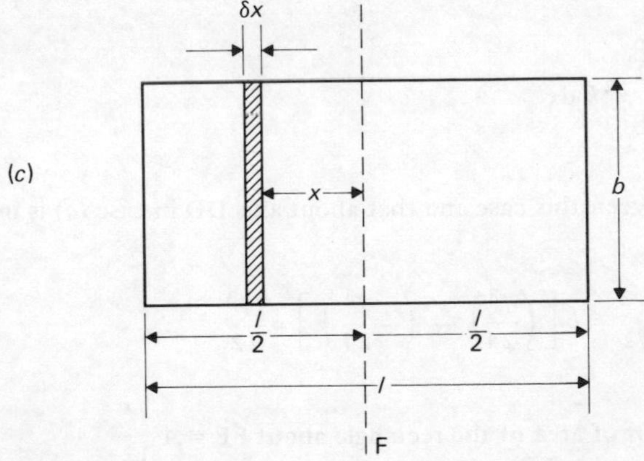

Figure 13

Second moment of area of strip about EE = $x^2 (l\delta x)$

Hence the total second moment of area of the rectangle about axis EE

$$= \sum_{x=0}^{x=b} x^2 l \delta x = \int_0^b x^2 l \, dx$$

$$= l \left[\frac{x^3}{3}\right]_0^b = \frac{lb^3}{3}$$

But the total area of the rectangle = lb

Hence **the second moment of area about axis EE** $= lb\left[\frac{b^2}{3}\right]$

$$= A \frac{b^2}{3}$$

Since $Ak_{EE}{}^2 = A \frac{b^2}{3}$

then $\quad k_{EE}{}^2 = \frac{b^2}{3}$ and $\boldsymbol{k_{EE}} = \frac{b}{\sqrt{3}}$

(c) To find the second moment of area of the rectangle about axis FF:

FF is an axis passing through the centre of the rectangle parallel to the breadth.

Consider, again, an elemental strip, width δx, parallel to, and distance x from, axis FF as shown in Fig. 13 (c).

Area of elemental strip = $b\delta x$

Second moment of area of the strip about FF = $x^2 (b\delta x)$

Hence the total second moment of area of the rectangle about axis FF

$$= \sum_{x=\frac{-l}{2}}^{x=\frac{l}{2}} x^2 b \delta x = \int_{\frac{-l}{2}}^{\frac{l}{2}} x^2 b \, dx$$

The only difference between this case and that about axis DD in case (*a*) is in the limits of integration.

$$\int_{\frac{-l}{2}}^{\frac{l}{2}} x^2 b \, dx = b\left[\frac{x^3}{3}\right]_{-l/2}^{l/2} = b\left[\left(\frac{l^3}{24}\right) - \left(-\frac{l^3}{24}\right)\right] = \frac{bl^3}{12}$$

Hence **the second moment of area of the rectangle about FF** $= A \frac{\boldsymbol{l^2}}{\boldsymbol{12}}$

$$(\text{radius of gyration})^2 = \frac{\text{second moment of area}}{\text{area}}$$

$$\text{i.e. } k_{FF}^2 = \frac{A\,\frac{l^2}{12}}{A} = \frac{l^2}{12}$$

$$k_{FF} = \frac{l}{\sqrt{12}} \text{ or } \frac{l}{2\sqrt{3}}$$

(ii) Second moment of area of a triangle about an edge

Consider triangle DEF shown in Fig. 14 (a). Let the edge EF be the base, of length b, and let the perpendicular height be h.

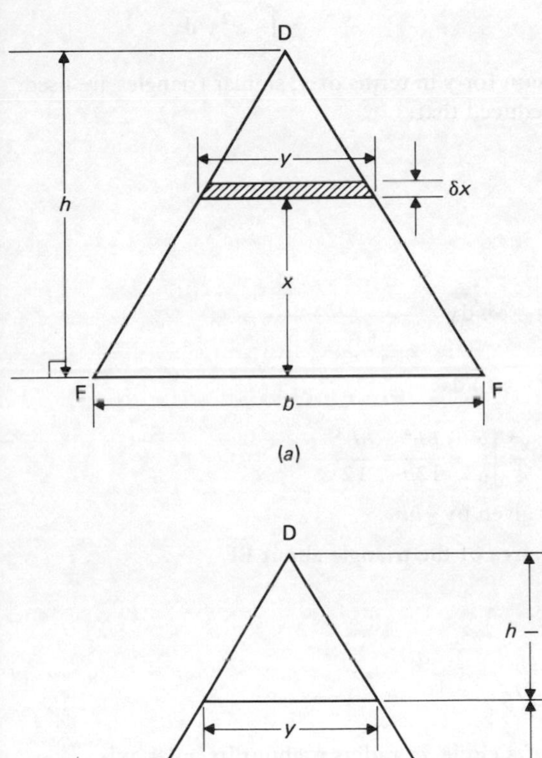

Figure 14

 To find the second moment of area of the triangle about edge EF:

Consider an elemental strip, width δx, parallel to and distance x from EF as shown in Fig. 14 (a). Let the length of the strip be y.

Area of strip $= y\delta x$

Second moment of area of the strip about EF $= x^2(y\delta x)$

Second moment of area of triangle DEF about EF $= \sum_{x=0}^{x=b} x^2 y\delta x$

$$= \int_0^b x^2 y \, dx$$

In order to obtain an expression for y in terms of x, similar triangles are used. From Fig. 14 (b) it may be deduced that:

$$\frac{h-x}{y} = \frac{h}{b},$$

from which $y = \frac{b}{h}(h - x)$

$$\text{Thus } \int_0^h x^2 y \, dx = \int_0^h x^2 \frac{b}{h}(h-x) \, dx$$

$$= \frac{b}{h}\int_0^h (x^2 h - x^3) \, dx$$

$$= \frac{b}{h}\left[\frac{x^3}{3}h - \frac{x^4}{4}\right]_0^h = \frac{bh^4}{12h} = \frac{bh^3}{12}$$

The area A of triangle DEF is given by $\frac{1}{2}bh$.

Hence **the second moment of area of the triangle about EF**

$$= (\tfrac{1}{2}bh)\left(\frac{h^2}{6}\right) = A\left(\frac{h^2}{6}\right)$$

Hence $k_{EF}{}^2 = \frac{h^2}{6}$ and $k_{EF} = \frac{h}{\sqrt{6}}$

(iii) Second moment of area of a circle, of radius r, about its polar axis

A polar axis is an axis through the centre of the circle perpendicular to the plane of the circle. This is shown as axis BB in Fig. 15 (a) and a plan view of the circle is shown in Fig. 15 (b).

Consider an elemental annulus of width δx (shown shaded in Fig. 15 (b)) and radius x from the centre of the circle (i.e. distance x from the polar axis BB). If δx is very small then the area of the elemental annulus equals the circumference times the width.

i.e. area of annulus $= (2\pi x)(\delta x)$.

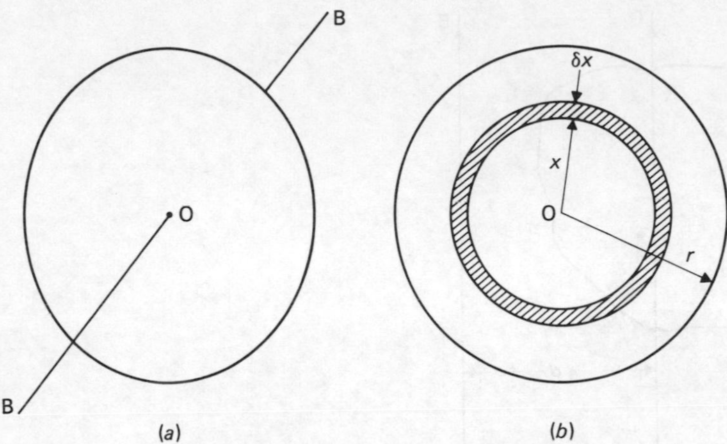

Figure 15

The second moment of area of the annulus about the centre of the circle

$$= (x^2)(2\pi x\delta x)$$

The total second moment of area of the circle about an axis through its centre, perpendicular to the plane

$$= \sum_{x=0}^{x=r} x^2(2\pi x\,\delta x)$$

$$= \int_0^r 2\pi x^3\,dx = 2\pi\left[\frac{x^4}{4}\right]_0^r = \frac{\pi r^4}{2}$$

But the total area A of the circle is given by πr^2.

Hence **the second moment of area of a circle about its polar axis**

$$= (\pi r^2)\left(\frac{r^2}{2}\right) = A\frac{r^2}{2}$$

Hence $k_{BB}{}^2 = \frac{r^2}{2}$ and $k_{BB} = \frac{r}{\sqrt{2}}$

There are two important theorems associated with second moments of areas which enable the second moment of area and the radius of gyration to be obtained about axes other than those used above.

3. Parallel-axis theorem

Let C be the centroid of any area A as shown in Fig. 16.

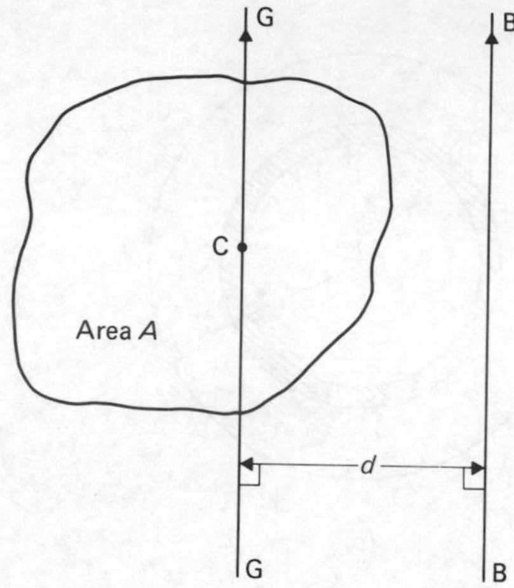

Figure 16

Let the second moment of area of A about an axis GG, passing through the centroid, be denoted by $Ak_{GG}{}^2$. Let BB be an axis parallel to axis GG and at perpendicular distance d from it. Axes GG and BB are in the same plane as area A. The parallel-axis theorem states:

$$Ak_{BB}{}^2 = Ak_{GG}{}^2 + Ad^2$$

or $$k_{BB}{}^2 = k_{GG}{}^2 + d^2$$

As an example, this may be shown to be true from the results obtained for the rectangle of length l, breadth b and area A.

The second moment of area of a rectangle about an axis through its centroid, parallel to the breadth (i.e. about axis FF of Fig. 13 (c)), is given by $A\dfrac{l^2}{12}$.

Let the second moment of area of the rectangle about an axis coinciding with the breadth (i.e. axis DD of Fig. 13 (a)) be denoted by $Ak_{DD}{}^2$. The distance between the parallel axes FF and DD is $\dfrac{l}{2}$. Hence, by the parallel-axis theorem:

$$Ak_{DD}{}^2 = Ak_{FF}{}^2 + Ad^2$$

$$= A\frac{l^2}{12} + A\left(\frac{l}{2}\right)^2$$

$$= A\left[\frac{l^2}{12} + \frac{l^2}{4}\right] = A\frac{l^2}{3}$$

This, of course, is the result previously obtained on p. 236

Second moment of area of a triangle about an axis through its centroid, parallel to the base, using the parallel-axis theorem

It was shown earlier that the second moment of area of a triangle about its base is $A\dfrac{h^2}{6}$, where h is the perpendicular height of the triangle. The centroid of a triangle is situated at a distance $\dfrac{h}{3}$ from the base.

Let the second moment of area of a triangle about an axis through its centroid, parallel to the base, be denoted by Ak_{CC}^2.

Thus, using the parallel-axis theorem:

$$A\frac{h^2}{6} = Ak_{CC}^2 + A\left(\frac{h}{3}\right)^2$$

i.e. $$Ak_{CC}^2 = A\frac{h^2}{6} - A\frac{h^2}{9} = A\frac{h^2}{18}$$

Since $A = \frac{1}{2}bh$ then $Ak_{CC}^2 = (\frac{1}{2}bh)\dfrac{h^2}{18}$ or $\dfrac{bh^3}{36}$

Hence $k_{CC}^2 = \dfrac{h^2}{18}$ and $k_{CC} = \dfrac{h}{\sqrt{18}}$ or $\dfrac{h}{3\sqrt{2}}$

Second moment of area of a triangle about an axis through its vertex, parallel with the base, using the parallel-axis theorem

If an axis is drawn through the vertex D of Fig. 14 (a), parallel to base EF, then:

$$Ak_D^2 = Ak_{CC}^2 + Ad^2$$

where d = distance between the centroid and the vertex of the triangle, i.e. $d = \frac{2}{3}h$.

$$\text{Hence } Ak_D^2 = A\frac{h^2}{18} + A(\tfrac{2}{3}h)^2$$

$$= A\left[\frac{h^2}{18} + \frac{4}{9}h^2\right] = A\frac{h^2}{2} \text{ or } \frac{bh^3}{4}$$

Thus $\quad k_D^2 = \dfrac{h^2}{2}$ and $k_D = \dfrac{h}{\sqrt{2}}$

4. Perpendicular-axis theorem

Consider a plane area A having three mutually perpendicular axes OX, OY and OZ, as shown in Fig. 17, with axes OX and OY lying in the plane of area A. The perpendicular-axis theorem states:

$$Ak_{OZ}^2 = Ak_{OX}^2 + Ak_{OY}^2$$

or $$k_{OZ}^2 = k_{OX}^2 + k_{OY}^2$$

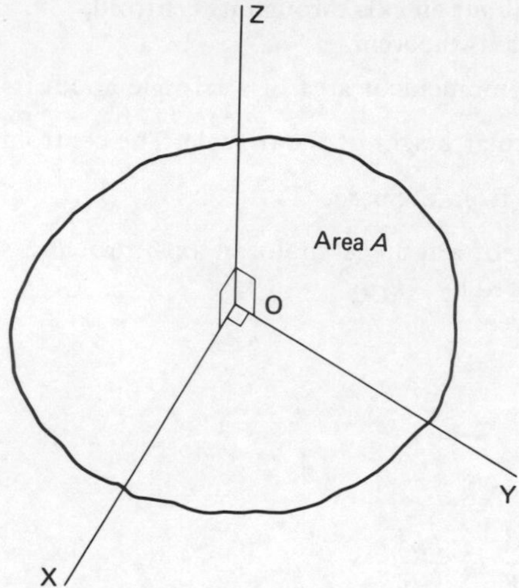

Figure 17

The second moment of area of a circle about a diameter, using the perpendicular-axis theorem

Consider a circle of radius r having three mutually perpendicular axes OX, OY and ZZ as shown in Fig. 18.

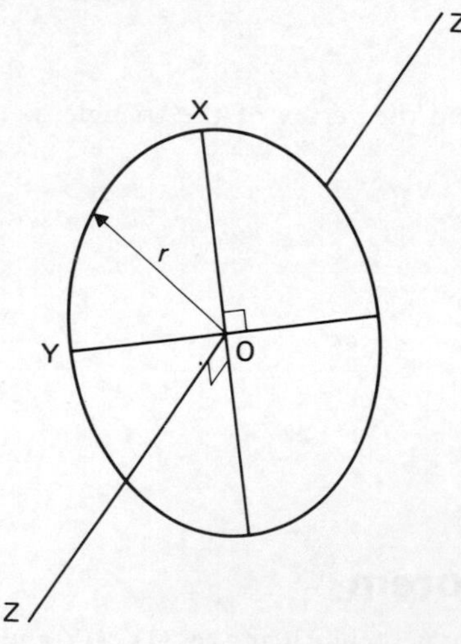

Figure 18

It was shown earlier that the second moment of area of a circle about its polar axis (i.e. axis ZZ of Fig. 18) is $A\dfrac{r^2}{2}$.

By symmetry, the second moment of area of the circle about axis OX will equal that about axis OY.

i.e. $Ak_{OX}^2 = Ak_{OY}^2$

By the perpendicular-axis theorem:

$$Ak_{ZZ}^2 = Ak_{OX}^2 + Ak_{OY}^2$$
$$= 2Ak_{OX}^2$$

i.e. $$A\frac{r^2}{2} = 2Ak_{OX}^2$$

$$Ak_{OX}^2 = A\frac{r^2}{4}$$

Hence **the second moment of area of a circle about a diameter** = $A\frac{r^2}{4}$ **or** $\frac{\pi r^4}{4}$.

Thus $k_{OX}^2 = \frac{r^2}{4}$ and $k_{OX} = \frac{r}{2}$

Second moment of area of a circle about a tangent using the parallel-axis theorem

Consider a circle of radius r having a tangent BB parallel to diameter XX as shown in Fig. 19.

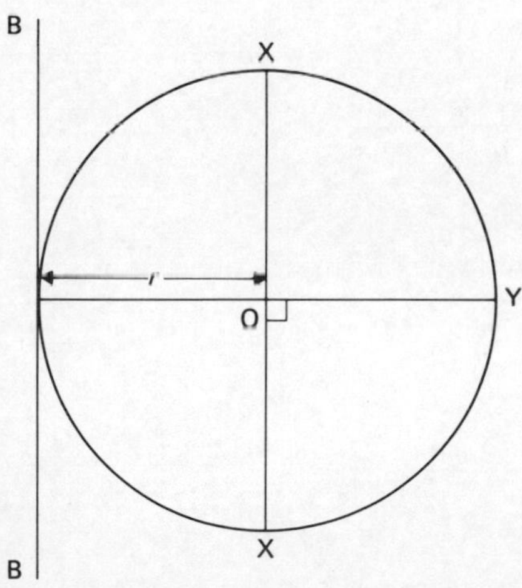

Figure 19

By the parallel-axis theorem:

$$Ak_{BB}^2 = Ak_{XX}^2 + Ar^2$$

But $Ak_{XX}^2 = A\frac{r^2}{4}$

Hence $Ak_{BB}{}^2 = A\frac{r^2}{4} + Ar^2 = A\left(\frac{r^2}{4} + r^2\right)$

Thus **the second moment of area of a circle about a tangent = $\frac{5}{4}Ar^2$.**

$$k_{BB}{}^2 = \tfrac{5}{4}r^2 \text{ and } k_{BB} = \frac{\sqrt{5}}{2}r$$

The second moment of area of a semicircle about its diameter

The second moment of area of a semicircle about its diameter is one-half of that for a circle about its diameter,

i.e. $\frac{1}{2}\left(\frac{\pi r^4}{4}\right)$ or $\frac{\pi r^4}{8}$

An alternative method of finding the second moment of area of a circle about a diameter is to take an elemental strip of width δx, distance x from axis OY as shown in Fig. 20.

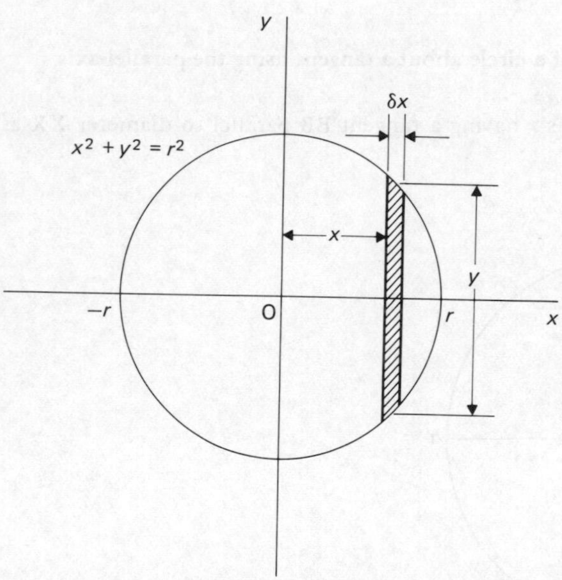

Figure 20

The second moment of area of the strip about OY = $x^2 y\delta x$

The second moment of area of the circle about OY

$$= \sum_{x=-r}^{x=r} x^2 y\delta x = \int_{-r}^{r} x^2 y \, dx \qquad \ldots (1)$$

If now the same procedure is adopted for a semicircle, the second moment of area of the semicircle about axis OY

$$= \sum_{x=0}^{x=r} x^2 y \delta x = \int_0^r x^2 y \, dx \qquad \ldots (2)$$

(Note that the evaluation of the integrals in equations (1) and (2) is too advanced for this text.)

Since $\int_{-r}^{r} x^2 y \, dx = 2\int_0^r x^2 y \, dx$ it follows that the second moment of area of a semicircle about its diameter is half of that for a circle about its diameter, i.e. $\frac{\pi r^4}{8}$.

Since area A of a semicircle $= \frac{\pi r^2}{2}$, the second moment of area about the diameter $= \left(\frac{\pi r^2}{2}\right)\left(\frac{r^2}{4}\right)$ or $A\frac{r^2}{4}$.

Thus **the radius of gyration k for a semicircle, about its diameter is $\frac{r}{2}$**, the same value as for the circle.

Summary of derived standard results

Shape	Position of axis	Second moment of area	Radius of gyration, k
Rectangle length l breadth b area A	1. Coinciding with b	$\frac{bl^3}{3}$ or $A\frac{l^2}{3}$	$\frac{l}{\sqrt{3}}$
	2. Coinciding with l	$\frac{lb^3}{3}$ or $A\frac{b^2}{3}$	$\frac{b}{\sqrt{3}}$
	3. Through centroid, parallel to b	$\frac{bl^3}{12}$ or $A\frac{l^2}{12}$	$\frac{l}{\sqrt{12}}$ or $\frac{l}{2\sqrt{3}}$
Triangle perpendicular height h base b area A	1. Coinciding with base	$\frac{bh^3}{12}$ or $A\frac{h^2}{6}$	$\frac{h}{\sqrt{6}}$
	2. Through centroid, parallel to base	$\frac{bh^3}{36}$ or $A\frac{h^2}{18}$	$\frac{h}{\sqrt{18}}$ or $\frac{h}{3\sqrt{2}}$
	3. Through vertex, parallel to base	$\frac{bh^3}{4}$ or $A\frac{h^2}{2}$	$\frac{h}{\sqrt{2}}$
Circle radius r area A	1. Through centre, perpendicular to plane (i.e. polar axis)	$\frac{\pi r^4}{2}$ or $A\frac{r^2}{2}$	$\frac{r}{\sqrt{2}}$
	2. Coinciding with diameter	$\frac{\pi r^4}{4}$ or $A\frac{r^2}{4}$	$\frac{r}{2}$
	3. About a tangent	$\frac{5\pi}{4}r^4$ or $\frac{5}{4}Ar^2$	$\frac{\sqrt{5}}{2}r$

Shape	Position of axis	Second moment of area	Radius of gyration, k
Semicircle radius r area A	Coinciding with diameter	$\frac{\pi r^4}{8}$ or $A\frac{r^2}{4}$	$\frac{r}{2}$

Worked problems on second moments of area

Problem 1. For each of the shapes shown in Fig. 21 find the second moment of area and the radius of gyration about the given axes.

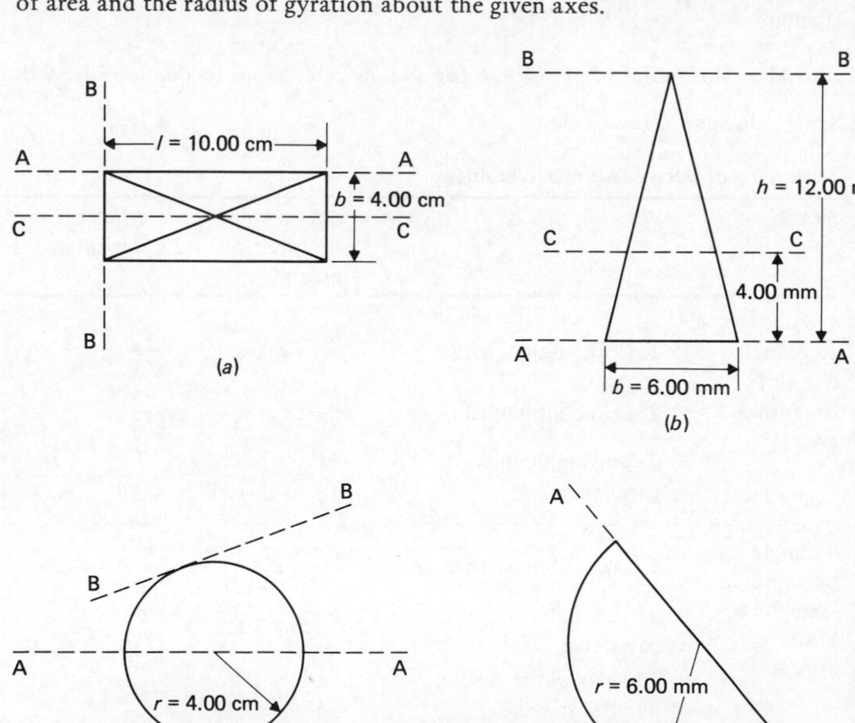

Figure 21

(a) For the rectangle:

Second moment of area about AA $= \frac{lb^3}{3} = \frac{10.00(4.00)^3}{3} =$ **213.3 cm⁴**

Radius of gyration, $k_{AA} = \dfrac{b}{\sqrt{3}} = \dfrac{4.00}{\sqrt{3}} = \mathbf{2.309\ cm}$

Second moment of area about BB $= \dfrac{bl^3}{3} = \dfrac{4.00(10.00)^3}{3} = \mathbf{1\ 333\ cm^4}$

Radius of gyration, $k_{BB} = \dfrac{l}{\sqrt{3}} = \dfrac{10.00}{\sqrt{3}} = \mathbf{5.774\ cm}$

The second moment of area about the centroid of a rectangle is $\dfrac{bl^3}{12}$ when the axis through the centroid is parallel with the breadth. In this case, the axis through the centroid is parallel with the length.

Hence the second moment of area about CC $= \dfrac{lb^3}{12} = \dfrac{10.00(4.00)^3}{12}$

$= \mathbf{53.33\ cm^4}$

Radius of gyration, $k_{CC} = \dfrac{b}{2\sqrt{3}} = \dfrac{4.00}{2\sqrt{3}} = \mathbf{1.155\ cm}$

(b) For the triangle:

Second moment of area about AA $= \dfrac{bh^3}{12} = \dfrac{(6.00)(12.00)^3}{12} = \mathbf{864\ mm^4}$

Radius of gyration, $k_{AA} = \dfrac{h}{\sqrt{6}} = \dfrac{12.00}{\sqrt{6}} = \mathbf{4.899\ mm}$

Second moment of area about BB $= \dfrac{bh^3}{4} = \dfrac{6.00(12.00)^3}{4} = \mathbf{2\ 592\ mm^4}$

Radius of gyration, $k_{BB} = \dfrac{h}{\sqrt{2}} = \dfrac{12.00}{\sqrt{2}} = \mathbf{8.485\ mm}$

Second moment of area about CC (i.e. through centroid) $= \dfrac{bh^3}{36}$

$= \dfrac{6.00(12.00)^3}{36} = \mathbf{288.0\ mm^4}$

Radius of gyration, $k_{CC} = \dfrac{h}{3\sqrt{2}} = \dfrac{12.00}{3\sqrt{2}} = \mathbf{2.828\ mm}$

(c) For the circle:

Second moment of area about AA (i.e. diameter) $= \dfrac{\pi r^4}{4} = \dfrac{\pi(4.00)^4}{4}$

$= \mathbf{201.1\ cm^4}$

Radius of gyration, $k_{AA} = \dfrac{r}{2} = \dfrac{4.00}{2} = \mathbf{2.000\ cm}$

Second moment of area about BB (i.e. tangent) $= \dfrac{5}{4}\pi r^4 = \dfrac{5}{4}\pi(4.00)^4$

$= \mathbf{1\ 005\ cm^4}$

Radius of gyration, $k_{BB} = \dfrac{\sqrt{5}}{2}r = \dfrac{\sqrt{5}}{2}(4.00) = \mathbf{4.472\ cm}$

(d) For the semicircle:

Second moment of area about AA (i.e. diameter) $= \frac{\pi r^4}{8} = \frac{\pi (6.00)^4}{8}$

$= \mathbf{508.9\ mm^4}$

Radius of gyration, $k_{AA} = \frac{r}{2} = \frac{6.00}{2} = \mathbf{3.000\ mm}$

Problem 2. Find the second moment of area of the rectangle of length l and breadth b shown in Fig. 22 about axis YY.

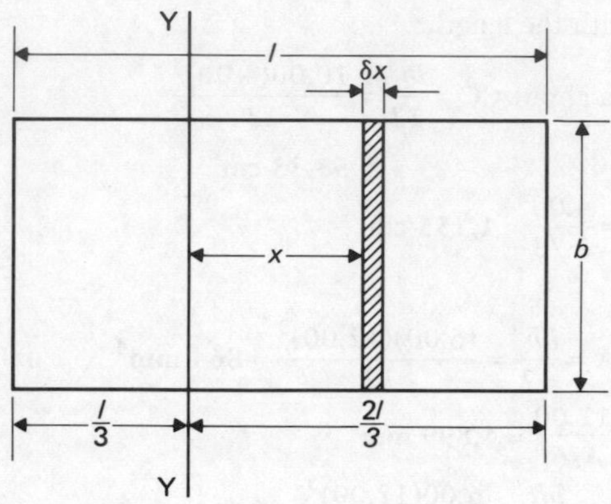

Figure 22

There are two methods of finding the second moment of area about axis YY.

Method 1. From first principles

Consider an elemental strip, width δx, parallel to and at distance x from axis YY as shown in Fig. 22.

Area of strip $= b\delta x$

Second moment of area of strip about YY $= x^2 (b\delta x)$

$$\text{Second moment of area of rectangle about YY} = \sum_{x=\frac{-l}{3}}^{x=\frac{2l}{3}} x^2 b\delta x$$

$$= \int_{\frac{-l}{3}}^{\frac{2l}{3}} x^2 b \, dx$$

$$= b\left[\frac{x^3}{3}\right]_{\frac{-l}{3}}^{\frac{2l}{3}}$$

$$= b\left[\frac{\left(\frac{2l}{3}\right)^3}{3} - \frac{\left(-\frac{l}{3}\right)^3}{3}\right]$$

$$= b\left[\frac{8l^3}{81} + \frac{l^3}{81}\right]$$

$$= \frac{bl^3}{9}\left(= A\frac{l^2}{9}\right)$$

Method 2. Using the parallel-axis theorem

The parallel-axis theorem states:

$$Ak_{YY}^2 = Ak_{CC}^2 + Ad^2$$

where Ak_{CC}^2 = second moment of area about the centroid, parallel with b, $= \frac{bl^3}{12}$, and d = perpendicular distance between axis YY and the centroid, i.e.

$$\frac{l}{2} - \frac{l}{3}, \text{ or } \frac{l}{6}$$

Hence $Ak_{YY}^2 = \frac{bl^3}{12} + bl\left(\frac{l}{6}\right)^2$

$$= \frac{bl^3}{12} + \frac{bl^3}{36} = \frac{bl^3}{9}$$

Problem 3. Find the second moment of area and the radius of gyration about axis XX for each of the shapes shown in Fig. 23. The broken lines indicate an axis through the centroid of the shape parallel to axis XX.

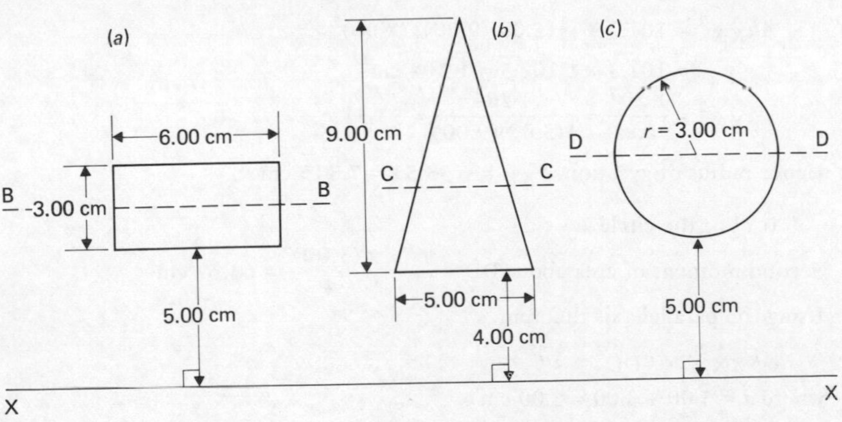

Figure 23

(a) For the rectangle:

Second moment of area about BB $= \dfrac{bl^3}{12}$ (where $l = 3.00$ cm, i.e. the length of the side at right angles to axis BB, and $b = 6.00$ cm)

$$= \frac{(6.00)(3.00)^3}{12}$$

$$= 13.50 \text{ cm}^4$$

Using the parallel-axis theorem:

$$Ak_{XX}^2 = Ak_{BB}^2 + Ad^2$$

where d = distance between BB and XX = 5.00 + 1.50 = 6.50 cm.

Hence the second moment of area of the rectangle about XX,

$$\boldsymbol{Ak_{XX}^2} = 13.50 + (18.00)(6.50)^2$$

$$= 13.50 + 760.5 = \mathbf{774.0\ cm^4}$$

$$k_{XX}^2 = \frac{774.0}{\text{area}} = \frac{774.0}{18.00} = 43.00$$

Hence radius of gyration, $\boldsymbol{k_{XX}} = \sqrt{43.00} = \mathbf{6.557\ cm}$

(b) For the triangle:

The second moment of area about CC $= \dfrac{bh^3}{36} = \dfrac{(5.00)(9.00)^3}{36}$

$$= 101.3 \text{ cm}^4$$

Using the parallel-axis theorem:

$$Ak_{XX}^2 = Ak_{CC}^2 + Ad^2$$

where $d = 4.00 + \frac{1}{3}(9.00) = 7.00$ cm

Hence the second moment of area of the triangle about XX,

$$\boldsymbol{Ak_{XX}^2} = 101.3 + [\tfrac{1}{2}(5.00)(9.00)]\,(7.00)^2$$

$$= 101.3 + 1\,102.5 = \mathbf{1\,204\ cm^4}$$

$$k_{XX}^2 = \frac{1\,204}{\text{area}} = \frac{1\,204}{\frac{1}{2}(5.00)(9.00)} = 53.51$$

Hence radius of gyration, $\boldsymbol{k_{XX}} = \sqrt{53.51} = \mathbf{7.315\ cm}$

(c) For the circle:

Second moment of area about DD $= \dfrac{\pi r^4}{4} = \dfrac{\pi(3.00)^4}{4} = 63.62$ cm^4

Using the parallel-axis theorem:

$$Ak_{XX}^2 = Ak_{DD}^2 + Ad^2$$

where $d = 5.00 + 3.00 = 8.00$ cm

Hence the second moment of area of the circle about XX,

$$Ak_{XX}^2 = 63.62 + [\pi(3.00)^2][8.00]^2$$
$$= 63.62 + 1\,810 = \mathbf{1\,874\ cm^4}$$
$$k_{XX}^2 = \frac{1\,874}{\text{area}} = \frac{1\,874}{\pi(3.00)^2} = 66.28$$

Hence radius of gyration, $\mathbf{k_{XX}} = \sqrt{66.28} = \mathbf{8.141\ cm}$

Problem 4. Calculate the second moment of area and the square of the radius of gyration of a rectangular lamina of length 56.0 cm and width 22.0 cm about an axis through one corner, perpendicular to the plane of the lamina.

Figure 24 shows the rectangle, having three mutually perpendicular axes at one corner.

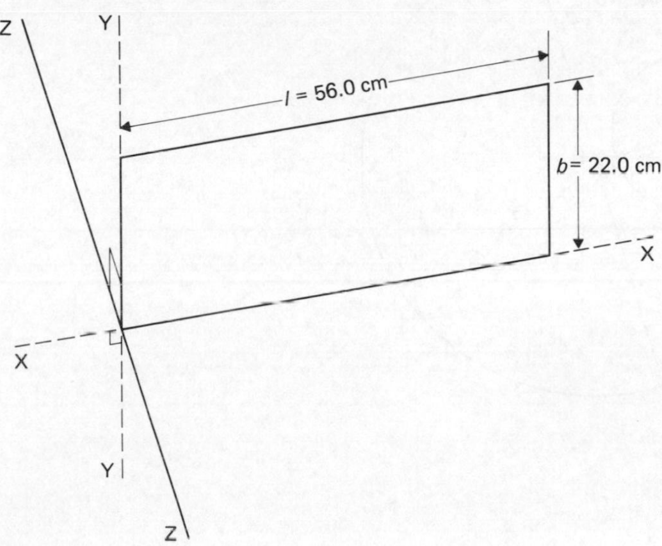

Figure 24

The perpendicular-axis theorem states:

$$Ak_{ZZ}^2 = Ak_{XX}^2 + Ak_{YY}^2$$

or $$k_{ZZ}^2 = k_{XX}^2 + k_{YY}^2$$

$$k_{XX}^2 = \frac{b^2}{3} = \frac{(22.0)^2}{3} = 161.3\ \text{cm}^2$$

$$k_{YY}^2 = \frac{l^2}{3} = \frac{(56.0)^2}{3} = 1\,045\ \text{cm}^2$$

Hence $\mathbf{k_{ZZ}^2} = 161.3 + 1\,045 = \mathbf{1\,206\ cm^2}$

$\mathbf{Ak_{ZZ}^2} = (56.0 \times 22.0)(1\,206) = \mathbf{1\,486\,000\ cm^4}$ or $\mathbf{1.486 \times 10^6\ cm^4}$

 Problem 5. A circular lamina, centre O, has a radius of 12.00 cm. A hole of radius 4.00 cm and centre A, where OA = 5.00 cm, is cut in the lamina. Find the second moment of area and the radius of gyration of the remainder about a diameter through O perpendicular to OA.

The circular lamina is shown in Fig. 25.

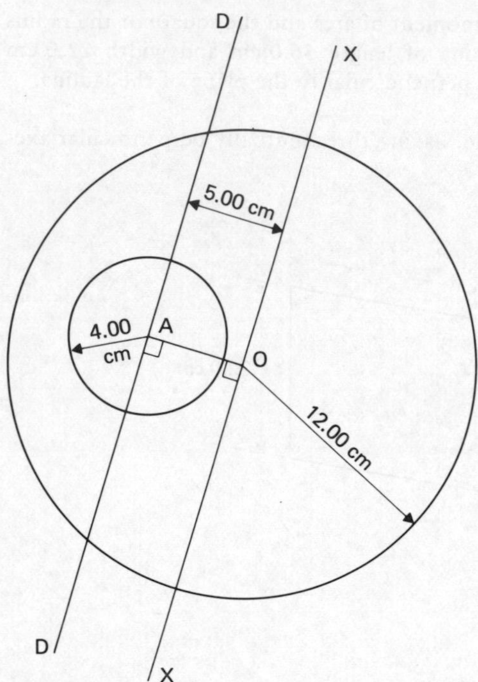

Figure 25

For the 12.00-cm-radius circle:

Second moment of area about XX (i.e. diameter) $= \dfrac{\pi r^4}{4} = \dfrac{\pi(12.00)^4}{4}$

$= 16\,290 \text{ cm}^4$

For the 4.00-cm-radius circle:
Second moment of area about diameter (i.e. axis DD), parallel to XX,

$= \dfrac{\pi r^4}{4} = \dfrac{\pi(4.00)^4}{4}$

$= 201.1 \text{ cm}^4$

Using the parallel-axis theorem, the second moment of area of the smaller circle about axis XX (denoted by $Ak_{XX}{}^2$) is given by:

$$Ak_{XX}^2 = Ak_{DD}^2 + (\text{area of smaller circle})(\text{perpendicular distance between DD and XX})^2$$
$$= 201.1 + [\pi(4.00)^2]\,[5.00]^2$$
$$= 1\,458\ \text{cm}^4$$

Hence **the second moment of area of the lamina remaining about its diameter XX is** 16 290 – 1 458, i.e. **14 832 cm^4**.

(Note that the second moment of area of the smaller circle is subtracted since that area has been removed from the lamina.)

Now $14\,832 = Ak_{XX}^2$

Area of lamina, $A = \pi(12.00)^2 - \pi(4.00)^2 = 402.1\ \text{cm}^2$

Hence $\quad k_{XX}^2 = \dfrac{14\,832}{\text{area}} = \dfrac{14\,832}{402.1}$

and **radius of gyration about XX,** $k_{XX} = \sqrt{\left(\dfrac{14\,832}{402.1}\right)}$
$= \mathbf{6.073\ cm}$

Problem 6. Determine the second moment of area about axis XX for the composite area shown in Fig. 26.

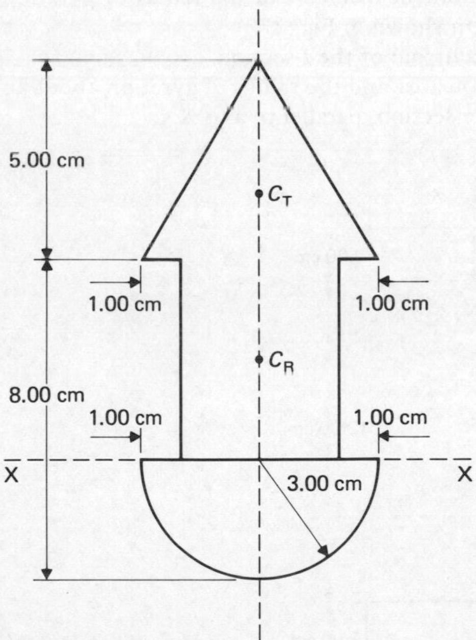

Figure 26

The composite area of Fig. 26 comprises:
a triangle, perpendicular height 5.00 cm, base 6.00 cm, centroid C_T,

a rectangle, length 5.00 cm, breadth 4.00 cm, centroid C_R and
a semicircle, radius 3.00 cm.

Second moment of area of triangle about an axis through C_T parallel to XX

$$= \frac{bh^3}{36} = \frac{(6.00)(5.00)^3}{36} = 20.83 \text{ cm}^4$$

By the parallel-axis theorem, the second moment of area of the triangle about

$$XX = 20.83 + [\tfrac{1}{2}(6.00)(5.00)]\,[5.00 + \tfrac{1}{3}(5.00)]^2$$

$$= \mathbf{687.5\ cm^4}$$

Second moment of area of rectangle about $XX = \dfrac{bl^3}{3} = \dfrac{(4.00)(5.00)^3}{3}$

$$= \mathbf{166.7\ cm^4}$$

Second moment of area of semicircle about $XX = \dfrac{\pi r^4}{8} = \dfrac{\pi(3.00)^4}{8}$

$$= \mathbf{31.81\ cm^4}$$

Total second moment of area about XX = 687.5 + 166.7 + 31.81

$$= \mathbf{886.0\ cm^4}$$

Problem 7. (a) Find the second moment of area and the radius of gyration about axis XX for the *I*-section beam shown in Fig. 27.
(b) Determine the position of the centroid of the *I*-section.
(c) Calculate the second moment of area and the radius of gyration about an axis NN through the centroid of the section, parallel to axis XX.

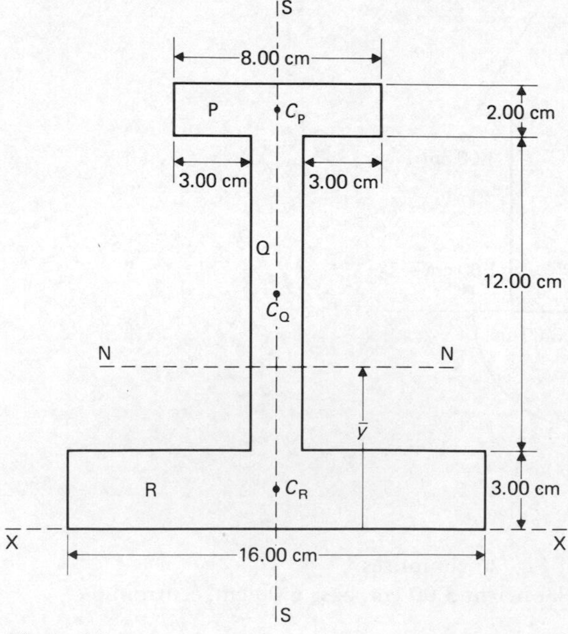

Figure 27

The *I*-section is divided into three rectangles, *P*, *Q* and *R* as shown in Fig. 27 with the centroid of each denoted by C_P, C_Q and C_R respectively.

(a) **For rectangle *P*:**

Second moment of area about C_P (an axis through C_P parallel to XX),

$$Ak_{C_P}^2 = \frac{bl^3}{12} \text{ (where } l = 2.00 \text{ cm and } b = 8.00 \text{ cm)}$$

$$= \frac{(8.00)(2.00)^3}{12} = 5.333 \text{ cm}^4$$

Using the parallel-axis theorem:

$$Ak_{XX}^2 = Ak_{C_P}^2 + Ad^2$$

where A = area of rectangle $P = 8.00 \times 2.00 = 16.00$ cm^2 and d = perpendicular distance between C_P and XX = 16.00 cm.

Hence $Ak_{XX}^2 = 5.333 + (16.00)(16.00)^2 = \mathbf{4\,101\ cm^4}$

For rectangle *Q*:

Similarly to above, $Ak_{C_Q}^2 = \dfrac{bl^3}{12} = \dfrac{(2.00)(12.00)^3}{12} = 288.0$ cm^4

$$Ak_{XX}^2 = 288.0 + [(2.00)(12.00)](9.00)^2 = \mathbf{2\,232\ cm^4}$$

For rectangle *R*:

$$Ak_{XX}^2 = \frac{lb^3}{3} = \frac{(16.00)(3.00)^3}{3} = \mathbf{144.0\ cm^4}$$

Total second moment of area for the *I*-section beam about axis XX
$= 4\,101 + 2\,232 + 144.0 = \mathbf{6\,477\ cm^4}$

Total area of *I*-section $= (8.00)(2.00) + (2.00)(12.00) + (16.00)(3.00)$
$= 88.00$ cm^2

$$Ak_{XX}^2 = 6\,477 \text{ cm}^4$$

Hence $$k_{XX}^2 = \frac{6\,477}{\text{area}} = \frac{6\,477}{88.00}$$

and radius of gyration about axis XX, $k_{XX} = \sqrt{\left(\dfrac{6\,477}{88.00}\right)} = \mathbf{8.579\ cm}$

(b) [Centroids of composite shapes such as that shown in Fig. 27 were discussed in *Technician Mathematics, Level 2*, Chapter 6.]

The centroid will lie on the axis of symmetry (shown as SS in Fig. 27).

Hence $\bar{x} = 0$.

Part	Area (a cm^2)	Distance of centroid from XX (i.e. y cm)	Moment about XX (i.e. ay cm^3)
P	16.00	16.00	256.00
Q	24.00	9.00	216.00
R	48.00	1.50	72.00
$\Sigma a = A = 88.00$			$\Sigma ay = 544.00$

$A\bar{y} = \Sigma ay$

Thus $\bar{y} = \frac{\Sigma ay}{A} = \frac{544.0}{88.00} = 6.182$ cm

The centroid is thus positioned on the axis of symmetry 6.182 cm from axis XX.

(c) $Ak_{XX}{}^2 = Ak_{NN}{}^2 + Ad^2$ from the parallel-axis theorem

$6\,477 = Ak_{NN}{}^2 + (88.00)(6.182)^2$

Hence $Ak_{NN}{}^2 = 6\,477 - 3\,363 = 3\,114$ cm^4

$$k_{NN}{}^2 = \frac{3\,114}{\text{area}} = \frac{3\,114}{88.00}\text{cm}^2$$

$$k_{NN} = \sqrt{\left(\frac{3\,114}{88.00}\right)} = 5.949 \text{ cm}$$

Thus the second moment of area about the centroid is 3 114 cm^4 and its radius of gyration is 5.949 cm.

Further problems on second moments of area may be found in the following section (5) (Problems 25–52).

5. Further problems

Centroids

1. Find the first moment of area about the axis XX for the shapes shown in Fig. 28.

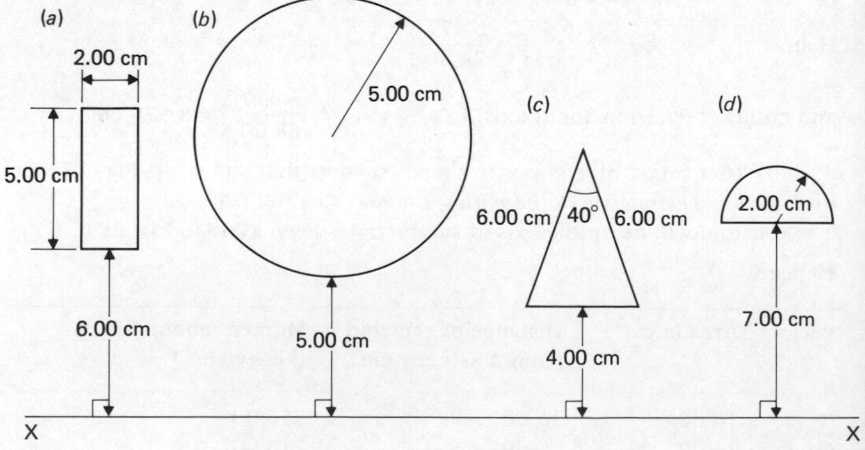

Figure 28

(a) [85 cm^3] (b) [250π cm^3] (c) [68.0 cm^3] (d) [49.3 cm^3]

In Problems 2–6 find the positions of the centroids of the areas bounded by the given curves, the x-axis and the given ordinates.

2. $y = 3x; x = 0, x = 4$ $[(2\frac{2}{3}, 4)]$
3. $y = 2x + 1; x = 0, x = 2$ $[(1\frac{2}{9}, 1\frac{13}{18})]$
4. $y = 4x^2; x = 1, x = 3$ $[(2.308, 11.17)]$
5. $y = x^3; x = 0, x = 1$ $[(\frac{4}{5}, \frac{2}{7})]$
6. $y = 2x(x + 1); x = -1, x = 0$ $[(-\frac{1}{2}, -\frac{1}{5})]$
7. Calculate the position of the centroid of the sheet of metal formed by the x-axis and the part of the curve $y = 5x - x^2$ which lies above the x-axis. $[(2\frac{1}{2}, 2\frac{1}{2})]$
8. Find the coordinates of the centroid of the plate which lies between the curve $\frac{y}{x} = x - 3$ and the x-axis. $[(1.5, -0.9)]$
9. Calculate the position of the centroid of the area of the metal plate bounded by the axes and the part of the curve $y = 4 - x^2$ which lies in the first quadrant. $[(0.75, 1.60)]$
10. A portion of the curve $y = 16 - x^2$ lies above the x-axis. Calculate the coordinates of the centroid of the area formed by the curve and the x-axis. $[(0, 6.4)]$
11. Find the position of the centroid of the area lying between $y = 3x^2$, the y-axis and the ordinates $y = 0$ and $y = 4$. $[(0.433, 2.4)]$
12. Determine the position of the centroid of the area enclosed by the curve $y = 3\sqrt{x}$, the x-axis and the ordinate $x = 4$. $[(2.4, 2.25)]$
13. Sketch the area enclosed by the curve $\frac{y}{\sqrt{3}} = \sqrt{x}$, the y-axis and the ordinate $y = 5$. Calculate the coordinates of the centroid of this area. $[(2.5, 3.75)]$
14. Sketch the curve $y^2 = 4x$ between limits of $x = 0$ and $x = 5$. Find the position of the centroid of this area. $[(3, 0)]$
15. Determine the position of the centroid of a metal template in the shape of a quadrant of a circle of radius r, given that the equation of a circle, centre O and radius r is $x^2 + y^2 = r^2$. Check your answer by using the theorem of Pappus.

    [On the centre line, distance $\frac{4\sqrt{2}r}{3\pi}$ (i.e. $0.6r$) from the centre.]
16. Find the coordinates of the centroid of the wooden lamina enclosed by the curve $y = 3 - \sqrt{x}$, the x-axis and the ordinate $x = 4$. $[(1.68, 0.9)]$
17. Find the points of interesection of the curves $y = 2x^2$ and $y = 4x$. Determine the coordinates of the centroid of the area enclosed by the two curves. $[(0, 0), (2, 8); (1, 3.2)]$
18. Determine the position of the centre of area of the metal plate enclosed between the curves $x^2 = y$ and $y^2 = x$. $[(0.45, 0.45)]$
19. Calculate the points of intersection of the curves $\frac{x^2}{2} = y$ and $y^2 = 2x$ and find the position of the centroid of the area enclosed by them. $[(0, 0), (2, 2); (0.9, 0.9)]$

20. Find the area of the sheet of thin cardboard bounded by the curve $y = 4x^2$ and the x-axis, between the limits $x = 0$ and $x = 3$. If this area is revolved about: (a) the x-axis; and (b) the y-axis, find the volume of the solid of revolution produced in each case. Determine the coordinates of the centroid of the cardboard: (i) by using integration; (ii) by using the theorem of Pappus.
[36 square units; (a) 777.6π cubic units; (b) 162π cubic units; (2.25, 10.8)]
21. Sketch the loop $y^2 = x(3 - x)^2$ which is the shape of a rudder. Calculate the coordinates of the centroid of the area enclosed by the loop.
[(1.286, 0)]
22. Determine the position of the centroid of the sheet of paper enclosed by the curve $y = 3x^2 + 2$, the y-axis and ordinates $y = 2$ and $y = 5$.
[(0.375, 3.80)]
23. Sketch the part of the curve $(x - 1)(4 - x)$ which lies above the x-axis. Find the area enclosed by the curve and the x-axis. If the area is revolved completely about the x-axis find the volume generated. Determine the position of the centroid of the area.
[$4\frac{1}{2}$ square units; 8.1π cubic units; (2.5, 0.9)]
24. Sketch the curves $y = 3x^2 + 2x - 1$ and $y + 5 = x(2x + 7)$ and determine the points of intersection. Calculate the coordinates of the centroid of the area enclosed by the two curves. [(1, 4), (4, 55); ($2\frac{1}{2}$, $25\frac{4}{7}$)]

Second moment of area

25. For the rectangle shown in Fig. 29 (a), find the second moment of area and the radius of gyration: (i) about axis AA; (ii) about axis BB; and (iii) about axis CC.
(i) [1 080 cm^4, 3.464 cm] (ii) [6 750 cm^4, 8.660 cm]
(iii) [1 688 cm^4, 4.331 cm]
26. For the triangle shown in Fig. 29 (b), find the second moment of area and the radius of gyration: (i) about axis AA; (ii) about axis BB; and (iii) about axis CC.
(i) [425.3 mm^4, 3.674 mm] (ii) [1 276 mm^4, 6.365 mm]
(iii) [141.8 mm^4, 2.122 mm]
27. For the circle shown in Fig. 29 (c), find the second moment of area and the radius of gyration: (i) about axis AA; (ii) about axis BB; and (iii) about axis CC.
(i) [490.9 cm^4, 2.500 cm] (ii) [2 454 cm^4, 5.590 cm]
(iii) [2 454 cm^4, 5.590 cm]
28. For the semicircle shown in Fig. 29 (d), find the second moment of area and the radius of gyration about axis AA. [100.5 cm^4, 2.000 cm]
29. Find the second moment of area of a rectangular wooden lamina having dimensions l and b, about an axis parallel to b at a distance $\frac{l}{4}$ from one end. [$\frac{7}{48}bl^3$]
30. Show that the radius of gyration of a semicircle of radius r about an axis through its centroid parallel with its diameter is given by $0.264\,3r$.

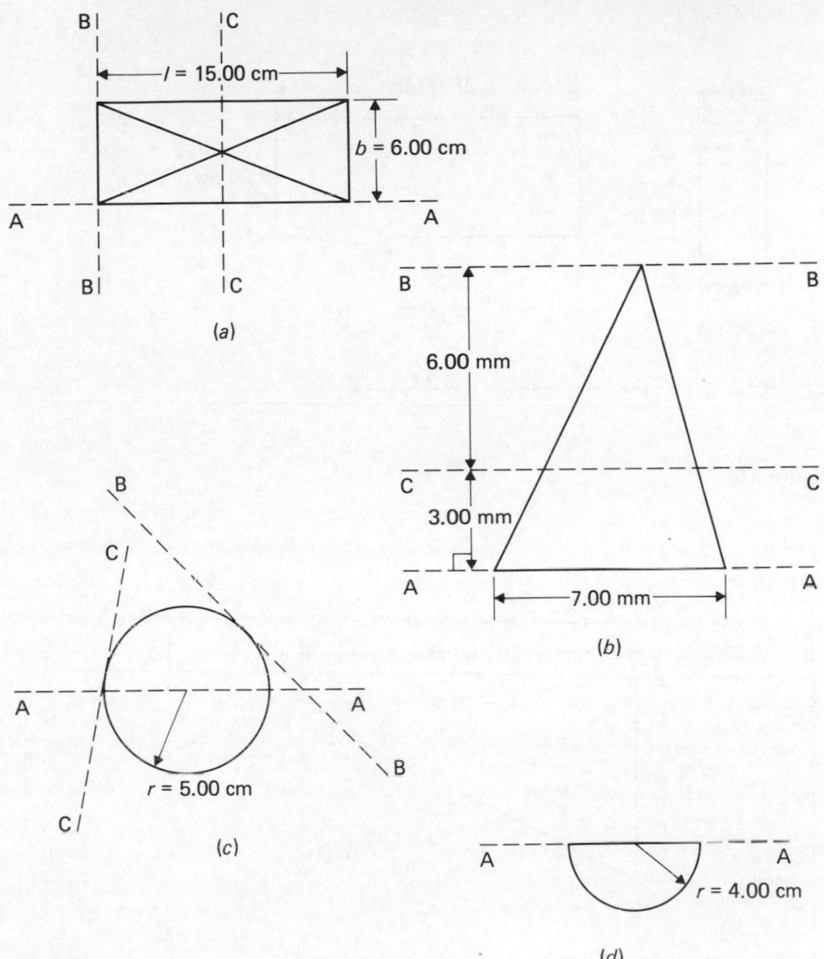

Figure 29

31. For each of the rectangles shown in Fig. 30, find the second moment of area and the radius of gyration about axis XX by using the parallel-axis theorem.
(a) [210.7 cm^4, 5.132 cm] (b) [20 200 cm^4, 12.66 cm]

32. For each of the triangles shown in Fig. 31, find the second moment of area and the radius of gyration about axis YY by using the parallel-axis theorem.
(a) [12 250 cm^4, 13.47 cm] (b) [1 695 cm^4, 8.069 cm]

33. For each of the circles shown in Fig. 32, find the second moment of area and the radius of gyration about axis AA by using the parallel-axis theorem.
(a) [53 410 mm^4, 11.85 mm] (b) [29 730 mm^4, 21.62 mm]

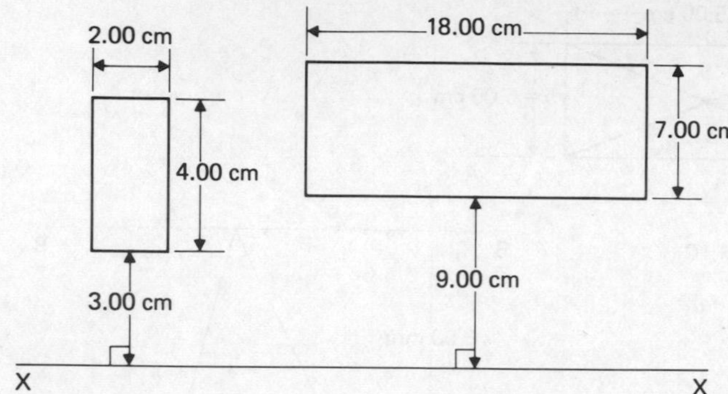

Figure 30

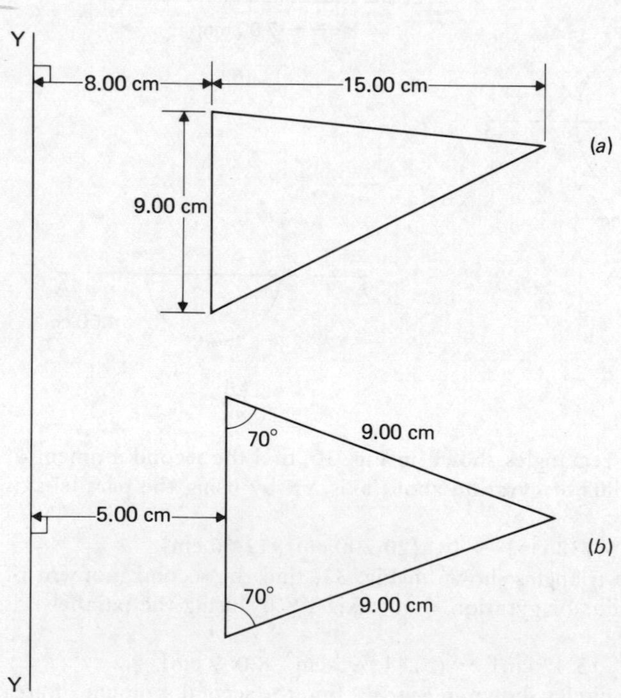

Figure 31

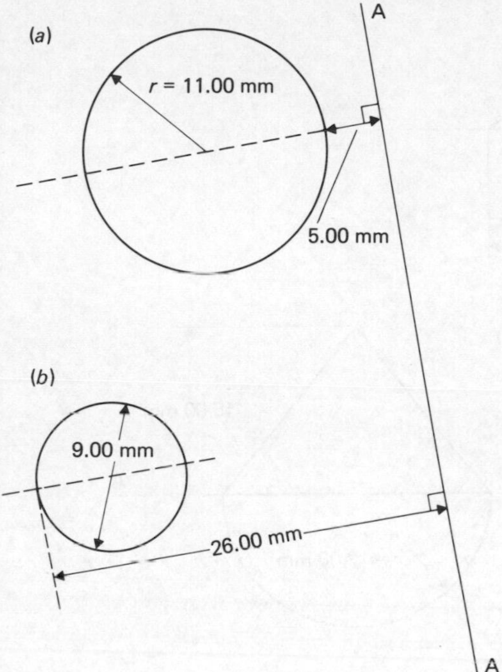

Figure 32

34. Calculate the second moment of area and the radius of gyration of a rectangular cover of length 32.0 mm and width 12.0 mm about an axis through one corner perpendicular to the plane of the lamina.
[149 500 mm^4, 19.73 mm]

35. Find the second moment of area and the radius of gyration of a plate in the shape of a quadrilateral PQRS about the side PQ, given that the lengths of PQ, PS and SR are 8.00, 10.00 and 12.00 cm respectively and that the angles QPS and PSR are right angles. [3 667 cm^4, 6.056 cm]

36. Determine the second moment of area and the square of the radius of gyration of a template in the shape of an isosceles triangle having sides of 5.00, 6.00 and 6.00 cm about an axis through its centroid and parallel to the 5.00-cm side. [38.14 cm^4, 2.347 cm^2]

37. Calculate the second moment of area and the radius of gyration of a circular metal cover of diameter 25.0 mm about an axis perpendicular to the plane of the cover and passing through the centre.
[38.350 mm^4, 8.839 mm]

38. For each of the composite shapes shown in Fig. 33, find the second moment of area and the radius of gyration about the axes AA.
(a) [285.8 cm^4, 2.210 cm] (b) [14 890 mm^4, 6.055 mm]
(c) [2 960 cm^4, 3.854 cm]

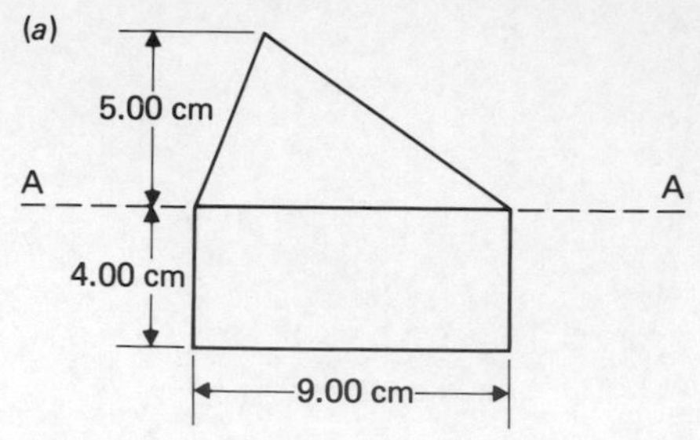

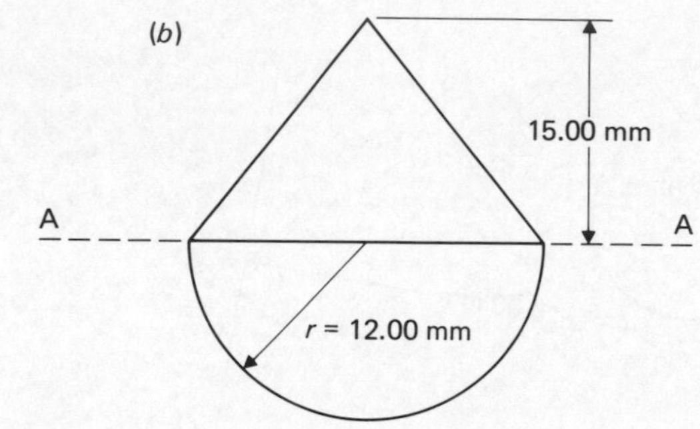

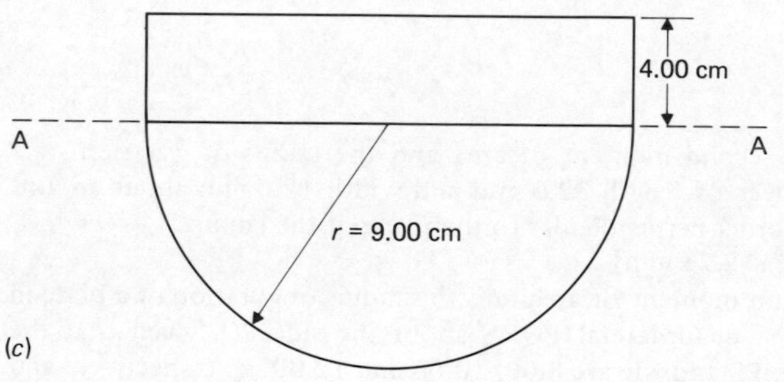

Figure 33

39. Determine the second moment of area of an equilateral triangular metal plate of side 8.00 cm about a pole passing through its centroid and perpendicular to the plane of the plate. [254.4 cm^4]

40. A circular cover, centre O, has a radius of 48.0 mm. A hole of radius 15.0 mm and centre B, where OB = 25.0 mm, is cut in the cover. Calculate the second moment of area (in cm^4) and the radius of gyration (in cm) of the remainder about a diameter through O perpendicular to OB. [368.8 cm^4, 2.376 cm]

41. Determine in terms of l, the second moment of area of the lamina shown in Fig. 34 about PQ, given that PQ = QR = ST = l and UT = $\frac{1}{3}$SR. [$\frac{10}{9}l^4$]

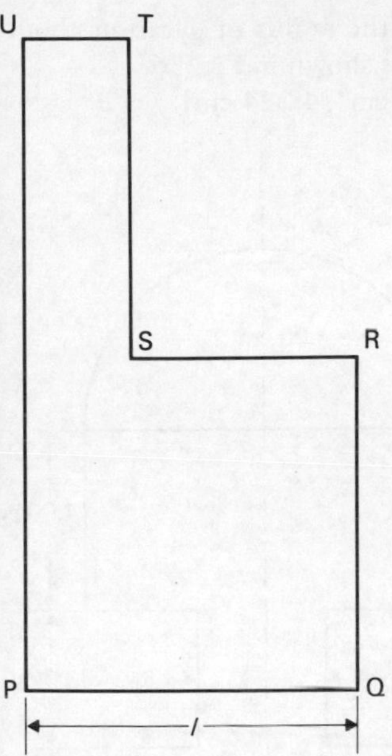

Figure 34

42. For each of the sections shown in Fig. 35, find the second moment of area and the radius of gyration about the axes AA.
(a) [3 184 cm^4, 8.145 cm] (b) [272 300 mm^4, 19.23 mm]

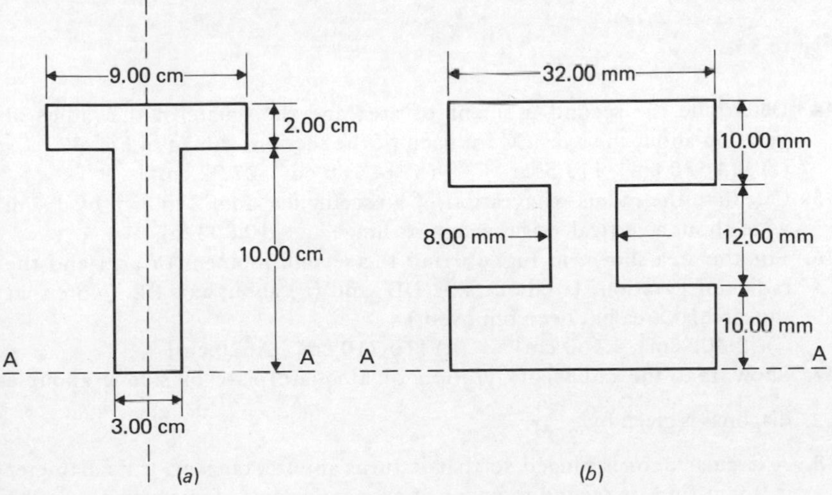

Figure 35

43. Calculate the second moment of area and the radius of gyration about the axes BB for each of the composite shapes shown in Fig. 36.
(a) [108 300 cm^4, 17.06 cm] (b) [1 394 cm^4, 4.224 cm]

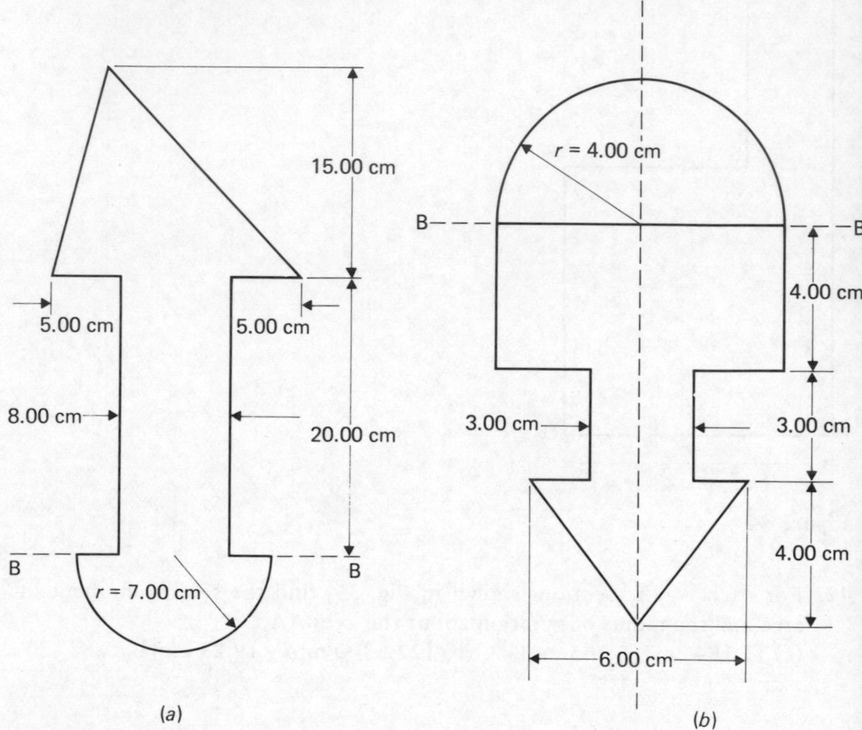

Figure 36

44. Determine the second moment of area and the square of the radius of gyration about the axes CC for each of the sections shown in Fig. 37.
(a) [14 530 cm^4, 113.5 cm^2] (b) [4 186 cm^4, 27.72 cm^2]

45. Calculate the radius of gyration of a rectangular door 2 m high by 1.4 m wide about a vertical axis through its hinge. [0.808 m]

46. For the area shown in Fig. 38 find the second moment of area and the radius of gyration: (i) about axis DD; and (ii) about axis EE. (Note that the circular area has been removed.)
(i) [2 302 cm^4, 4.460 cm] (ii) [10 710 cm^4, 9.620 cm]

47. Show that the radius of gyration of a square plate of side x about a diagonal is given by $\dfrac{x}{2\sqrt{3}}$.

48. A circular door is hinged so that it turns about a tangent. If its diameter is 0.9 m find its second moment of area and radius of gyration about the hinge. [0.161 m^4; 0.503 m]

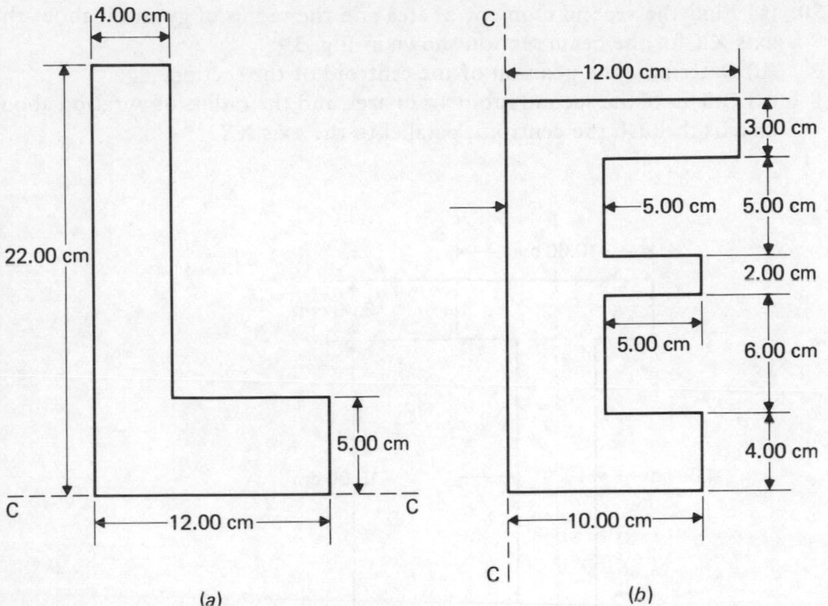

Figure 37

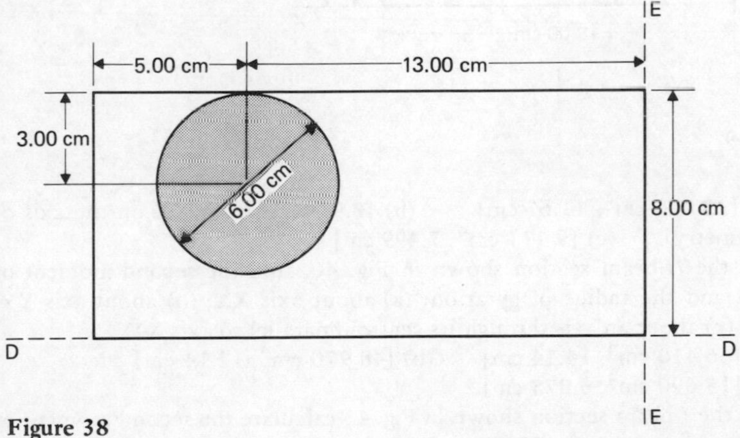

Figure 38

49. A uniform rectangular template of dimensions 32.0 cm and 24.0 cm has a circular hole of radius 10.0 cm removed. The centre of the hole coincides with the centroid of the rectangle. Calculate the second moment of area and the radius of gyration about an axis through the centroid perpendicular to the template. [86 690 cm^4, 13.82 cm]

50. (a) Find the second moment of area and the radius of gyration about the axis XX for the beam section shown in Fig. 39.
(b) Determine the position of the centroid of the section.
(c) Calculate the second moment of area and the radius of gyration about an axis through the centroid, parallel to the axis XX.

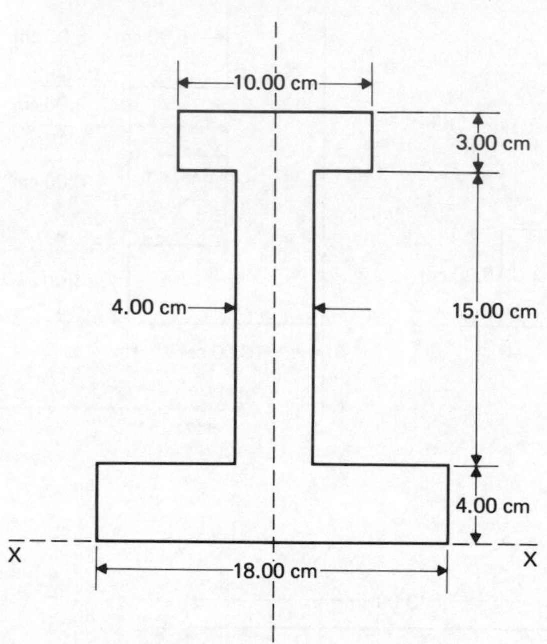

Figure 39

(a) [22 020 cm^4, 11.67 cm] (b) [8.944 cm from XX on the axis of symmetry] (c) [9 111 cm^4, 7.499 cm]

51. For the *H*-beam section shown in Fig. 40, find the second moment of area and the radius of gyration: (a) about axis XX; (b) about axis YY; and (c) about an axis through its centroid parallel to axis XX.
(a) [66 410 cm^4, 14.14 cm] (b) [48 970 cm^4, 12.14 cm]
(c) [15 690 cm^4, 6.875 cm]

52. For the *I*-beam section shown in Fig. 41 calculate the second moment of area and the square of the radius of gyration: (i) about axis AA; (ii) about axis BB; (iii) about the axis of symmetry CC; (iv) about an axis through the centroid parallel to axis AA; (v) about an axis through the centroid perpendicular to the place of the section.
(i) [24 450 cm^4, 103.6 cm^2] (ii) 45 810 cm^4, 194.1 cm^2]
(iii) [7 995 cm^4, 33.88 cm^2] (iv) [10 320 cm^4, 43.73 cm^2]
(v) [18 315 cm^4, 77.61 cm^2]

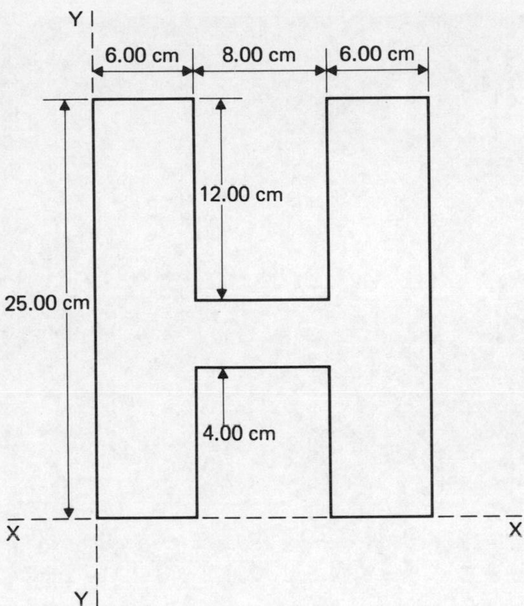

Figure 40

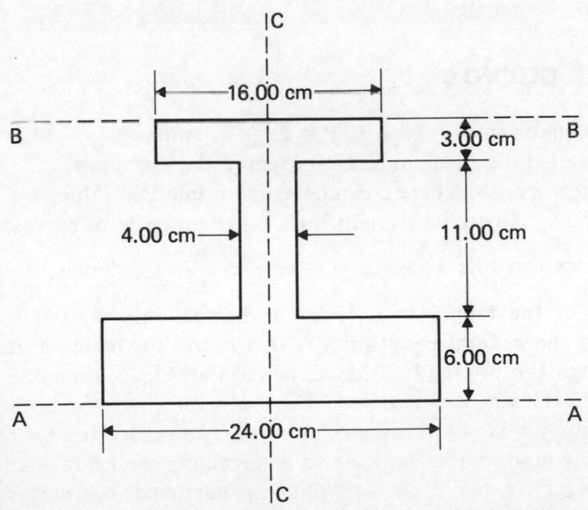

Figure 41

Chapter 9

Differential equations

1. Families of curves

A graph depicting the equations $y = 3x + 1$, $y = 3x + 2$, and $y = 3x - 4$ is shown in Fig. 1, and three parallel straight lines are seen to be the result. Equations of the form $y = 3x + c$, where c can have any numerical value, will produce an infinite number of parallel straight lines called a **family of curves.** A few of these can be seen in Fig. 1. Since $y = 3x + c$, $\frac{dy}{dx} = 3$, that is, the slope of every member of the family is 3. When additional information is given, for example, both the **general equation** $y = 3x + c$, and the member of the family passing through the point (2, 2), as shown as P in Fig. 1, then one particular member of the family is identified. The only line meeting both these conditions is the line $y = 3x - 4$. This is established by substituting $x = 2$ and $y = 2$ in the general equation $y = 3x + c$ and determining the value of c. Then, $2 = 3(2) + c$, giving $c = -6 + 2$, i.e. -4. Thus the **particular solution** is $y = 3x - 4$. Similarly, at point Q having coordinates $(-2, -4)$, the particular member of the family meeting both the conditions that it belongs to the family $y = 3x + c$ and that it passes through Q is the line $y = 3x + 2$. The additional information given, to enable a particular member of a family to be selected, is called the **boundary conditions.**

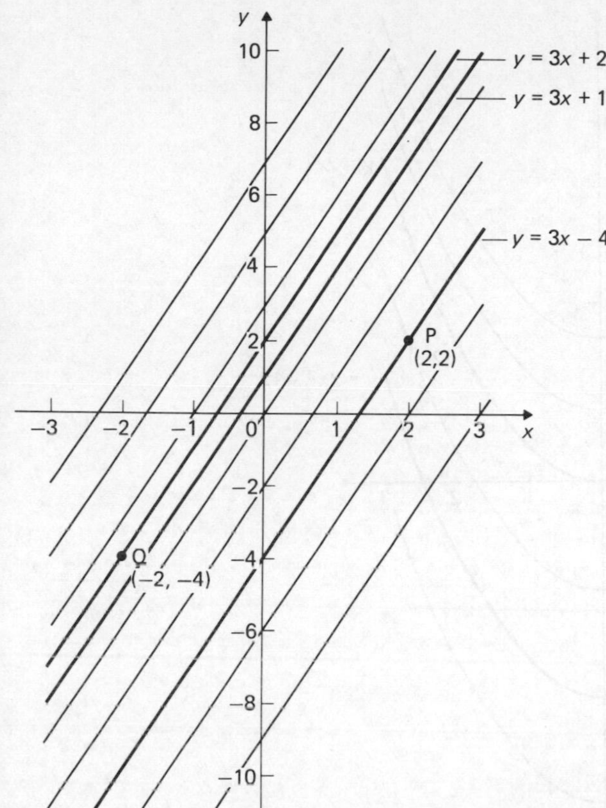

Figure 1 Some members of the family of curves satisfying the equation $y = 3x + c$

The equation, $\frac{dy}{dx} = 3$, is called a **differential equation** since it contains a differential coefficient. It is also called a **first-order** differential equation, since it contains the first differential coefficient only, and has no differential coefficients such as $\frac{d^2y}{dx^2}$ or higher orders.

Another family of an infinite number of curves is produced by drawing a graph depicting the equations $y = 2x^2 + c$. Two of the curves in the family are $y = 2x^2$ (when $c = 0$) and $y = 2x^2 - 12$ (when $c = -12$) and these curves, together with others belonging to the family, are shown in Fig. 2.

The slope at any point of these curves is found by differentiating $y = 2x^2 + c$ and is given by $\frac{dy}{dx} = 4x$, i.e. the gradient of all of the curves is given by 4 times the value of the abscissa at every point. When boundary con-

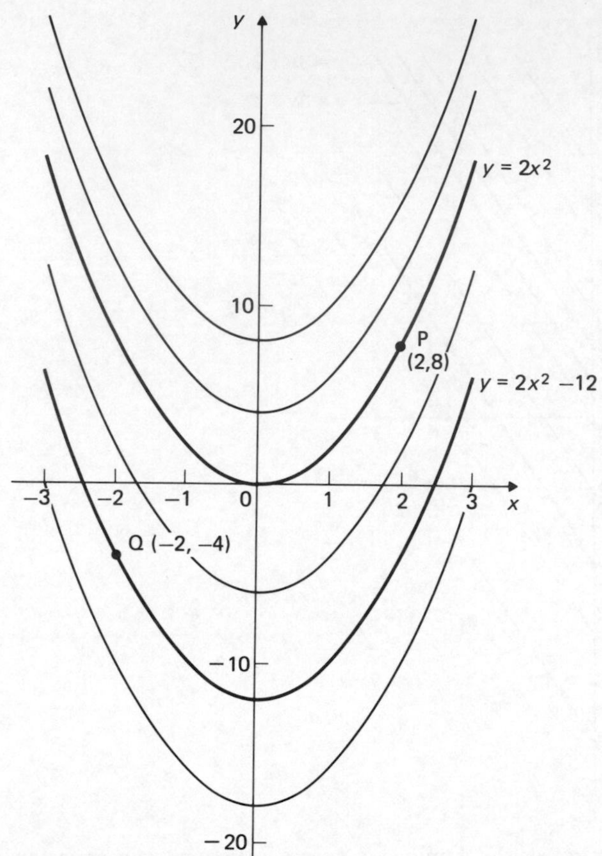

Figure 2 Some members of the family of curves satisfying the equation $y = 2x^2 + c$

ditions are stated, particular curves can be identified. For example, the curve belonging to the family of curves $y = 2x^2 + c$ and which passes through point P, having coordinates (2, 8), is obtained by substituting $x = 2$ and $y = 8$ in the general equation. This gives $8 = 2(2)^2 + c$, i.e. $c = 0$, and hence the curve is $y = 2x^2$. Similarly, the curve satisfying the general equation and passing through the point Q = (−2, −4) is $y = 2x^2 - 12$.

2. The solution of differential equations of the form $\dfrac{dy}{dx} = f(x)$

In Section 1, it was stated that a differential equation is an equation containing a differential coefficient. These equations occur frequently in both technology and science, since many systems can be represented by them. As a simple example, many systems can be represented by a first-order differential equation depicting the law of natural growth or decay, introduced in Chapter 3.

This law applies where the rate of change of a variable is directly proportional to the variable itself, i.e.

$$\frac{dQ}{dt} = kQ, \text{ where } k \text{ is a constant.}$$

One of the many occurrences governed by this law is the rate of cooling of a body. The rate at which it cools depends on such things as the mass of the body, its surface area, the nature of its surface, the temperature difference between the body and its surroundings and so on. However, by measuring the temperature of a body and remeasuring its temperature a known time later, it is possible to predict completely the way in which the body will cool.

Since for the body, $\frac{d\theta}{dt} = -k\theta$, it is shown in Section 3 that $\theta = Ae^{-kt}$, and from the measurements of time and temperature, the constants A and k for a particular body can be determined by using simultaneous equations. Thus by measuring time and the change in temperature, the complete cooling curve for the body can be predicted without any measurements being taken of mass, surface area, nature of surface, and so on.

There are many types of differential equation and possibly the simplest type is of the form $\frac{dy}{dx} = f(x)$.

The solution of any first-order differential equation of the form $\frac{dy}{dx} = f(x)$ involves eliminating the differential coefficient, i.e., forming an equation which does not contain $\frac{dy}{dx}$. This is achieved by integrating, because $\int \frac{dy}{dx}\,dx$ is equal to y.

For example, when $\frac{dy}{dx} = 2x + 3$, by integrating:

$$\int \frac{dy}{dx}\,dx = \int (2x + 3)\,dx$$

giving $\quad y = x^2 + 3x + c$

It was shown in Section 1 that the solution of an equation of this form produces a family of curves. A solution to a differential equation containing an arbitrary constant of integration is called the **general solution** of the differential equation and is representative of a family of curves.

When additional information is given, so that a particular curve from the family can be identified, the solution is called a **particular solution** of the differential equation or just a **particular integral.**

For example, when determining the particular solution to the equation $\frac{dy}{dx} = x^2 + 5$, given the boundary conditions that x is equal to 2 when y is equal to 5, the first step is to integrate to obtain the general solution.

When $\frac{dy}{dx} = x^2 + 5$, then $y = \frac{x^3}{3} + 5x + c$, the general solution.

Using the information given, i.e. substituting for x and y in the general solution, gives:

$$5 = \frac{(2)^3}{3} + 5(2) + c$$

and so $c = 5 - \frac{8}{3} - 10 = -7\frac{2}{3}$

Hence the particular solution is

$$y = \frac{x^3}{3} + 5x - 7\tfrac{2}{3}.$$

Worked problems on the solution of differential equations of the form $\frac{dy}{dx} = f(x)$

Problem 1. Find the general solutions to the equations:

(a) $\frac{dy}{dx} + \frac{5}{x} = 4x$

(b) $x\frac{dy}{dx} = 3x^3 - 4x^2 + 5x$

(c) $5\frac{dM}{d\theta} = 3e^{\theta} - 4e^{-\theta}$

(d) $\frac{ds}{dt} = u + at$, where u and a are constants

(e) $\frac{di}{dt} = \omega I_m \cos \omega t$, where ω and I_m are constants

Each of these equations is of the form $\frac{dy}{dx} = f(x)$ and the general solution can be obtained by integration.

(a) $\frac{dy}{dx} = 4x - \frac{5}{x}$

Integrating: $\int \frac{dy}{dx}\,dx = \int \left(4x - \frac{5}{x}\right) dx$

$$y = \frac{4x^2}{2} - 5 \ln x + c$$

i.e. $\mathbf{y = 2x^2 - 5 \ln x + c}$

(b) $x\frac{dy}{dx} = 3x^3 - 4x^2 + 5x$

Dividing throughout by x gives:

$$\frac{dy}{dx} = 3x^2 - 4x + 5$$

Integrating gives: $y = \frac{3x^3}{3} - \frac{4x^2}{2} + 5x + c$

i.e. $\quad$ $\mathbf{y = x^3 - 2x^2 + 5x + c}$

(c) $5\frac{dM}{d\theta} = 3e^{\theta} - 4e^{-\theta}$

$$\frac{dM}{d\theta} = \tfrac{1}{5}(3e^{\theta} - 4e^{-\theta})$$

Integrating gives: $\mathbf{M = \tfrac{1}{5}(3e^{\theta} + 4e^{-\theta}) + c}$

(d) $\frac{ds}{dt} = u + at$

Integrating gives: $\mathbf{s = ut + \tfrac{1}{2}at^2 + c}$

(e) $\frac{di}{dt} = \omega I_m \cos \omega t$

Integrating gives: $i = \frac{\omega I_m}{\omega} \sin \omega t + c$

i.e. $\quad$ $\mathbf{i = I_m \sin \omega t + c}$

Problem 2. Find the particular solutions of the following equations satisfying the given boundary conditions:

(a) $\frac{dy}{dx} + x = 2$ and $y = 3$ when $x = 1$

(b) $x\frac{dy}{dx} = 3 - x^3$ and $y = 3\frac{2}{3}$ when $x = 1$

(c) $3\frac{dr}{d\theta} + \cos\theta = 0$ and $r = 5$ when $\theta = \frac{\pi}{2}$

(d) $\frac{dy}{dx} = e^x - 2\sin 2x$ and $y = 2$ when $x = \frac{\pi}{4}$

(a) $\frac{dy}{dx} = 2 - x$

Integrating gives: $y = 2x - \frac{x^2}{2} + c$, the general solution.

Substituting the boundary conditions $y = 3$ and $x = 1$ to evaluate c gives:

$$3 = 2(1) - \frac{(1)^2}{2} + c$$

i.e. $c = 1\frac{1}{2}$

Hence the particular solution is $\mathbf{y = 2x - \frac{x^2}{2} + 1\frac{1}{2}}$

(b) $x\frac{dy}{dx} = 3 - x^3$

Dividing throughout by x to express the equation in the form $\frac{dy}{dx} = f(x)$ gives:

$$\frac{dy}{dx} = \frac{3}{x} - x^2$$

Integrating gives: $y = 3 \ln x - \frac{x^3}{3} + c$, the general solution.

Substituting the boundary conditions gives: $3\frac{2}{3} = 3 \ln 1 - \frac{(1)^3}{3} + c$

i.e. $c = 4$

Hence the particular solution is $\boldsymbol{y = 3 \ln x - \frac{x^3}{3} + 4}$

(c) $3\frac{dr}{d\theta} + \cos\theta = 0$

$$\frac{dr}{d\theta} = -\tfrac{1}{3}\cos\theta$$

Integrating gives: $r = -\frac{1}{3}\sin\theta + c$, the general solution.

Substituting the boundary conditions gives: $5 = -\frac{1}{3}\sin\frac{\pi}{2} + c$

i.e. $c = 5\frac{1}{3}$

Hence the particular solution is: $\boldsymbol{r = -\frac{1}{3}\sin\theta + 5\frac{1}{3}}$

(d) $\frac{dy}{dx} = e^x - 2\sin 2x$

Integrating gives: $y = e^x + \cos 2x + c$, the general solution.

Substituting the boundary conditions gives: $2 = e^{\pi/4} + \cos\frac{\pi}{2} + c$

i.e. $c = 2 - e^{\pi/4}$

Hence the particular solution is $\boldsymbol{y = e^x + \cos 2x + 2 - e^{\pi/4}}$

Expressed in this form, the true value of y is stated. The value of $e^{\pi/4}$ is 2.193 3 correct to 4 decimal places and the result can be expressed as $y = e^x + \cos 2x - 0.193\,3$, correct to 4 decimal places. However, when a result can be accurately expressed in terms of e or π, then it is usually better to leave the result in this form, unless the problem specifies the accuracy required.

Further problems on the solution of differential equations of the form $\frac{dy}{dx} = f(x)$ *may be found in Section 4 (Problems 1–30).*

3. The solution of differential equations of the form $\dfrac{dQ}{dt} = kQ$

Chapter 3 dealt with the natural laws of growth and decay, which are of the form $y = Ae^{kx}$ where A and k are constants. For such a law to apply, the rate of change of a variable must be proportional to the variable itself. This can be shown by differentiation. Since

$$y = Ae^{kx}$$

$$\frac{dy}{dx} = Ake^{kx}, \text{ i.e. } \frac{dy}{dx} = kAe^{kx}$$

But Ae^{kx} is equal to y. Hence, $\dfrac{dy}{dx} = ky$

Three of the natural laws stated in Chapter 3 are shown below and all such laws can be shown to be of a similar form.

(i) For linear expansion, the amount by which a rod expands when heated depends on the length of the rod, that is, the increase of length with respect to temperature is proportional to the length of the rod. Thus, mathematically:

$$\frac{dl}{d\theta} = kl \text{ and the law is } l = l_0 e^{k\theta}$$

(ii) For Newton's law of cooling, the fall of temperature with respect to time is proportional to the excess of its temperature above that of its surroundings, i.e.

$$\frac{d\theta}{dt} = -k\theta \text{ and the law is } \theta = \theta_0 e^{-kt}$$

(iii) In electrical work, when current decays in a circuit containing resistance and inductance connected in series, the change of current with respect to time is proportional to the current flowing at any instant, i.e.

$\dfrac{di}{dt} = ki$ and the law is $i = Ae^{kt}$ where $k = -\dfrac{1}{T}$ and t is the time constant of the circuit.

These are just some of many examples of natural or exponential laws. In general, differential equations of the form $\dfrac{dQ}{dt} = kQ$ depict natural laws and the solutions are always of the form $Q = Ae^{kt}$. This can be shown as follows:

Since $\dfrac{dQ}{dt} = kQ$, $\dfrac{dQ}{Q} = k\,dt$

Integrating: $\displaystyle\int \frac{dQ}{Q} = \int k\,dt$

i.e. $\qquad \ln Q = kt + c$

 By the definition of a logarithm, if $y = e^x$ then $x = \ln y$, i.e. if $\ln y = x$, then $y = e^x$. It follows that when $\ln q = kt + c$

$$Q = e^{(kt+c)}$$

By the laws of indices, $e^a e^b = e^{(a+b)}$ and applying this principle, gives:

$$Q = e^{kt} e^c$$

But e^c is a constant, say A, thus

$$Q = Ae^{kt}$$

Checking by differentiation:

when $Q = Ae^{kt}$

$$\frac{dQ}{dt} = kAe^{kt} = kQ$$

Hence Ae^{kt} is a solution to the differential equation $\frac{dQ}{dt} = kQ$.

Thus the general solution of any differential equation of the form

$$\frac{dQ}{dt} = kQ \textbf{ is } Q = Ae^{kt}$$

and when boundary conditions are given, the particular solution can be obtained, as shown in the worked problems following.

Worked problems on the solution of equations of the form $\frac{dQ}{dt} = kQ$

Problem 1. Solve the equation $\frac{dy}{dx} = 6y$ given that $y = 3$ when $x = 0.5$.

Since $\frac{dy}{dx} = 6y$ is of the form $\frac{dQ}{dt} = kQ$, the solution to the general equation will be of the form $Q = Ae^{kt}$, i.e. $y = Ae^{6x}$

Substituting the boundary conditions gives: $3 = Ae^{6(0.5)}$

i.e. $$A = \frac{3}{e^3}$$

Hence the particular solution is $\boldsymbol{y = \frac{3}{e^3} e^{6x} = 3e^{3(2x-1)}}$

Problem 2. Determine the particular solutions of the following equations and their given boundary conditions, expressing the values of the constants correct to 4 significant figures:

(a) $\frac{dM}{di} - 4M = 0$ and $M = 5$ when $i = 1$

(b) $\frac{1}{15}\frac{dl}{dm} + \frac{m}{4} = 0$ and $l = 15.41$ when $m = 0.714\,3$

(a) Rearranging the equation into the form $\frac{dQ}{dt} = kQ$ gives:

$$\frac{dM}{di} = 4i$$

The general solution is of the form $Q = Ae^{kt}$, giving

$$M = Ae^{4i}$$

Substituting the boundary conditions gives: $5 = Ae^{(4)(1)}$

i.e. $A = 0.091\ 58$

Hence the particular solution is $\boldsymbol{M = 0.091\ 58\, e^{4i}}$

(b) Writing the equation in the form $\frac{dQ}{dt} = kQ$ gives:

$$\frac{dl}{dm} = -\frac{15}{4}\, l$$

The general solution is $l = Ae^{-\frac{15}{4}m}$

Substituting the boundary conditions gives: $15.41 = Ae^{(-\frac{15}{4})(0.714\ 3)}$

i.e. $A = 224.4$

Hence the particular solution is $\boldsymbol{l = 224.4\, e^{-3.750m}}$

Problem 3. The decay of current in an electrical circuit containing resistance R ohms and inductance L henrys in series is given by $L\frac{di}{dt} + Ri = 0$, where i is the current flowing at time t seconds. Determine the general solution of the equation. In such a circuit, R is 5 kΩ, L is 3 henrys and the current falls to 5 A in 0.7 ms. Determine how long it will take for the current to fall to 2 amperes. Express your answer correct to 2 significant figures.

Since $\frac{di}{dt} = -\frac{R}{L}\,i$, then the general solution of the equation is

$$i = Ae^{-\frac{Rt}{L}}$$

By substituting the given values of R, L, i and t, the value of constant A is determined, i.e.

$$5 = Ae^{\left(\frac{-5\times 10^3\times 0.7\times 10^{-3}}{3}\right)}$$

$$= Ae^{-\frac{3.5}{3}}$$

giving $A = 16.056$

To determine the time for i to fall to 2 A, substituting in the general solution for i, A, R and L gives

$$2 = 16.056\, e^{\left(\frac{-5\times 10^3\times t}{3}\right)}$$

$$= 16.056\, e^{\left(\frac{-5t}{3}\right)}$$ when t is stated in milliseconds.

Thus, $\quad e^{\left(-\frac{5t}{3}\right)} = \dfrac{2}{16.056} = 0.124\ 56,$

and taking natural logarithms, gives:

$$-\frac{5t}{3}\ \ln e = \ln 0.124\ 56$$

But $\ln e = 1$, hence $t = -\frac{3}{5} \ln 0.124\ 56 = 1.25$ ms

i.e. **the time for i to fall to 2 A is 1.3 ms,** correct to 2 significant figures.

Problem 4. A copper conductor heats up to 50°C when carrying a current of 200 A. If the temperature coefficient of linear expansion, α_0, for copper is 17×10^{-6}/°C at 0°C and the equation relating temperature θ with length l is $\dfrac{dl}{d\theta} = \alpha l$, find the increase in length of the conductor at 50°C correct to the nearest centimetre, when l is 1 000 m at 0°C.

Since $\dfrac{dl}{d\theta} = \alpha l$, then $l = Ae^{\alpha\theta}$, the general solution.

But l is 1 000 when θ is 0°C, and substituting these values in the general solution of the equation gives:

$1\ 000 = Ae^0$, i.e. $A = 1\ 000$

Substituting for A, α and θ in the general equation, gives

$l = 1\ 000\ e^{(17 \times 10^{-6} \times 50)} = 1\ 000.850$ m

i.e. **the increase in length of the conductor at 50°C is 85 cm,** correct to the nearest centimetre.

Problem 5. The rate of cooling of a body is proportional to the excess of its temperature above that of its surrounding, θ°C.

The equation is: $\dfrac{d\theta}{dt} = k\theta$, where k is a constant.

A body cools from 90°C to 70°C in 3.0 minutes at a surrounding temperature of 15°C. Determine how long it will take for the body to cool to 50°C.

The general solution of the equation $\dfrac{d\theta}{dt} = k\theta$ is $\theta = Ae^{kt}$.

Letting the temperature 90°C correspond to a time t of zero gives an excess of body temperature above the surroundings of (90 – 15).

Hence, $(90 - 15) = Ae^{(k)(0)}$, i.e. $A = 75$.

3.0 minutes later, the general solution becomes:

$(70 - 15) = 75\ e^{(k)(3)}$

i.e. $\quad e^{3k} = \dfrac{55}{75}$

Taking natural logarithms,

$$3k = \ln\frac{55}{75},\ k = \frac{1}{3}\ln\frac{55}{75}$$

i.e. $k = -0.103\ 39$

At 50°C, $(50 - 15) = 75\ e^{-0.103\ 39t}$

$$\frac{35}{75} = e^{-0.103\ 39t}$$

Taking natural logarithms gives:

$$t = -\frac{1}{0.103\ 39}\ln\frac{35}{75}$$

$$= 7.37$$

That is, **the time for the body to cool to 50°C is 7.37 minutes,** or 7 minutes 22 seconds, correct to the nearest second.

Problem 6. The rate of decay of a radioactive material is given by $\frac{dN}{dt} = -\lambda N$ where λ is the decay constant and λN the number of radioactive atoms disintegrating per second. Determine the half-life of a zinc isotope, taking the decay constant as 2.22×10^{-4} atoms per second.

The half-life of an element is the time for N to become one-half of its original value. Since $\frac{dN}{dt} = -\lambda N$, then applying the general solution to this equation gives:

$N = Ae^{-\lambda t}$, where the constant A represents the original number of radioactive atoms present. For half-life conditions, the ratio $\frac{N}{A}$ is $\frac{1}{2}$, hence

$$\tfrac{1}{2} = e^{-\lambda t} = e^{-2.22\times10^{-4}t}$$

Thus, $\ln\frac{1}{2} = -2.22 \times 10^{-4}t.$

i.e. $$t = -\frac{1}{2.22 \times 10^{-4}}\ln 0.5$$

$$= 3\ 122 \text{ seconds or } 52 \text{ minutes } 2 \text{ seconds.}$$

Thus, **the half-life is 52 minutes,** correct to the nearest minute.

Further problems on the solution of equations of the form $\frac{dQ}{dt} = kQ$ *may be found in the following section (4) (Problems 31–41).*

4. Further problems

Solution of equations of the form $\frac{dy}{dx} = f(x)$

In Problems 1–15, find the general solutions of the equations.

1. $\frac{dy}{dx} = 3x - \frac{4}{x^2}$ $\left[y = \frac{3x^2}{2} + \frac{4}{x} + c\right]$
2. $\frac{dy}{dx} + 3 = 4x^2$ $\left[y = \frac{4x^3}{3} - 3x + c\right]$
3. $3\frac{dy}{dx} + \frac{2}{\sqrt{x}} = 5\sqrt{x}$ $[y = \frac{2}{3}\sqrt{x}(\frac{5}{3}x - 2) + c]$
4. $\frac{du}{dV} - \frac{1}{V} = 4$ $[u = 4V + \ln V + c]$
5. $6 - 5\frac{dy}{dx} = \frac{1}{x - 2}$ $[y = \frac{1}{5}(6x - \ln(x - 2)) + c]$
6. $2x^2 - \frac{3}{x} + 4\frac{dy}{dx} = 0$ $\left[y = \frac{1}{4}\left(3\ln x - 2\frac{x^3}{3}\right) + c\right]$
7. $\frac{di}{d\theta} = \cos\theta$ $[i = \sin\theta + c]$
8. $6\frac{dV}{dt} = 4\sin\left(100t + \frac{\pi}{6}\right)$ $\left[V = -\frac{1}{150}\cos\left(100t + \frac{\pi}{6}\right) + c\right]$
9. $\frac{di}{dt} - \frac{t}{10} + 140 = 0$ $\left[i = \frac{t^2}{20} - 140t + c\right]$
10. $3\frac{dv}{dt} + 0.7t^2 - 1.4 = 0$ $[v = \frac{1}{9}(4.2t - 0.7t^3) + c]$
11. $\frac{dy}{d\theta} = 3e^{\theta} - \frac{4}{e^{2\theta}}$ $\left[y = 3e^{\theta} + \frac{2}{e^{2\theta}} + c\right]$
12. $\frac{dV}{dx} = 3x - \frac{5}{x} - \sec^2 x$ $\left[V = \frac{3x^2}{2} - 5\ln x - \tan x + c\right]$
13. $\frac{1}{2}\frac{dy}{dx} + 2x^{\frac{1}{2}} = e^{\frac{x}{2}}$ $\left[y = 4\left(e^{\frac{x}{2}} - \frac{2x^{\frac{3}{2}}}{3}\right) + c\right]$
14. $x\frac{dy}{dx} = 2 - 3x^2$ $\left[y = 2\ln x - \frac{3x^2}{2} + c\right]$
15. $\frac{dM}{d\theta} = \frac{1}{2}\sin 3\theta - \frac{1}{3}\cos 2\theta$ $[M = -\frac{1}{6}(\cos 3\theta + \sin 2\theta) + c]$

In Problems 16–25, determine the particular solutions of the differential equations for the boundary conditions given.

16. $x\frac{dy}{dx} - 2 = x^3$ and $y = 1$ when $x = 1$. $\left[y = 2\ln x + \frac{x^3}{3} + \frac{2}{3}\right]$
17. $x\left(x - \frac{dy}{dx}\right) = 3$ and $y = 2$ when $x = 1$. $\left[y = \frac{x^2}{2} - 3\ln x + 1\frac{1}{2}\right]$
18. $\frac{ds}{dt} - 4t^2 = 9$ and $s = 27$ when $t = 3$. $\left[s = 9t + \frac{4t^3}{3} - 36\right]$
19. $e^{-p}\frac{dq}{dp} = 5$ and $q = 2.718$ when $p = 0$. $[q = 5e^{p} - 2.282]$

20. $3 - \frac{dy}{dx} = e^{2x} - 2e^x$ and $y = 7$ when $x = 0$.
$[y = 3x - \frac{1}{2}e^{2x} + 2e^x + 5\frac{1}{2}]$

21. $\frac{dy}{d\theta} - \sin 3\theta = 5$ and $y = \frac{5\pi}{6}$ when $\theta = \frac{\pi}{6}$. $[y = 5\theta - \frac{1}{3}\cos 3\theta]$

22. $3 \sin\left(2\theta - \frac{\pi}{3}\right) + 4\frac{dv}{d\theta} = 0$ and $v = 3.7$ when $\theta = \frac{2\pi}{3}$.
$\left[v = \frac{3}{8}\left(\cos\left(2\theta - \frac{\pi}{3}\right)\right) + 4.075\right]$

23. $\frac{1}{6}\frac{dM}{d\theta} + 1 = \sin\theta$ and $M = 3$ when $\theta = \pi$.
$[M = 3\{(2\pi - 1) - 2\cos\theta - 2\theta\}]$

24. $\frac{2}{(u+1)^2} = 4 - \frac{dz}{du}$ and $z = 14$ when $u = 5$. $\left[z = 4u + \frac{2}{u+1} - 6\frac{1}{3}\right]$

25. $\frac{1}{2e^x} + 4 = x - 3\frac{dy}{dx}$ and $y = 3$ when $x = 0$.
$\left[y = \frac{1}{3}\left\{\frac{x^2}{2} + \frac{1}{2e^x} - 4x\right\} + 2\frac{5}{6}\right]$

26. The bending moment of a beam, M, and shear force F are related by the equation $\frac{dM}{dx} = F$, where x is the distance from one end of the beam. Determine M in terms of x when $F = -w(l - x)$ where w and l are constants, and $M = \frac{1}{2}wl^2$ when $x = 0$. $[M = \frac{1}{2}w(l - x)^2]$

27. The angular velocity ω of a flywheel of moment of inertia I is given by $I\frac{d\omega}{dt} + N = 0$, where N is a constant. Determine ω in terms of t given that $\omega = \omega_0$ when $t = 0$. $\left[\omega = \omega_0 - \frac{Nt}{I}\right]$

28. The gradient of a curve is given by $\frac{dy}{dx} = 2x - \frac{x^2}{3}$. Determine the equation of the curve if it passes through the point $x = 3, y = 4$.
$\left[y = x^2 - \frac{x^3}{9} - 2\right]$

29. The acceleration of a body a is equal to its rate of change of velocity, $\frac{dv}{dt}$. Determine an equation for v in terms of t given that the velocity is u when $t = 0$. $[v = u + at]$

30. The velocity of a body v, is equal to its rate of change of distance, $\frac{dx}{dt}$. Determine an equation for x in terms of t given $v = u + at$, where u and a are constants and $x = 0$ when $t = 0$. $[x = ut + \frac{1}{2}at^2]$

Solution of equations of the form $\frac{dQ}{dt} = kQ$

In Problems 31–33 determine the general solutions to the equations.

31. $\dfrac{dp}{dq} = 9p \qquad [p = Ae^{9q}]$

32. $\dfrac{dm}{dn} + 5m = 0 \qquad [m = Ae^{-5n}]$

33. $\frac{1}{6}\dfrac{dw}{dx} + \frac{3}{5}w = 0 \qquad [w = Ae^{-\frac{18}{5}x}]$

In Problems 34–36 determine the particular solutions to the equations, expressing the values of the constants correct to 3 significant figures.

34. $\dfrac{dQ}{dt} = 15.0Q$ and $Q = 7.3$ when $t = 0.015$. $\qquad [Q = 5.83\,e^{15.0t}]$

35. $\frac{1}{7}\dfrac{dl}{dm} - \frac{1}{3}l = 0$ and $l = 1.7 \times 10^4$ when $m = 3.4 \times 10^{-2}$.
$[l = 1.57 \times 10^4\,e^{2.33m}]$

36. $0.741\dfrac{dy}{dx} + 0.071\,y = 0$ and $y = 73.4$ when $x = 15.7$
$[y = 330\,e^{-0.0958x}]$

37. The difference in tension, T newtons, between two sides of a belt when in contact with a pulley over an angle of θ radians and when it is on the point of slipping, is given by $\dfrac{dT}{d\theta} = \mu T$, where μ is the coefficient of friction between the material of the belt and that of the pulley at the point of slipping. When $\theta = 0$ radians, the tension is 170 N and the coefficient of friction as slipping starts is 0.31. Determine the tension at the point of slipping when θ is $\dfrac{5\pi}{6}$ radians. Also determine the angle of lap in degrees, to give a tension of 340 N just before slipping starts.
[383 N, 128°]

38. The charge Q coulombs at time t seconds for a capacitor of capacitance C farads when discharging through a resistance of R ohms is given by:
$$R\frac{dQ}{dt} + \frac{Q}{C} = 0$$
A circuit contains a resistance of 500 kilohms and a capacitance of 8.7 microfarads, and after 147 milliseconds the charge falls to 7.5 coulombs. Determine the initial charge and the charge after one second, correct to 3 significant figures. $\qquad$ [7.76 C, 6.17 C]

39. The rate of decay of a radioactive substance is given by $\dfrac{dN}{dt} = -\lambda N$, where λ is the decay constant and λN the number of radioactive atoms disintegrating per second. Determine the half-life of radium in years (i.e. the time for N to become one-half of its original value) taking the decay constant for radium as 1.36×10^{-11} atoms per second and assuming a '365-day' year. $\qquad$ [1 616 years]

40. The variation of resistance, R ohms, of a copper conductor with temperature, θ°C, is given by $\dfrac{dR}{d\theta} = \alpha R$, where α is the temperature coefficient of

resistance of copper. Taking α as 39×10^{-4} per °C, determine the resistance of a copper conductor at 30°C, correct to 4 significant figures, when its resistance at 80°C is 57.4 ohms. [47.23 ohms]

41. The rate of growth of bacteria is directly proportional to the amount of bacteria present. Form a differential equation for the rate of growth when n is the number of bacteria at time t seconds. If the number of bacteria present at $t = 0$ is n_0, solve the equation. When the number of bacteria doubles in one hour, determine by how many times it will have increased in twelve hours. [$n = n_0 e^{kt}$, 2^{12}]

Chapter 10

Probability

1. Definitions and simple probability

Probability

The probability of an event occurring means the likelihood or chance of it occurring. It is measured on a scale extending from a minimum of zero to a maximum of one. On this scale, zero corresponds to an absolute impossibility and one corresponds to an absolute certainty. When one red and one black marble are concealed in a bag and one of these is drawn from it, the probability of drawing a red marble is $\frac{1}{2}$ (or 1 in 2) and that of drawing a black marble is $\frac{1}{2}$ (or 1 in 2). When three red and one black marble are placed in the bag, the probability of drawing one red marble is $\frac{3}{4}$ (or 3 in 4) and that of drawing one black marble is $\frac{1}{4}$ (or 1 in 4).

For a bag containing m red marbles and n black marbles, the probability p of drawing a red marble is given by the ratio:

$$p = \frac{\text{number of red marbles}}{\text{the total number of marbles}}, \text{ i.e. } p = \frac{m}{m+n}$$

Also the probability q of not drawing a red marble, i.e. drawing a black marble, is the ratio:

$$q = \frac{\text{number of black marbles}}{\text{the total number of marbles}}, \text{ i.e. } q = \frac{n}{m+n}$$

In general, when p is the probability of an event happening and q is the probability of an event not happening, then

$q = 1 - p$ and the total probability, $p + q$ is unity.

Expectation

Generally, expectation can be defined as the product of the probability of success and the number of attempts. Thus if I buy 10 tickets in a raffle and the probability of winning a prize is $\frac{1}{200}$, the expectation of success, i.e. winning a prize, is $10 \times \frac{1}{200}$ or $\frac{1}{20}$. Expectation of success can be considered also in the following way.

If p is the probability of success and M is the reward for success, then the product is defined as the expectation of success, i.e.

expectation, $E = pM$

For example, if the probability p of winning a prize is $\frac{1}{50}$ and the prize, M, is £100, the expectation is pM, i.e. $\frac{1}{50} \times$ £100 or £2.

Independent and dependent events

When the occurrence of one event does not affect the probability of the occurrence of another, then the events are called **independent**. For example, a bag contains six red and four white balls. A ball is drawn at random from the bag and then replaced (a process termed '**with replacement**'), after its colour is noted. The probability of drawing a red ball is $\frac{6}{10}$. A second ball is now drawn from the bag and its colour noted, and again the probability of drawing a red ball is $\frac{6}{10}$. Since the probability of drawing a red ball on the second draw is in no way affected by the first draw, these two events are independent.

Conversely, when the probability of one event occurring does effect the probability of another event occurring, then the events are called **dependent.** For the bag containing six red and four white balls, on the first draw the probability of drawing a red ball is $\frac{6}{10}$. If another ball is withdrawn without putting the first ball back (a process termed '**without replacement**'), then the probability of drawing a red ball on the second draw is either $\frac{5}{9}$, when a red ball is removed as a result of the first draw, or $\frac{6}{9}$, when a white ball is removed as a result of the first draw.

Mutually exclusive events

Two or more events are called **mutually exclusive** when the occurrence of one of them excludes the occurrence of the others, i.e. when not more than one event can happen at the same time. An example is provided by the tossing of a coin, since the appearance of a head mutually excludes the appearance of a tail.

Worked problems on simple probability

Problem 1. Determine the probabilities of:
(a) drawing a black ball from a bag containing 7 red and 20 black balls;
(b) selecting at random a male from a group of 15 males and 29 females;
(c) winning a prize in a raffle by buying 5 tickets when a total of 450 tickets are sold; and
(d) winning a prize in a raffle by buying 10 tickets when there are 6 prizes and a total of 800 tickets are sold.

(a) Applying $p = \frac{n}{m+n}$, where p is the probability of drawing a black ball, m is the number of red balls and n the number of black balls, gives:

$$p = \frac{20}{7+20} = \frac{20}{27}$$

i.e. **the probability of drawing a black ball is $\frac{20}{27}$.**

(b) Applying $p = \frac{m}{m+n}$, where p is the probability of selecting a male, m is the number of males and n is the number of females, gives:

$$p = \frac{15}{15+29} = \frac{15}{44}$$

i.e. **the probability of selecting a male is $\frac{15}{44}$.**

(c) Applying $p = \frac{m}{m+n}$, where p is the probability of winning a prize, m is the number of tickets bought and $m + n$ the total number of tickets sold, gives:

$$p = \frac{m}{m+n} = \frac{5}{450} = \frac{1}{90}.$$

i.e. **the probability of winning a prize is $\frac{1}{90}$.**

(d) The probability of winning a prize by buying one ticket is the ratio:

$\frac{\text{number of prizes}}{\text{number of tickets sold}}$, i.e. $\frac{6}{800}$ or $\frac{3}{400}$.

That is, the probability of success is $\frac{3}{400}$. Hence the probability of winning a prize by buying ten tickets is (the probability of success) × (the number of attempts),

i.e. $10 \times \frac{3}{400}$ or $\frac{3}{40}$.

Thus, **the probability of winning a prize is $\frac{3}{40}$.**

Problem 2.

(a) The probability of an event happening is $\frac{3}{19}$. Determine the probability of it not happening.
(b) A bench can seat 7 people. Determine the probability of any one person: (i) sitting in the middle; and (ii) sitting at the end.
(c) A person's chance of winning a raffle is $\frac{1}{18}$. If he buys 15 tickets determine the total number sold.
(d) A bag contains 8 red, 5 blue and 4 white balls. Determine the probabilities of drawing: (i) a red ball; (ii) a blue ball; and (iii) a white ball.

(a) When p is the probability of an event happening and q the probability of the event not happening, then $p + q = 1$. Hence $q = 1 - p$. The value of p is given as $\frac{3}{19}$, hence

$$q = 1 - \frac{3}{19} = \frac{16}{19}$$

i.e. **the probability of the event not happening is $\frac{16}{19}$.**

(b) There are 7 seats on the bench and the probability of one person sitting on any particular seat is $\frac{1}{7}$. Hence:
(i) the probability of one person sitting in the middle of the bench is $\frac{1}{7}$; and
(ii) the probability of one person sitting on either of the two ends of the bench is $\frac{1}{7}$ for the left end and $\frac{1}{7}$ for the right end, giving a total probability of $\frac{2}{7}$.

(c) Let p be the probability of the person winning the raffle, m be the number of tickets the person bought and $m + n$ the total number of tickets sold. Then:

$$p = \frac{m}{m+n}, \text{ i.e. } m + n = \frac{m}{p}$$

$$m + n = \frac{15}{\frac{1}{18}} = 270$$

i.e. **the total number of tickets sold is 270.**

(d) Let p_R be the probability of drawing a red ball, given by the ratio:

$$p_R = \frac{\text{number of red balls}}{\text{the total number of balls}}$$

i.e. $p_R = \dfrac{8}{8+5+4} = \dfrac{8}{17}$

Similarly, $p_B = \dfrac{5}{17}$ and $p_W = \dfrac{4}{17}$

i.e. **the probability of drawing a red ball is $\frac{8}{17}$, a blue ball $\frac{5}{17}$ and a white ball $\frac{4}{17}$.**

Further problems on simple probability may be found in Section 3 (Problems 1–8).

2. Laws of probability

The addition law of probability

This law applies to mutually exclusive events and is usually recognised by the words **'either . . . or'**. It states that when two events are mutually exclusive, if the probability of the first event happening is p_1 and the probability of the second event happening is p_2, then the probability of **either** the first **or** the second event happening is $p_1 + p_2$. For example, fifteen cards are marked from 1 to 15 and one is drawn at random. It is required to determine the probability of the card selected being a multiple of 2 or 3. The probability of the card being a multiple of 2 is $\frac{7}{15}$ (given by the numbers 2, 4, 6, 8, 10, 12 and 14). The probability of the card being a multiple of 3 is $\frac{3}{15}$ (given by the numbers 3, 9 and 15, the numbers 6 and 12 already having been selected). Hence the probability of the card being a multiple of **either 2 or 3 is**

$$\frac{7}{15} + \frac{3}{15}, \text{ that is, } \frac{10}{15} \text{ or } \frac{2}{5}.$$

Similarly for n mutually exclusive events which can be related by the addition law, where n is a positive integer, the total probability is $p_1 + p_2 + p_3 + \ldots + p_n$, the events being linked by the words **or.**

The multiplication law of probability

This law applies to both dependent and independent events and is usually recognised by the words **'both . . . and'**. It states that when the probability of one event happening is p_1 and the probability of a second event happening is p_2, then the probability of **both** the first **and** the second events happening is $p_1 \times p_2$. This law is derived as follows. Assuming the first event can happen in a total of n_1 ways, of which a_1 are successful, and the second event can happen in a total of n_2 ways, of which a_2 are successful, then the probability of both the first and the second events happening is given by combining each successful first event with each successful second event, i.e. $a_1 \times a_2$. For example, if there are two successful first events, say $a_1 = 1$ and 2, and three successful second events, say $a_2 = 3$, 4 and 5, then the probability of both the first and second events happening is given by the combinations 1 and 3, 1 and 4, 1 and 5, 2 and 3, 2 and 4, 2 and 5, i.e. 6 combinations given by $a_1 \times a_2$. Also, the total possible number of occurrences is $n_1 \times n_2$, i.e. each of the number of ways the first event can happen combined with each of the number of ways the second event can happen. Hence, the probability of both the first and the second event happening is

$$p = \frac{a_1 \times a_2}{n_1 \times n_2} = p_1 \times p_2.$$

As an example of the multiplication law, it is required to determine the probability of drawing two white balls in succession from a bag containing six red and four white balls without replacement. Since the problem can be expressed as drawing **both** one white ball **and** another white ball, the multiplication law is indicated. The probability p_1 of drawing one white ball on the first draw is

$$p_1 = \frac{\text{number of white balls}}{\text{the total number of balls}} = \frac{4}{10} = \frac{2}{5}$$

The number of white balls is now reduced by 1, to 3, and the total number of balls reduced by 1, to 9. If p_2 is the probability of drawing one white ball on the second draw, then

$$p_2 = \frac{\text{number of white balls}}{\text{the total number of balls}} = \frac{3}{9} = \frac{1}{3}.$$

Applying the multiplication law gives:

$$\text{probability of drawing two white balls} = p_1 \times p_2 = \frac{2}{5} \times \frac{1}{3} = \frac{2}{15}.$$

Similarly for n events which can be related by the multiplication law, the total probability is $p_1 \times p_2 \times p_3 \times \ldots \times p_n$, the events being linked by the words **and.**

Problem 1. Two balls are drawn in turn with replacement from a bag containing 8 red balls, 15 white balls, 24 black balls and 17 orange balls. Determine the probabilities of having:

(a) two red balls;
(b) a red and a white ball;
(c) no orange balls;
(d) a black and red or a black and orange ball;
(e) at least one black ball;
(f) at most one orange ball; and
(g) a white ball on the first draw but the second ball not white.

(a) The probability of drawing a red ball on the first draw is the ratio: $\frac{\text{number of red balls}}{\text{total number of balls}}$, i.e. $\frac{8}{8+15+24+17} = \frac{8}{64} = \frac{1}{8}$.

Since drawing is with replacement, the red ball is now returned to the bag and a second draw made. The probability of drawing a red ball on the second draw is again $\frac{1}{8}$. Thus the probability of drawing **both** a red ball on the first draw **and** a red ball on the second draw is $\frac{1}{8} \times \frac{1}{8} = \frac{1}{64}$, the **'both . . . and'** indicating the muliplication law. That is, the probability of selecting two red balls is $\mathbf{\frac{1}{64}}$, with replacement.

(b) The probability of drawing a red ball on the first draw is $\frac{1}{8}$ (see part (a)). The probability of drawing a white ball on the second draw is $\frac{15 \text{ (white balls)}}{64 \text{ (total number of balls)}}$.

Hence the probability of drawing **both** a red ball on the first draw **and** a white ball on the second draw is $\frac{1}{8} \times \frac{15}{64} = \frac{15}{512}$.

The probability of having a red and white ball can also be achieved by drawing a white ball first and a red ball second. Hence the probability of drawing **both** a white ball on the first draw **and** a red ball on the second draw is $\frac{15}{64} \times \frac{1}{8} = \frac{15}{512}$.

The probability of drawing **either** a red then white ball **or** a white then red ball is $\frac{15}{512} + \frac{15}{512}$ (the 'either . . . or' indicating the addition law), i.e. $\frac{15}{256}$. That is, the probability of having a red ball and a white ball is $\mathbf{\frac{15}{256}}$, with replacement.

(c) The probability of having no orange balls really means the probability of drawing a red, white or black ball on both the first and second draws. The probability of drawing a red, white or black ball is $\frac{8 + 15 + 24}{64}$, that is $\frac{47}{64}$.

Hence the probability of drawing red, white and black balls on **both** the first **and** the second draws is $\frac{47}{64} \times \frac{47}{64}$, i.e. $\frac{2\,209}{4\,096}$ since drawing is with replacement. That is, the probability of drawing no orange balls is $\mathbf{\frac{2\,209}{4\,096}}$, with replacement.

(d) The probability of having a black and red is $\frac{24}{64} \times \frac{8}{64} + \frac{8}{64} \times \frac{24}{64}$, that is, $\frac{192}{2\,048}$ (see part (b)), or $\frac{3}{32}$.

The probability of having a black and orange is $\frac{24}{64} \times \frac{17}{64} + \frac{17}{64} \times \frac{24}{64} = \frac{408}{2\,048}$ (see part (b)), or $\frac{51}{256}$.

Hence the probability of having **either** a black and red **or** a black and orange is $\frac{3}{32} + \frac{51}{256}$, i.e. $\frac{75}{256}$. Thus, the probability of having a black and red or a black and orange ball is $\mathbf{\frac{75}{256}}$, with replacement.

(e) The outcome of at least one black ball can be achieved by drawing a black and a non-black ball, a non-black ball and a black ball, or by drawing two black balls.

The probability of drawing a black and a non-black is $\frac{24}{64} \times \frac{40}{64}$, i.e. $\frac{15}{64}$. Since drawing is with replacement, the probability of drawing a non-black and a black ball is also $\frac{15}{64}$.

The probability of drawing **either** a black and non-black **or** a non-black and black ball is $\frac{15}{64} + \frac{15}{64}$, i.e. $\frac{15}{32}$. The probability of drawing two black balls is $\frac{24}{64} \times \frac{24}{64}$, i.e. $\frac{9}{64}$. Thus the probability of drawing **either** (a black and non-black or a non-black and black) **or** two black balls is $\frac{15}{32} + \frac{9}{64}$, i.e. $\frac{39}{64}$. Thus, the probability of drawing at least one black ball is $\mathbf{\frac{39}{64}}$, with replacement.

(f) The possibilities for at most one orange ball are: (i) an orange ball on the first draw and a non-orange ball on the second draw; (ii) a non-orange ball on the first draw and an orange ball on the second draw; or (iii) no orange balls at all.

The probability of (i) is $\frac{17}{64} \times \frac{47}{64}$, i.e. $\frac{799}{4\,096}$

The probability of (ii) is $\frac{47}{64} \times \frac{17}{64}$, i.e. $\frac{799}{4\,096}$

The probability of (iii) is $\frac{47}{64} \times \frac{47}{64}$, i.e. $\frac{2\,209}{4\,096}$

The probability of (i) **or** (ii) **or** (iii) is $\frac{799 + 799 + 2\,209}{4\,096}$, i.e. $\frac{3\,807}{4\,096}$. That is, the probability of having at most one orange ball is $\mathbf{\frac{3\,807}{4\,096}}$, with replacement.

(g) The probability of having a white ball on the first draw and one non-white ball on the second is $\frac{15}{64} \times \frac{49}{64}$, i.e. $\mathbf{\frac{735}{4\,096}}$.

Problem 2. The probability of three events happening are $\frac{1}{8}$ for event A, $\frac{1}{5}$ for event B, and $\frac{2}{7}$ for event C. Determine:

(a) the probability of all three events happening;
(b) the probability of event A and B but not C happening;
(c) the probability of only event B happening; and
(d) the probability of event A or event B happening but not event C.

Let p_A, p_B and p_C be the probabilities of events A, B and C happening respectively. Let $\overline{p_A}$, $\overline{p_B}$ and $\overline{p_C}$ be the probabilities of those events not happening. Then:

$$p_A = \frac{1}{8}, p_B = \frac{1}{5}, p_C = \frac{2}{7}, \overline{p_A} = \frac{7}{8}, \overline{p_B} = \frac{4}{5} \text{ and } \overline{p_C} = \frac{5}{7},$$

since the probability of an event happening plus the probability of it not happening must be unity, i.e. $p_A + \overline{p_A} = 1$, etc.

(a) The probability of all three events happening is the same as the probability of A **and** B **and** C, i.e. the multiplication law is indicated. Thus:

$p_A \times p_B \times p_C = \frac{1}{8} \times \frac{1}{5} \times \frac{2}{7} = \frac{1}{140}$.

That is, the probability that all three events will happen is $\mathbf{\frac{1}{140}}$. An example of this is that if there are three commonly occurring faults during the manufacture of an article and these occur with the probabilities shown as p_A, p_B and p_C, then the probability of all three of these faults occurring is any one article is $\mathbf{\frac{1}{140}}$.

(b) The probability of **both** A **and** B happening is $p_A \times p_B = \frac{1}{8} \times \frac{1}{5} = \frac{1}{40}$.

The probability of **both** (A and B) happening **and** C not happening is $p_A \times p_B \times \overline{p_C}$, i.e. $\frac{1}{8} \times \frac{1}{5} \times \frac{5}{7}$, i.e. $\frac{1}{56}$.

That is, the probability of A and B but not C happening is $\mathbf{\frac{1}{56}}$.

(c) The probability of only event B happening means that **both** event A and event C are not happening **and** event B is happening.

The probability of this occurring is given by $\overline{p_A} \times p_B \times \overline{p_C}$, i.e. $\frac{7}{8} \times \frac{1}{5} \times \frac{5}{7}$, that is, $\frac{1}{8}$.

Thus, the probability of only event B happening is $\mathbf{\frac{1}{8}}$.

(d) The probability of event A **or** event B happening is $(p_A + p_B)$.

The probability of (event A or event B happening) **and** event C not happening is $(p_A + p_B) \times \overline{p_C}$. That is, $\overline{p_C}(p_A + p_B)$. Then,

$$\overline{p_C}(p_A + p_B) = \frac{5}{7}\left(\frac{1}{8} + \frac{1}{5}\right)$$

$$= \frac{5}{7} \times \frac{13}{40}$$

$$= \frac{13}{56}$$

i.e. the probability of event A or event B but not event C happening is $\mathbf{\frac{13}{56}}$.

Problem 3. One bag contains 3 red and 5 black marbles and a second bag contains 4 green and 7 white marbles. One marble is drawn from the first bag and two marbles from the second bag, without replacement. Determine the probability of having:

(a) one red and two white marbles;

(b) no green marbles; and

(c) either one black and two green or one black and two white marbles.

(a) The probability of having one red marble on the first draw is $\frac{3}{8}$.

The probability of having one white marble on the second draw is $\frac{7}{11}$.

The probability of having one white marble on the third draw is $\frac{6}{10}$, since there is no replacement. Hence the probability of having one red **and** one white **and** one white marble on successive draws is $\frac{3}{8} \times \frac{7}{11} \times \frac{6}{10} = \frac{63}{440}$.

i.e. the probability of having one red and two white marbles is $\mathbf{\frac{63}{440}}$, without replacement.

(b) The probability of having no green marbles on the first draw is unity, since the first bag contains no green marbles.

The probability of drawing no green marbles on the second draw is the probability of drawing a white marble (since the second bag only contains green and white marbles), i.e. $\frac{7}{11}$.

The probability of drawing no green marbles on the third draw is again the probability of drawing a white marble, i.e. $\frac{6}{10}$, since there is no replacement.

Hence the probability of drawing no green marbles on the first draw **and** no green marbles on the second draw **and** no green marbles on the third draw is $\frac{1}{1} \times \frac{7}{11} \times \frac{6}{10}$, i.e. $\frac{21}{55}$.

That is, the probability of having no green marbles is $\frac{21}{55}$, without replacement.

(c) The probability of one black and two green marbles, without replacement, is $\frac{5}{8} \times \frac{4}{11} \times \frac{3}{10}$ (see part (a)), i.e. $\frac{3}{44}$.

The probability of one black and two white marbles, without replacement, is $\frac{5}{8} \times \frac{7}{11} \times \frac{6}{10}$ (see part (a)), i.e. $\frac{21}{88}$.

Thus, the probability of **either** one black and two green marbles **or** one black and two white is $\frac{3}{44} + \frac{21}{88}$, i.e. $\frac{27}{88}$.

Further problems on the laws of probability may be found in the following section (3) (Problems 9–21).

3. Further problems

Simple probability

1. A box contains 132 rivets of which 23 are undersized, 47 are oversized and 62 are satisfactory. Determine the probability of drawing at random: (a) one undersized; (b) one oversized; and (c) one satisfactory rivet from the box. (a) $\left[\frac{23}{132}\right]$ (b) $\left[\frac{47}{132}\right]$ (c) $\left[\frac{31}{66}\right]$
2. Four hundred resistors are examined and 6 per cent are found to be defective. Determine the probability that one selected at random will be defective and also the probability that it will not be defective. $\left[\frac{3}{50}, \frac{47}{50}\right]$
3. A purse contains 7 copper and 13 silver coins. Determine the probability of selecting a copper coin when one is taken at random. $\left[\frac{7}{20}\right]$
4. Determine the probability of winning a prize in a raffle by buying 3

tickets, when there are 7 prizes and a total of 450 tickets are sold. $\left[\dfrac{7}{150}\right]$

5. Determine the probability of drawing a multiple of 3 from 50 cards marked from 1 to 50. $\left[\dfrac{8}{25}\right]$
6. Determine the probability of an event not happening when the probability of it happening is $\dfrac{7}{93}$. $\left[\dfrac{86}{93}\right]$
7. The probability of winning a prize in a raffle by buying 4 tickets is $\dfrac{1}{300}$. Find out how many tickets were sold in a raffle having 5 prizes. [6 000]
8. A batch of 700 components contains 16 defective ones. Determine the probability of selecting at random one defective item. $\left[\dfrac{4}{175}\right]$

Laws of probability

Problems 9–14 refer to a box that contains 131 similar transistors, of which 70 are satisfactory, 43 give too high a gain under normal operating conditions and 18 give too low a gain. Determine the probabilities stated.

9. The probability when drawing two transistors in turn, at random, with replacement, of having: (a) two satisfactory; (b) none with low gain; (c) one satisfactory and one with high gain; (d) one with low gain and none satisfactory.

   (a) $\left[\dfrac{4\,900}{17\,161}\right]$ (b) $\left[\dfrac{12\,769}{17\,161}\right]$ (c) $\left[\dfrac{6\,020}{17\,161}\right]$ (d) $\left[\dfrac{1\,548}{17\,161}\right]$
10. Determine the probabilities required in Problem 9, but with no replacement between draws.

    (a) $\left[\dfrac{483}{1\,703}\right]$ (b) $\left[\dfrac{6\,328}{8\,515}\right]$ (c) $\left[\dfrac{602}{1\,703}\right]$ (d) $\left[\dfrac{1\,089}{8\,515} \text{ or } \dfrac{2\,117}{17\,030}\right]$
11. The probability, when drawing two transistors at random, with replacement, of having: (a) two satisfactory or two with high gain; (b) the second draw being not low gain; and (c) at least one with high gain.

    (a) $\left[\dfrac{6\,749}{17\,161}\right]$ (b) $\left[\dfrac{113}{131}\right]$ (c) $\left[\dfrac{9\,417}{17\,161}\right]$
12. Determine the probabilities required in Problem 11, but with no replacement between draws. (a) $\left[\dfrac{3\,318}{8\,515}\right]$ (b) either $\left[\dfrac{112}{130} \text{ or } \dfrac{113}{130}\right.$, depending on the gain of the first transistor drawn$\left.\right]$ (c) $\left[\dfrac{4\,687}{8\,515}\right]$
13. The probability, when drawing two transistors at random, with replacement, of having: (a) two satisfactory or one with high gain and one with low gain; (b) at most one with low gain; and (c) one satisfactory and one

with low gain or one satisfactory and one with high gain.

(a)$\left[\dfrac{6\,448}{17\,161}\right]$ (b)$\left[\dfrac{16\,837}{17\,161}\right]$ (c)$\left[\dfrac{8\,540}{17\,161}\right]$

14. Determine the probabilities required in Problem 13, but with no replacement between draws. . (a)$\left[\dfrac{3\,189}{8\,515}\right]$ (b)$\left[\dfrac{8\,362}{8\,515}\right]$ (c)$\left[\dfrac{854}{1\,703}\right]$

15. A box contains 13 tickets numbered from 1 to 13. Two tickets are drawn from the box, one at a time, with replacement. Determine the probabilities that they are: (a) odd then even; and (b) even then odd numbers.

(a)$\left[\dfrac{42}{169}\right]$ (b)$\left[\dfrac{42}{169}\right]$

16. Determine the probabilities required in Problem 15, but with no replacement between draws. (a)$\left[\dfrac{7}{26}\right]$ (b)$\left[\dfrac{7}{26}\right]$

17. The probabilities of an engine failing are given by: p_1, failure due to overheating; p_2, failure due to ignition problems; p_3, failure due to fuel blockage. When $p_1 = \dfrac{1}{7}$, $p_2 = \dfrac{2}{9}$ and $p_3 = \dfrac{3}{11}$, determine the probabilities of:

(a) both p_1 and p_2 happening;
(b) either p_2 or p_3 happening; and
(c) both p_1 and either p_2 or p_3 happening.

(a) $\left[\dfrac{2}{63}\right]$ (b)$\left[\dfrac{49}{99}\right]$ (c)$\left[\dfrac{7}{99}\right]$

18. Acturial tables show that the life expectancy of three men, A, B and C, over a twenty-year period depends on their age and is given by $p_A = \dfrac{4}{15}$, $p_B = \dfrac{11}{15}$ and $p_C = \dfrac{14}{15}$. Determine the probabilities that in twenty years:

(a) all three men will be alive;
(b) A will be alive but B and C will be dead;
(c) at least one man will be alive.

(a)$\left[\dfrac{616}{3\,375}\right]$ (b)$\left[\dfrac{16}{3\,375}\right]$ (c)$\left[\dfrac{3\,331}{3\,375}\right]$

19. Bag A contains 13 white and 15 black marbles. Bag B contains 17 green and 19 red marbles. Three marbles are drawn at random, one from bag A and two from bag B, with replacement. Determine the probabilities of having:

(a) one white, one green and one red marble;
(b) no black marbles and two green marbles;
(c) either one white and two green or one white, one red and one green marble.

(a)$\left[\dfrac{8\,398}{36\,288}\right]$ (b)$\left[\dfrac{3\,757}{36\,288}\right]$ (c)$\left[\dfrac{12\,155}{36\,288}\right]$

20. Determine the probabilities required in Problem 19, but with no replacement between draws. (a)$\left[\dfrac{8\,398}{35\,280}\right]$ (b)$\left[\dfrac{221}{2\,205}\right]$ (c)$\left[\dfrac{5\,967}{17\,640}\right]$

21. Three types of seed are planted with a chance of growth of $\frac{1}{8}, \frac{1}{5}, \frac{1}{10}$. What are the probabilities that they all grow, that only one type grows, and that at least one type grows? $\left[\frac{1}{400}, \frac{127}{400}, \frac{37}{100}\right]$

Chapter 11

The binomial and poisson probability distributions

1. Probability distributions

Frequency distributions are discussed in *Technician Mathematics, Level 2*, Chapter 8, in which it is shown how data can be grouped into classes containing the variable and its frequency. These distributions can be readily analysed to give a better understanding of the data by means of histograms, frequency polygons and numerical values, like measures of central tendency and deviation. The relative frequency of a class is the ratio of the frequency of the class to the total frequency of all classes, and hence probability and relative frequency are closely linked. It follows that information on probabilities can be presented in a similar way to information on frequencies. These data then form a **probability distribution** and contain information on the variable and its probability. A probability distribution can also be represented pictorially by means of a histogram and frequency polygon. A **discrete** probability distribution is one in which the values of probability can have certain values only and will have 'steps' in any pictorial representation. A **continuous** probability distribution can have any value between certain limits and will be a smooth curve when represented pictorially.

Three of the principal theoretical probability distributions are introduced in this book. Two discrete probability distributions called the binomial distribution and the Poisson distribution are introduced in this chapter and a continuous distribution called the normal distribution is introduced in Chapter 12.

2. The binomial distribution

As the name implies, the binomial distribution deals with 'two numbers' only and these are often taken as the probability that an event will happen, p (called the probability of success), and the probability that an event will not happen, q (called the probability of failure). In this context, $p + q$ must be equal to unity. The binomial distribution is the basis of much of the statistical work done in industrial inspection and in research.

Suppose that a large number of balls are placed in a bag, 10 per cent of them being red and the remainder black. Let the number be large enough so that if drawing takes place with replacement or without replacement, the result is almost the same. The probability p of drawing a red ball is $\frac{1}{10}$. The probability of drawing two red balls (**both** a red ball **and** another red ball) is given by the multiplication law of probability and is $\frac{1}{10} \times \frac{1}{10}$, i.e. $\frac{1}{100}$. It follows that the probability of drawing n red balls is $\left(\frac{1}{10}\right)^n$. Similarly, the probability of drawing n black balls is $\left(\frac{9}{10}\right)^n$. These probabilities can be readily determined, but to determine the probability that when 7 balls are drawn, 3 will be red and the remainder black is far more difficult and it is this type of problem which can be solved using the binomial distribution.

Let R signify a red ball and B a black ball. The various possibilities which exist when drawing two balls from the bag are: two reds, a red and a black, a black and red and two blacks, i.e.

R R, R B, B R, B B

The probability of drawing two red balls is p^2, a red and black ball is pq, and so on, and summarising the results when two balls are drawn at random:

Result	R R	R B or B R	B B
Probability	p^2	$2pq$	q^2

When three balls are drawn at random from the bag, the various possibilities are as shown below, together with their probabilities.

R R R	p^3
R R B	p^2q
R B R	p^2q
R B B	pq^2
B R R	p^2q
B R B	pq^2
B B R	pq^2
B B B	q^3

 Summary:

Result	3R	2R, 1B	1R, 2B	3B
Probability	p^3	$3p^2q$	$3pq^2$	q^3

When the same procedure is carred out for four balls drawn at random, the summary of results is:

Result	4R	3R, 1B	2R, 2B	1R, 3B	4B
Probability	p^4	$4p^3q$	$6p^2q^2$	$4pq^3$	q^4

Comparing these summaries with the results obtained in Chapter 2 (the binomial expansion) for expansions of $(a+b)^2$, $(a+b)^3$ and $(a+b)^4$, we can see that the terms obtained in the probability summaries are of the same form as those obtained when carrying out a binomial expansion.

In practical statistics, for example, sampling in industry for inspection purposes, red balls in a bag correspond to defective items in a large batch of items and the black balls correspond to items which are not defective. In this case the binomial probability distribution is defined as:

The probability that 0, 1, 2, 3, . . . defective items in a sample of n *items drawn at random from a large population, whose probability of defective items is* p *and whose probability of non-defective items is* q, *is given by the successive terms of the expansion of* (q + p)n, *taking terms in succession from left to right.*

For example, a certain machine is producing, say, 1 000 components per hour, and batches of 10 are taken at random for inspection purposes every 30 minutes. By inspecting these components, it is possible to predict over a period of time, the defect rate for the machine. Let this be, say, 15 per cent, then p is $\frac{15}{100}$ or $\frac{3}{20}$. The probability q of a non-defective item, is $1 - \frac{3}{20}$, i.e. $\frac{17}{20}$. If a sample of, say, 4 components is selected at random from the output of the machine, the probability of having 0, 1, 2, 3 or 4 defective components is given by the successive terms of the expansion of $(q+p)^4$ taken from left to right. Applying the binomial expansion to $(q+p)^4$ gives:

$$(q+p)^4 = q^4 + 4q^3p + 6q^2p^2 + 4qp^3 + p^4,$$

and substituting $p = \frac{3}{20}$ and $q = \frac{17}{20}$ gives:

$$\left(\frac{17}{20}+\frac{3}{20}\right)^4 = \left(\frac{17}{20}\right)^4 + 4\left(\frac{17}{20}\right)^3\left(\frac{3}{20}\right) + 6\left(\frac{17}{20}\right)^2\left(\frac{3}{20}\right)^2 + 4\left(\frac{17}{20}\right)\left(\frac{3}{20}\right)^3 + \left(\frac{3}{20}\right)^4$$

$$= 0.522 + 0.368 + 0.098 + 0.011 + 0.001$$

when these values are determined correct to 3-decimal-place accuracy. This result shows that:

Number defective	0	1	2	3	4
Probability	0.522	0.368	0.098	0.011	0.001

This result means that when drawing random samples of 4 components from the output of the machine when it has a defect production rate of 15 per cent, it is likely that when drawing, say 100 such samples, in 52 samples there will be no defective items, in 37 samples there will be one defective item, in 10 samples there will be two defective items, and so on. Providing these probabilities remain reasonably constant, it can be predicted that the production defect rate for the machine is remaining reasonably constant.

A more general statement of the binomial probability distribution is:

If p *is the probability that an event will happen and* q *the probability that it will not happen, then the probability that the event will happen 0, 1, 2, 3, . . . times in* n *trials is given by the successive terms of the expansion of* (q + p)n, *taken from left to right.*

It is shown in Chapter 2 that the terms comprising the expansion of $(q + p)^n$ can either be obtained using Pascal's triangle or by using the general binomial expansion:

$$(q + p)^n = q^n + nq^{n-1}p + \frac{n(n-1)}{2!}q^{n-2}p^2 + \ldots$$

An example of this more general application is to determine the probability of a three-child family having, say, 2 boys, assuming the probability of birth of a boy and girl are the same. The probability of a boy being born is $\frac{1}{2}$ and the probability of a girl being born is also $\frac{1}{2}$. The probability of 0, 1, 2 or 3 boys being born to a family of 3 children is given by the successive terms of the expansion of $(q + p)^3$ taken from left to right.

$$(q + p)^3 = q^3 + 3q^2p + 3qp^2 + p^3$$

and substituting $q = \frac{1}{2}$ and $p = \frac{1}{2}$ gives:

$$\left(\frac{1}{2} + \frac{1}{2}\right)^3 = \left(\frac{1}{2}\right)^3 + 3\left(\frac{1}{2}\right)^2\left(\frac{1}{2}\right) + 3\left(\frac{1}{2}\right)\left(\frac{1}{2}\right)^2 + \left(\frac{1}{2}\right)^3$$

$$= \frac{1}{8} \quad + \quad \frac{3}{8} \quad + \quad \frac{3}{8} \quad + \quad \frac{1}{8}$$

and these terms give the probabilities of	0 boys	1 boy	2 boys	3 boys

i.e. the probability of there being 2 boys in a 3-child family is $\frac{3}{8}$.

The results of a binomial probability distribution can be represented pictorially by drawing a histogram, and the histograms for the machine having 15 per cent defective items and for the number of boys in a 3-child family are shown in Figs 1 (a) and 1 (b) respectively.

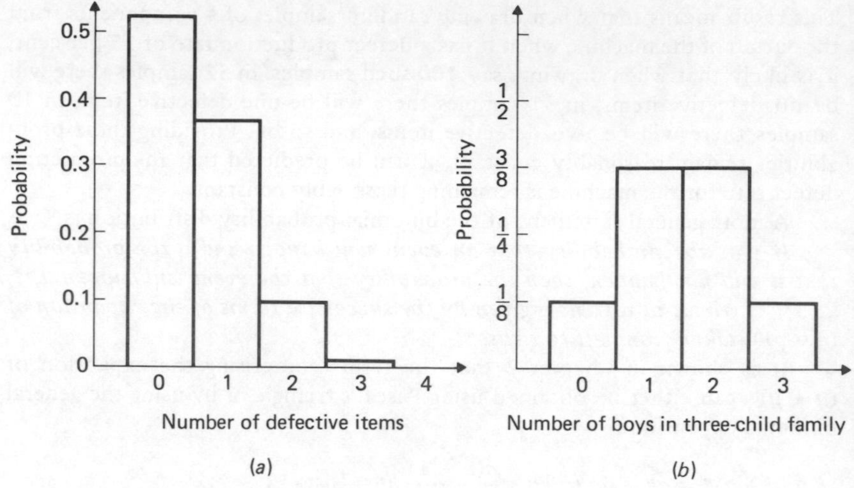

Figure 1 Histograms of binomial probability distribution

Worked problems on the binomial distribution

Problem 1. A bag contains a large number of marbles of which 25 per cent are white and the remainder green. Five marbles are drawn from the bag at random. Determine the probability of having:

(a) 2 white and 3 green marbles;
(b) at least 3 white marbles; and
(c) not more than 3 green marbles.

Let p be the probability of having a white marble and q be the probability of not having a white marble, i.e. the probability of having a green marble. The sample number n is 5. Then the probability of having 0, 1, 2, 3, 4 or 5 white marbles is given by the successive terms of the expansion of $(q + p)^5$ taken from left to right. Using Pascal's triangle to obtain the expansion of $(q + p)^5$ (see Chapter 2) gives:

$$(q + p)^5 = q^5 + 5q^4p + 10q^3p^2 + 10q^2p^3 + 5qp^4 + p^5.$$

Substituting $q = \frac{3}{4}$ and $p = \frac{1}{4}$ gives:

$$\left(\frac{3}{4} + \frac{1}{4}\right)^5 = \left(\frac{3}{4}\right)^5 + 5\left(\frac{3}{4}\right)^4\left(\frac{1}{4}\right) + 10\left(\frac{3}{4}\right)^3\left(\frac{1}{4}\right)^2 + 10\left(\frac{3}{4}\right)^2\left(\frac{1}{4}\right)^3 + $$
$$+ 5\left(\frac{3}{4}\right)\left(\frac{1}{4}\right)^4 + \left(\frac{1}{4}\right)^5$$

and summarising:

Term	1st	2nd	3rd	4th	5th	6th
Marbles	0W, 5G	1W, 4G	2W, 3G	3W, 2G	4W, 1G	5W, 0G
Probability	$\left(\frac{3}{4}\right)^5$	$5\left(\frac{3}{4}\right)^4\left(\frac{1}{4}\right)$	$10\left(\frac{3}{4}\right)^3\left(\frac{1}{4}\right)^2$	$10\left(\frac{3}{4}\right)^2\left(\frac{1}{4}\right)^3$	$5\left(\frac{3}{4}\right)\left(\frac{1}{4}\right)^4$	$\left(\frac{1}{4}\right)^5$

(a) The probability of having 2 white and 3 green marbles is given by the 3rd term of the expansion, i.e. $10\left(\frac{3}{4}\right)^3\left(\frac{1}{4}\right)^2$, i.e. $\frac{10 \times 3^3}{4^5}$ or $\frac{135}{512}$.

(b) The probability of having at least 3 white marbles is made up of the sum of the 4th, 5th and 6th terms.

The 4th term $= 10\left(\frac{3}{4}\right)^2\left(\frac{1}{4}\right)^3 = \frac{10 \times 3^2}{4^5} = \frac{90}{1\,024}$

The 5th term $= 5\left(\frac{3}{4}\right)\left(\frac{1}{4}\right)^4 = \frac{5 \times 3}{4^5} = \frac{15}{1\,024}$

The 6th term $= \left(\frac{1}{4}\right)^5 = \frac{1}{1\,024}$

Hence the sum of these terms is $\frac{90 + 15 + 1}{1\,024} = \frac{106}{1\,024} = \frac{53}{512}$.

(c) The probability of having not more than 3 green marbles is given by the sum of the 3rd, 4th, 5th and 6th terms. Since the sum of all the terms is unity and the values of the 3rd and subsequent terms have been previously calculated, then the required value is

$\frac{53}{512} + \frac{135}{512}$, these terms being the solutions to parts (b) and (a).

i.e. $\frac{188}{512}$ or $\frac{47}{128}$.

Problem 2. A machine produces 20 per cent defective components. In a sample of 6, drawn at random, determine the probability that:
(a) there will be 4 defective items;
(b) there will be not more than 3 defective items; and
(c) all the items will be non-defective.

Let p be the probability of a component being defective, then $p = \frac{1}{5}$. Also, let q be the probability of a component not being defective, then $q = \frac{4}{5}$. The probability of 0, 1, 2, 3, 4, 5 or 6 defective items in a random sample of 6 items is given by the successive terms of the expansion of $(q + p)^6$, taken from left to right. Using Pascal's triangle to expand $(q + p)^6$ gives:

$$(q+p)^6 = q^6 + 6q^5p + 15q^4p^2 + 20q^3p^3 + 15q^2p^4 + 6qp^5 + p^6$$

and corresponds to: 0 defectives, 1 defective, 2 defectives, 3 defectives, 4 defectives, 5 defectives, 6 defectiv[es]

The problem requires only information on the first 5 terms of the expansion, so only these will be considered.

Substituting for q and p gives:

$$\left(\frac{4}{5}+\frac{1}{5}\right)^6 = \left(\frac{4}{5}\right)^6 + 6\left(\frac{4}{5}\right)^5\left(\frac{1}{5}\right) + 15\left(\frac{4}{5}\right)^4\left(\frac{1}{5}\right)^2 + 20\left(\frac{4}{5}\right)^3\left(\frac{1}{5}\right)^3 + 15\left(\frac{4}{5}\right)^2\left(\frac{1}{5}\right)^4 + \ldots$$

Summarising:

Term	1st	2nd	3rd	4th	5th
Defectives	0	1	2	3	4
Probability	$\left(\frac{4}{5}\right)^6$	$6\left(\frac{4}{5}\right)^5\left(\frac{1}{5}\right)$	$15\left(\frac{4}{5}\right)^4\left(\frac{1}{5}\right)^2$	$20\left(\frac{4}{5}\right)^3\left(\frac{1}{5}\right)^3$	$15\left(\frac{4}{5}\right)^2\left(\frac{1}{5}\right)^4$

(a) The probability of 4 defective items is given by the 5th term and is

$$15\left(\frac{4}{5}\right)^2\left(\frac{1}{5}\right)^4, \text{ i.e. } \frac{15 \times 4^2}{5^6} \text{ or } \frac{240}{15\,625} \text{ or } \frac{48}{3\,125}.$$

(b) The probability of not more than 3 defective items is given by the sum of the first 4 terms.

$$\text{1st term} = \left(\frac{4}{5}\right)^6 = \frac{4\,096}{15\,625}$$

$$\text{2nd term} = 6\left(\frac{4}{5}\right)^5\left(\frac{1}{5}\right) = \frac{6 \times 4^5}{5^6} = \frac{6\,144}{15\,625}$$

$$\text{3rd term} = 15\left(\frac{4}{5}\right)^4\left(\frac{1}{5}\right)^2 = \frac{3\,840}{15\,625}$$

$$\text{4th term} = 20\left(\frac{4}{5}\right)^3\left(\frac{1}{5}\right)^3 = \frac{1\,280}{15\,625}$$

Hence the sum of the first 4 terms is $\dfrac{4\,096 + 6\,144 + 3\,840 + 1\,280}{15\,625}$

that is, $\dfrac{15\,360}{15\,625}$, or $\dfrac{3\,072}{3\,125}$.

(c) The probability of all items non-defective is given by the first term, and is $\dfrac{4\,096}{15\,625}$ from part (b).

Problem 3. Four hundred families have 4 children each. Assuming equal probabilities for boy and girl births, determine how many families will have: (a) 3 boys; (b) 2 girls; and (c) either 2 boys and 2 girls or 3 boys and 1 girl.

Let the probability of a boy being born be p and that of a girl being born be q. Then $p = q = \frac{1}{2}$. The probability of 0, 1, 2, 3 or 4 boys in a family having 4 children is given by the successive terms of the expansion of $(q + p)^4$, taken from left to right.

$$(q + p)^4 = q^4 + 4q^3p + 6q^2p^2 + 4qp^3 + p^4$$

Number of boys 0 1 2 3 4

(a) The probability of having 3 boys is given by the 4th term.

$$4qp^3 = 4\left(\frac{1}{2}\right)\left(\frac{1}{2}\right)^3 = \frac{4}{16} = \frac{1}{4}.$$

(b) The probability of having 2 girls is the same as the probability of having 2 boys and is given by the 3rd term of the expansion.

$$6q^2p^2 = 6\left(\frac{1}{2}\right)^2\left(\frac{1}{2}\right)^2 = \frac{6}{16} = \frac{3}{8}.$$

(c) The probability of having 2 boys and 2 girls or 3 boys and 1 girl will be the sum of the 3rd and 4th terms. From parts (a) and (b),

this is $\frac{1}{4} + \frac{3}{8}$, i.e. $\frac{5}{8}$.

Further problems on the binomial probability distribution may be found in Section 4 (Problems 1–15).

3. The Poisson distribution

The calculations associated with the binomial distribution become very laborious when the sample number n becomes larger than about 10. When n is large and the probability p is small, so that the expectation np is less than 5, then a very good approximation to the binomial distribution can be obtained by using another probability distribution called the Poisson distribution, in which the calculations are normally far easier. In addition, the Poisson distribution is also used in its own right to determine probabilities associated with events which cannot be resolved by using a binomial distribution.

The binomial expansion of $(q + p)^n$ is:

$$q^n + nq^{n-1}p + \frac{n(n-1)}{2!}q^{n-2}p^2 + \frac{n(n-1)(n-2)}{3!}q^{n-3}p^3 + \ldots$$

Also in the binomial distribution $q = 1 - p$. Hence when p is small $q \simeq 1$. Making this approximation, the binomial expansion of $(q + p)^n$ becomes:

$$1 + np + \frac{n(n-1)}{2!}p^2 + \frac{n(n-1)(n-2)}{3!}p^3 + \ldots$$

Also, when n is large, terms such as $(n - 1)$, $(n - 2)$, $(n - 3)$, . . . are approximately equal to n, and applying this approximation to the binomial expansion of $(q+p)^n$ gives:

$$1 + np + \frac{n^2p^2}{2!} + \frac{n^3p^3}{3!} + \ldots$$

Writing the expectation, np, as λ gives:

$$1 + \lambda + \frac{\lambda^2}{2!} + \frac{\lambda^3}{3!} + \ldots$$

but it is stated in Chapter 3 that the power series for

$$e^{\lambda} \text{ is } 1 + \lambda + \frac{\lambda^2}{2!} + \frac{\lambda^3}{3!} + \ldots$$

One of the requirements of any probability distribution is that the sum of all the probabilities is equal to unity. Since n and p are in no way related, this is not necessarily the case for the terms making up the expansion of e^{λ}. To meet this condition, both sides of the equation are divided by e^{λ}, giving:

$$\frac{e^{\lambda}}{e^{\lambda}} = 1 = \frac{1}{e^{\lambda}}\left(1 + \lambda + \frac{\lambda^2}{2!} + \frac{\lambda^3}{3!} + \ldots\right)$$

The successive terms are $e^{-\lambda}, \lambda e^{-\lambda}, \dfrac{\lambda^2 e^{-\lambda}}{2!}, \dfrac{\lambda^3 e^{-\lambda}}{3!}, \ldots$

These successive terms taken from left to right give a very good approximation to the binomial distribution when n is large, p is small and λ is the expectation np.

For example, a machine produces 3 per cent defective items. We can determine the probability that there will be, say, two defective items in a sample of 15 items selected at random from the output of the machine by using either: (a) the binomial distribution; or (b) the Poisson approximation to the binomial distribution.

(a) Using the binomial distribution, $p = 0.03$, $q = 1 - p = 0.97$ and $n = 15$. Also

$$(q+p)^{15} = q^{15} + 15q^{14}p + \frac{15 \times 14}{2!}q^{13}p^2 + \ldots$$

these terms giving the probabilities of having 0, 1 or 2 defective items respectively. Taking the third term, the probability of having two defective items is

$\dfrac{15 \times 14}{2!}q^{13}p^2$, and substituting for q and p gives

$\dfrac{15 \times 14}{2!}(0.97)^{13}(0.03)^2$, i.e. $105 \times 0.673\,0 \times 0.000\,9$,

or 0.063 6.

(b) For the Poisson approximation to the binomial distribution, $n = 15$, $p = 0.03$, so $\lambda = np = 0.45$. The probability of having 0, 1, 2, . . . defective items is given by the terms $e^{-\lambda}$, $\lambda e^{-\lambda}$, $\frac{\lambda^2 e^{-\lambda}}{2!}$, . . . respectively. Taking the third term, the probability of two defective items is

$$\frac{\lambda^2 e^{-\lambda}}{2!}, \text{ i.e. } \frac{0.45^2 e^{-0.45}}{2}, \text{ or } 0.064\,6.$$

These results differ by less than 2 per cent and when n becomes 50 or more and np is less than 5, the difference between the binomial and the Poisson approximation to the binomial distribution is barely detectable. It can be seen that the calculations in (b) are easier than those in (a) and the difference in the ease of calculations becomes more noticeable as n becomes larger.

The principal use of the Poisson distribution is to determine the probability of an event happening where the probability of the event not happening is not known. For example, the probability of a particular machine breaking down can be determined by noting the number of times it breaks down in a certain period (this is the expectation). It is not known how many times it did not break down in that period.

The Poisson distribution can be stated as follows:

If the chance of an event occurring at any instant is constant and the expectation of the event occurring in a period of time is λ, then the probability of the event occurring 0, 1, 2, 3, . . . times is given by the successive terms of the expansion of

$$e^{-\lambda}\left(1 + \lambda + \frac{\lambda^2}{2!} + \frac{\lambda^3}{3!} + \ldots\right), \textit{ taken from left to right.}$$

For example, if between 10 and 11 o'clock in the morning, the average number of telephone calls received by the switchboard of a company is 4 per minute, the Poisson distribution can be used to determine the probability that in any particular minute, say, 3 calls will arrive. The number of calls which did not arrive is not known, hence the binomial distribution cannot be used. The probability of receiving 0, 1, 2 or 3 calls is given by the terms $e^{-\lambda}$, $\lambda e^{-\lambda}$, $\frac{\lambda^2 e^{-\lambda}}{2!}$ or $\frac{\lambda^3 e^{-\lambda}}{3!}$, repsectively, where λ is the expectation of a call arriving and is 4 per minute. The probability that there will be 3 calls is given by the 4th term, i.e. $\frac{\lambda^3 e^{-\lambda}}{3!}$. Substituting $\lambda = 4$ gives: $\frac{\lambda^3 e^{-\lambda}}{3!} = \frac{4^3 e^{-4}}{3!} = 0.195$. That is, the probability of 3 calls arriving in any particular minute is 0.195. This type of information can be used to determine the number of lines required by a switchboard in order to keep the probability of a person getting an engaged tone, when ringing the switchboard, to within certain limits.

Worked problems on the Poisson distribution

Problem 1. If 2 per cent of the electric light bulbs produced by a company are defective, determine the probability that in a sample of 60 bulbs: (a) 3 bulbs; (b) not more than 3 bulbs; and (c) at least 2 bulbs, will be defective.

Since n is large and p is small, the Poisson approximation to the binomial distribution can be used. The expectation λ is np, i.e. 60×0.02 or 1.2. The probability of having 0, 1, 2, 3, . . . defective bulbs is given by the terms $e^{-\lambda}$, $\lambda e^{-\lambda}, \frac{\lambda^2 e^{-\lambda}}{2!}, \frac{\lambda^3 e^{-\lambda}}{3!}, \ldots$ respectively.

(a) The probability of having 3 defective bulbs is given by $\frac{\lambda^3 e^{-\lambda}}{3!}$. Substituting for λ gives

$$\frac{\lambda^3 e^{-\lambda}}{3!} = \frac{1.2^3 e^{-1.2}}{3 \times 2} = 0.086\,7$$

That is, **the probability of having 3 defective bulbs is 0.086 7.**

(b) The probability of not more than 3 bulbs being defective is the probability of there being no bulbs, 1 bulb, 2 bulbs and 3 bulbs defective, that is, the sum of the first 4 terms of the Poisson distribution. Now

$$e^{-\lambda} = e^{-1.2} = 0.301\,2$$

$$\lambda e^{-\lambda} = 1.2 \times 0.301\,2 = 0.361\,4$$

$$\frac{\lambda^2 e^{-\lambda}}{2!} = \frac{1.2^2 \times 0.301\,2}{2} = 0.216\,9$$

and $\frac{\lambda^3 e^{-\lambda}}{3!} = 0.086\,7$ from part (a).

Thus, the probability of having not more than 3 bulbs defective is given by the sum of these probabilities, i.e.

$$0.301\,2 + 0.361\,4 + 0.216\,9 + 0.086\,7 \text{ or } 0.966\,2.$$

That is, **the probability of having not more than 3 bulbs defective is 0.966 2.**

(c) The probability that at least 2 bulbs will be defective means two or more of the sample being defective. Since the total probability is unity, the probability of 2 or more being defective is the total probability less the probability of having no defective bulbs and the probability of having one defective bulb, i.e.

$$1 - e^{-\lambda} - \lambda e^{-\lambda}.$$

From part (b) this is $1 - (0.301\,2 + 0.361\,4)$, i.e. 0.337 4.

Thus, **the probability of having two or more defective bulbs is 0.337 4.**

Problem 2. A team scores an average of 3 goals per match in 45 matches. Determine in how many matches they would expect to score: (a) 4 goals; and (b) less than 2 goals, assuming a Poisson distribution.

The probability of scoring 0, 1, 2, 3 or 4 goals is given by the successive terms of the expansion of $e^{-\lambda}\left(1+\lambda+\frac{\lambda^2}{2!}+\frac{\lambda^3}{3!}+\ldots\right)$ taken from left to right.

Hence:

Number of goals scored	0	1	2	3	4
Probability	$e^{-\lambda}$	$\lambda e^{-\lambda}$	$\frac{\lambda^2 e^{-\lambda}}{2!}$	$\frac{\lambda^3 e^{-\lambda}}{3!}$	$\frac{\lambda^4 e^{-\lambda}}{4!}$

The expectation of scoring, λ, is 3 per match, giving the value of $e^{-\lambda}$ as 0.049 8.

(a) The probability of scoring 4 goals in one match is given by $\frac{\lambda^4 e^{-\lambda}}{4!}$,

i.e. $\frac{3^4 \times 0.049\,8}{24}$ or 0.168 1.

Thus the expectation of scoring 4 goals in any of the 45 matches is 0.168 1 × 45, i.e. 7.565, which is 'rounded-up' to **8 matches.**

(b) The probability of scoring less than two goals in one match is the sum of the probabilities of scoring no goals and one goal, i.e. $e^{-\lambda} + \lambda e^{-\lambda}$. Substituting $\lambda = 3$ gives

$$
\begin{aligned}
e^{-\lambda} + \lambda e^{-\lambda} &= 0.049\,8 + 3 \times 0.049\,8 \\
&= 0.199\,2
\end{aligned}
$$

The expectation of scoring less than two goals in any of the 45 matches is 0.199 2 × 45, i.e. 8.964, which is 'rounded-up' to **9 matches.**

Problem 3. Special drills of a certain diameter are kept in a machine shop store. A survey reveals that they are drawn from the store, on average, 1.5 times a day. Determine for a period of 200 working days:
(a) the number of days when none of the drills are used; and
(b) the number of days when four of the drills are in use, assuming the demand follows a Poisson distribution.

The probability of the demand being for 0, 1, 2, 3, 4, . . . drills is given by the successive terms of the expansion of $e^{-\lambda}\left(1+\lambda+\frac{\lambda^2}{2!}+\frac{\lambda^3}{3!}+\frac{\lambda^4}{4!}+\ldots\right)$, taken from left to right. Hence:

Demand for drills	0	1	2	3	4
Probability	$e^{-\lambda}$	$\lambda e^{-\lambda}$	$\frac{\lambda^2 e^{-\lambda}}{2!}$	$\frac{\lambda^3 e^{-\lambda}}{3!}$	$\frac{\lambda^4 e^{-\lambda}}{4!}$

The daily expectation of there being a demand, λ, is 1.5, giving a value of $e^{-\lambda}$ of 0.223 1.

(a) The probability of there being a demand for none of the drills in a day is given by $e^{-\lambda}$, i.e. 0.223 1. In 200 days, it is probable that no drills are used on 200 × 0.223 1 = 44.62, i.e. **45 days** when 'rounded-up'.

(b) The probability of there being a demand for four drills on any day is given by $\frac{\lambda^4 e^{-\lambda}}{4!}$, i.e.

$$\frac{(1.5^4)(0.223\ 1)}{(4)(3)(2)(1)} = 0.047\ 06$$

In 200 days, it is probable that four drills are used on 200 × 0.047 06 = 9.412, i.e. **10 days** when 'rounded-up'.

Further problems on the Poisson distribution may be found in the following section (4) (Problems 16–28).

4. Further problems

Binomial distribution

Problems 1–5 refer to a box containing a large number of capacitors of which 70 per cent are within given tolerance values and the remainder are not. Determine the probabilities stated.

1. When three capacitors are drawn at random, determine the probability that: (a) two are not within; and (b) two are within the given tolerance values. (a) [0.189] (b) [0.441]
2. When six capacitors are drawn at random, determine the probability that: (a) there are three not within; and (b) there are not more than two within the given tolerance values. (a) [0.185 2] (b) [0.479 7]
3. When nine capacitors are drawn at random, determine the probability that: (a) less than three are not within; and (b) six are within the given tolerance values. (a) [0.462 8] (b) [0.266 8]
4. When five capacitors are drawn at random, find the probability that: (a) more than four are not within; and (b) there are two not within the given tolerance values. (a) [0.002 4] (b) [0.308 7]
5. When seven capacitors are drawn at random, find the probability that: (a) at least two are within; and (b) at most three are not within the given tolerance values. (a) [0.996 2] (b) [0.874 0]
6. A machine produces 10 per cent defective components. In a sample of five drawn at random, find the probability of having: (a) three defective; (b) two defective; and (c) no defective components.
   (a) [0.008] (b) [0.073] (c) [0.591]
7. A target is hit by a marksman once in four shots, on average. When firing seven shots, determine the probability of: (a) obtaining three hits; and (b) obtaining at least two hits. (a) [0.173 0] (b) [0.558 9]
8. A box containing 75 components and when inspected, five are found to be defective. When six are chosen at random from the box, determine the probability of: (a) having no defective components; and (b) having more than three defective components in the sample.
   (a) [0.661 0] (b) [2.656×10^{-4}]

9. The probability of winning a prize at a fair is once in each eight tries, on average. Determine the probability of winning three prizes in nine tries. [0.073 6]
10. The probability of passing an examination is 0.65. Determine the probability that out of eight students: (a) just three; (b) just five; and (c) just seven will pass the examination.
(a) [0.080 8] (b) [0.278 6] (c) [0.137 3]
11. Six people are all the same age and in good health. Actuarial tables show that the probability that they will all be alive in 25 years is $\frac{3}{7}$. Find the probability that in 25 years: (a) all six; (b) two; and (c) at least three will be alive. (a) [0.006 2] (b) [0.293 8] (c) [0.643 3]
12. The output of a machine has, on average, 94 per cent perfect components. Determine the probability that in a sample of four components, more than one will be imperfect. [0.019 9]
13. Resistors are packed in packets of ten and there are, on average, 2 per cent defective. Determine the probability of finding two defective resistors in any packet. [0.015 3]
14. A large consignment of eggs is delivered to a shop and on average there is one broken egg in every four boxes of six eggs delivered. Determine the probability of a shopper having a box with two broken eggs in it. [0.022 0]
15. A sampling inspection scheme samples ten components from each batch supplied. If any defective items are found, the batch is returned to the supplier. If the supplier's batches have, on average, 5 per cent defective components, determine the percentage of batches returned to the supplier. [40.13 per cent]

The Poisson distribution

16. The output of an automatic machine is inspected by taking samples of 60 items. If the probability of a defective item is 0.001 5, find the probability of having: (a) two defective items; (b) more than two defective items in a sample. (a) [0.003 7] (b) [1.35×10^{-4}]
17. If 4 per cent of the tyres produced by a company are defective, find the probability that in a sample of 40 tyres there will be no defective tyres. [0.201 9]
18. In the previous problem, find the probability that the sample will contain one or two or three defective tyres. [0.719 3]
19. Inspection shows that 90 per cent of the components produced by a process are perfect. Find the probability that in a sample of ten components chosen at random, two will be defective by using: (a) the binomial distribution; and (b) the Poisson distribution.
(a) [0.193 7] (b) [0.183 9]
20. The probability of a person having an accident in a certain period of time is 0.001. Determine the probability that out of 2 000 people: (a) just three; and (b) more than two will have an accident.
(a) [0.180 4] (b) [0.323 3]

21. In a two-hour period an average of $2\frac{1}{2}$ telephone calls per minute arrive at a switchboard. Determine the probabilities that: (a) one; (b) three; (c) less than five; and (d) more than six calls will arrive in any particular minute.
(a) [0.205 2] (b) [0.213 8] (c) [0.891 2] (d) [0.042 0]
22. Of 40 lathes in a machine shop, the outage due to breakdowns averages 0.8 per week. Determine the probability that in a given week, more than two lathes will break down. [0.047 4]
23. In 30 games a team scores on average two goals per game. Determine the probability that they will score two goals in their next game.
[0.270 7]
24. A 600-page book contains an average of one error per page, distributed at random. Find the probability that there will be two or more errors on a page. [0.264 2]
25. A company has, on average, 120 man-days lost due to sickness every 100 working days. Determine on how many days over a period of 50 working days they may expect: (a) no absence due to sickness; and (b) three or more men to be absent. (a) [16] (b) [7 days]
26. The deposition of grit particles from the atmosphere was measured by counting the numbers of particles deposited on 200 prepared cards in a specified time, and the following table was drawn up:

No. of particles	0	1	2	3	4	5	6
No. of cards	45	65	52	24	11	3	0

Calculate the mean and variance of this distribution, each correct to one place of decimals, and hence show that it is reasonable to assume that the deposition of grit particles is according to a Poisson distribution. Using the mean calculated above, calculate also the theoretical probabilities of obtaining 0, 1, 2, 3, 4, or 5 or more particles on any one card, and hence prepare a table similar to the experimental one above for the expected frequencies of grit particles on 200 cards placed at random.

[mean = variance = 1.5;

probabilities	0.223 1	0.334 7	0.251 0	0.126 0	0.047 0	0.014 1
	45	67	50	25	10	3

]

27. If the probability of suffering side-effects from an efection is 10^{-3}, what is the expectation that out of 2 000 people: (a) three; (b) not more than two patients will suffer side-reactions? (a) [0.180 4] (b) [0.323 3]
28. The mean number of breakdowns in a distillation plant is 2.5 per 5-day week. What is the probability of no breakdowns on a particular day? If the factory works 50 weeks in a year, on how many days may two or more breakdowns be expected? [0.606 5, 23 days]

Chapter 12

The normal probability distribution

1. The normal probability curve

The binomial and Poisson distributions deal with the occurrence of distinct events, that is, they are discrete probability distributions. Many instances occur where distributions are continuous, particularly when data relates to measured quantities such as length, mass, time, electric current, temperature, luminous intensity and their derived units. There are many instances where data from these sources approximate to a distribution called the **normal** distribution.

When the sample number, n, in the binomial distribution is made very large and a histogram is drawn of the results, it is found that the shape of the diagram approaches a symmetrical mathematical curve. This curve has an equation of the form: $y = \frac{1}{\sigma} e^{-\frac{x^2}{2\sigma^2}}$, where σ is the standard deviation of the data and x and y are the coordinates of points on the curve. This curve is called the **normal probability curve**, the y-values the **probability density** and the x-values the **variable** or **variate.** Figure 1 shows that the normal probability curve is a symmetrical curve.

It is the most important of the various distribution curves. When a large number of values are taken of, say, peoples' heights, masses, intelligence quotients, or the sizes of parts produced by a machine or the marks gained in an examin-

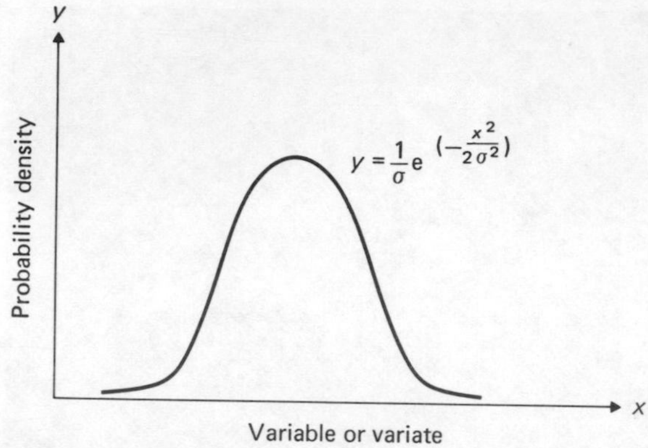

Figure 1 The normal probability curve

ation, it is found that when the data are plotted as a frequency curve, a curve of similar shape to the normal probability curve is produced in each case.

If any normal probability curve is drawn and the standard deviation, σ, of the data calculated, then the area under the curve and the value of the standard deviation are related. It is found that if a vertical line is drawn to pass through the maximum value of the normal probability curve (the mean value since the curve is symmetrical) and two other vertical lines are drawn at a distance of one standard deviation on either side of this line, then in all cases:

> the area enclosed by the curve and the vertical lines at distances $\pm 1\sigma$ from the mean value is about $66\frac{2}{3}\%$ of the total area under the curve (the true value being 68.26%).

Similarly, it is found that:

> the area enclosed by the curve and the vertical lines at distances $\pm 2\sigma$ from the mean is about 95% of the total area under the curve (the true value being 95.44%).

Finally, it is found that:

> the area enclosed by the curve and the vertical lines at distances $\pm 3\sigma$ from the mean is about $99\frac{3}{4}\%$ of the total area under the curve (the true value being 99.74%).

A normal probability curve showing these areas can be seen in Fig. 2.

The relationship between a normal probability curve and the area contained between the vertical lines drawn at various distances from the mean value forms the basis for such techniques as sampling and quality control. Because of this, it is worth becoming familiar with and committing to memory the approximate areas quoted in the last paragraph.

In histograms and distribution curves the area under the curve is proportional to frequency. In a normal probability curve, since the area between the curve and the vertical lines drawn at ± 1 standard deviation from the mean value is roughly $66\frac{2}{3}\%$ of the total area, it follows that $66\frac{2}{3}\%$ of the total fre-

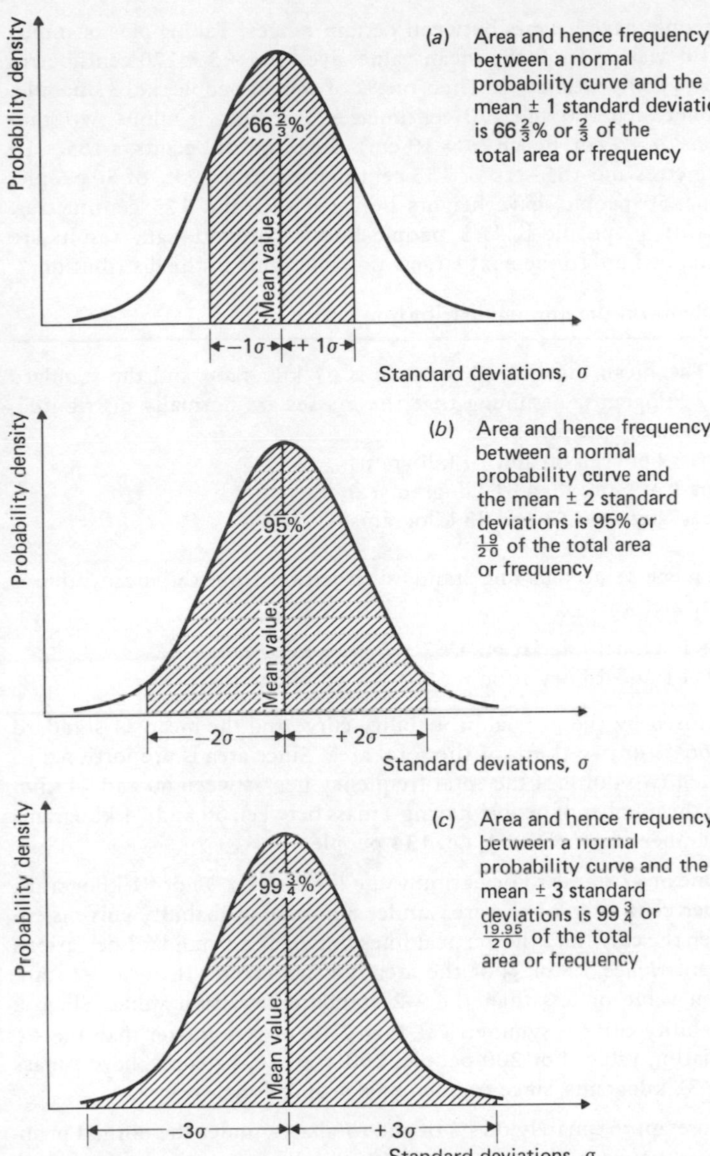

Figure 2

quency also lies between these values of standard deviation for normally distributed data.

For example, if the mean value of the heights of 50 people is 165 centimetres and the standard deviation is 5 centimetres, it is possible to predict how many people have heights between certain ranges. Taking plus or minus one standard deviation from the mean value gives 165 + 5 = 170 centimetres and 165 − 5 = 160 centimetres. Then $66\frac{2}{3}$% of the 50 people, i.e. 33 people, have heights between 160 and 170 centimetres. For plus or minus two standard deviations (σ = 5 cm, hence 2σ = 10 cm), the range of heights is 165 + 10 or 175 centrimetres and 165 − 10 or 155 centimetres. Thus, 95% of 50 people, i.e. 48 of the 50 people, have heights between 155 and 175 centimetres. Strictly, 95% of 50 people is 47.5 people but for discrete data, results are invariably 'rounded-up' to the next largest possible value in the distribution.

Worked problems on the normal distribution curve

Problem 1. The mean mass of 200 people is 67 kilograms and the standard deviation is 7 kilograms. Assuming that the masses are normally distributed, determine how many people:
(a) have a mass between 60 and 74 kilograms;
(b) have a mass of more than 81 kilograms; and
(c) have a mass between 53 and 88 kilograms.

(a) The mean value plus one standard deviation from the mean value is usually abbreviated to:

mean + 1 standard deviation = 67 + 7 = 74 kilograms.
Also, mean − 1 standard deviation = 67 − 7 = 60 kilograms.

The area enclosed by the normal probability curve and the mean ±1 standard deviation is $66\frac{2}{3}$% or two-thirds of the total area. Since area is proportional to frequency, then two-thirds of the total frequency lies between 60 and 74 kilograms. Hence the number of people having a mass between 60 and 74 kilograms is 200 (the number of people) × $\frac{2}{3}$, i.e. **134 people.**

(b) The mean +2 standard deviation value is 67 + (2 × 7) or 81 kilograms. Ninety-five per cent or $\frac{19}{20}$ of the area under a normal probability curve is enclosed between the curve and the vertical lines drawn at ±2 standard deviations from the mean. Hence 5% or $\frac{1}{20}$ of the area is either greater than the +2 standard deviation value or less than the −2 standard deviation value. Since a normal probability curve is symmetrical, then 2½% or $\frac{1}{40}$ is greater than the +2 standard deviation value. For 200 people, 200 × $\frac{1}{40}$, i.e. **5 people,** have a mass of more than 81 kilograms, since frequency is proportional to area.

(c) Because approximately $66\frac{2}{3}$% of the total area under the normal probability curve is between the mean and ±1 standard deviation, about 95% between the mean and ±2 standard deviations, and about $99\frac{3}{4}$% between the mean and ±3 standard deviations, the areas under a normal curve (and hence the frequencies) can be subdivided as shown in Fig. 3.

The areas marked (A) are each half of $66\frac{2}{3}$%, i.e. $33\frac{1}{3}$%. The areas marked (B) are half of 95%, that is, $47\frac{1}{2}$%, less the $33\frac{1}{3}$% already shown as area (A). Thus the area marked (B) is ($47\frac{1}{2} - 33\frac{1}{3}$)% or $14\frac{1}{6}$%. The areas marked (C) are half of $99\frac{3}{4}$%, i.e. $49\frac{7}{8}$% less the areas marked (A) and (B), that is ($49\frac{7}{8} - 33\frac{1}{3}$ −

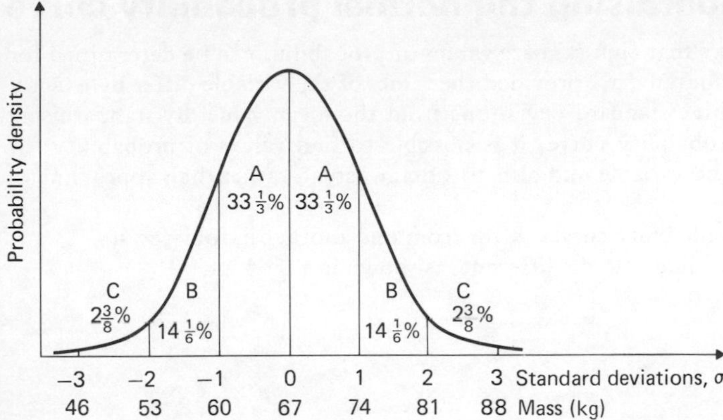

Figure 3 Areas (or frequencies) between the normal probability curve and the mean ± 1, 2 or 3 standard deviations

$14\frac{1}{6}$)%, or $2\frac{3}{8}$%. Using these values, the total area (and hence frequency) between 53 and 88 kilograms is obtained by adding the appropriate areas in Fig. 3, i.e.

$$\left(14\tfrac{1}{6} + 33\tfrac{1}{3} + 33\tfrac{1}{3} + 14\tfrac{1}{6} + 2\tfrac{3}{8}\right)\% = 97\tfrac{3}{8}\%.$$

(Alternatively, $99\frac{3}{4} - 2\frac{3}{8} = 97\frac{3}{8}$%.)

The number of people is $\dfrac{97\frac{3}{8}}{100} \times 200 = 194\frac{3}{4}$.

As this is a discrete distribution, this must be rounded up to the next possible value, giving a result of **195 people** having a mass of between 53 and 88 kilograms.

Problem 2. A sample of 60 bolts produced by a machine was measured and the mean diameter was found to be 0.402 centimetres. The standard deviation of the sample was 0.000 5 centimetres. If only bolts having diameters between 0.401 and 0.403 centimetres were accepted, determine how may of the sample were rejected, assuming a normal distribution.

The range of bolts being accepted is 0.403–0.401, i.e. 0.002 centimetres. Since the standard deviation is 0.000 5 centimetres, this range corresponds to ±2 standard deviations.

The area and also the frequency between a normal distribution curve and the mean ±2 standard deviations is 95%. Hence the frequency outside of the range is 5% or $\frac{1}{20}$. Thus the number of bolts rejected is:

$60 \times \frac{1}{20}$ = **3 bolts.**

Further problems on the normal probability curve may be found in Section 4 (Problems 1–6).

2. Standardising the normal probability curve

Section 1 shows that approximate values of probability can be determined for normally distributed data, provided the values of the variable differ by exactly one, two or three standard deviations from the mean value. By standardising the normal probability curve, it is possible to find values of probability for any value of the variable and also to obtain actual, rather than approximate, results.

Normal probability curves differ from one another in four ways:

(i) The mean values can be different, as shown in Fig. 4.

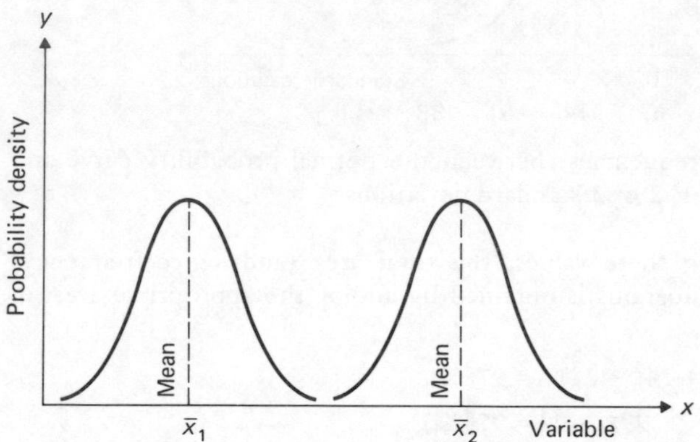

Figure 4 Normal probability curves with the same standard deviation but having different mean values

(ii) The values of their standard deviations can be different, as shown in Fig. 5.

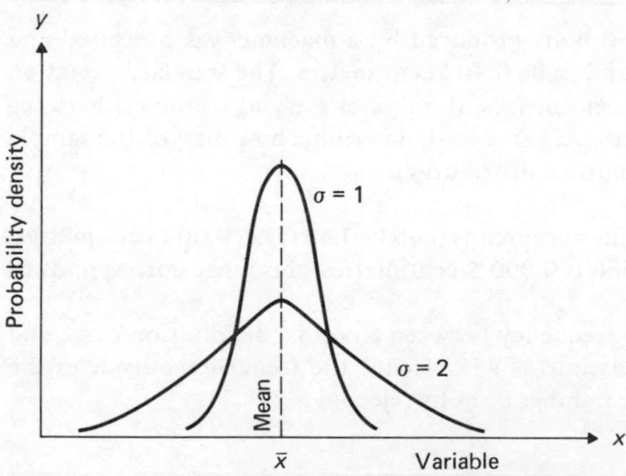

Figure 5 Normal probability curves having the same mean value but with different values of standard deviation

(iii) The magnitude and units of the variables can be different.
(iv) The areas under their normal probability curves can be different.

The process of standardisation is to alter the normal probability curve to make it independent of these four items. This is achieved as follows:

(a) by writing $(x - \bar{x})$ for x in the equation of the normal probability curve, where $\bar{x}$ is the mean value, the origin of the graph is moved to $\bar{x}$. The equation now becomes

$$y = \frac{1}{\sigma} e^{\left(-\frac{(x-\bar{x})^2}{2\sigma^2}\right)}$$

All normal probability curves amended in this way have their mean values as the origin and hence the two curves shown in Fig. 4 would now sit on top of one another.

(b) The horizontal axis is scaled in standard deviations instead of the variable, x. This is achieved by using the relationship: $z = \frac{x - \bar{x}}{\sigma}$, where z is called the **normal standard variate.** For example, if the mean value of a distribution is, say, 147 centimetres and the value of the standard deviation is, say, 5 centimetres, then the variable value, say, 152 centimetres is 147 + 5, that is the mean value plus one standard deviation. Similarly, 137 centimetres is the mean value minus two standard deviations. Thus the x-axis is rescaled as the z-axis and some of the values are as shown:

x-axis (centimetres)	137	142	147	152	157
z-axis (standard deviations)	−2	−1	0	1	2

By introducing the normal standard variate, the origin of all normal probability curves is $\sigma = 0$ and hence they all have the same origin. In addition, the scaling of all normal probability curves is in standard deviations and all normal probability curves have the same scaling. Also, since standard deviation is a numerical value and can be independent of units, all normal probability curves can have the same units. Thus the introduction of the normal standard variate gives all normal probability curves a common value of unity for their standard deviation and can make them independent of units. Since $z = \frac{x - \bar{x}}{\sigma}$, then $\frac{(x - \bar{x})^2}{2\sigma^2}$ becomes $\frac{z^2}{2}$ and since on the z-scale, σ has the value unity, the equation of the normal probability curve now becomes:

$$y = e^{\left(-\frac{z^2}{2}\right)}$$

The area between the normal probability curve and the z-axis is given by:

$$\int_{-\infty}^{\infty} e^{\left(-\frac{z^2}{2}\right)} dz$$

To evaluate this integral requires techniques not yet introduced (double integrals and the use of polar coordinates), giving its value as $\sqrt{(2\pi)}$. One of the requirements of a probability distribution curve is that the area beneath the curve is equal to unity, since the total area represents the total probability. To make the area under the normal probability curve unity, it is therefore necessary to divide the value of y by $\sqrt{(2\pi)}$. This gives the equation of the **standarised normal curve** as:

$$y = \frac{1}{\sqrt{(2\pi)}} e^{\left(-\frac{z^2}{2}\right)}$$

and by using it, any normally distributed data is represented by the same curve.

The area under the standardised normal curve represents probability and the area under the curve between limits, say z_1 and z_2 can be found by determining the value of:

$$\int_{z_2}^{z_1} \frac{1}{\sqrt{(2\pi)}} e^{\left(-\frac{z^2}{2}\right)} dz$$

Tables are available giving the values of this integral for z-values between 0.00 and 3.99 and one such table is Table 1, shown on page 333.

Worked problems on determining probabilities using a table of partial areas beneath the standardised normal curve

---

Problem 1. A certain machine produces components having a mean length of 15 centimetres. As a result of measuring samples of these components, it is found that the standard deviation is 0.2 centimetres. A test is carried out on a sample to check whether the data on the lengths of the components is normally distributed and it is found that this is so.
(a) Determine the number of components likely to have a length of less than 14.95 centimetres in a batch of 1 000 components.
(b) Determine the number of components likely to be between 14.95 and 15.15 centimetres long in a batch of 1 000 components.
(c) Determine the number of components likely to be larger than 15.43 centimetres long.

(a) The z-value is given by $z = \frac{x - \bar{x}}{\sigma}$, where x is the variable value (14.95 cm), $\bar{x}$ the mean value (15 cm) and σ the value of the standard deviation (0.2 cm). Then

$$z = \frac{14.95 - 15}{0.2} = -0.25 \text{ standard deviations.}$$

Using Table 1, the partial area beneath the standardised normal curve corresponding to a z-value of 0.25 is 0.098 7, the minus sign showing that it lies to the left of the zero z-value ordinate. This area is shown by the shaded area in Fig. 6 (a), and is 0.098 7 of the total under the curve, the total area being unity.

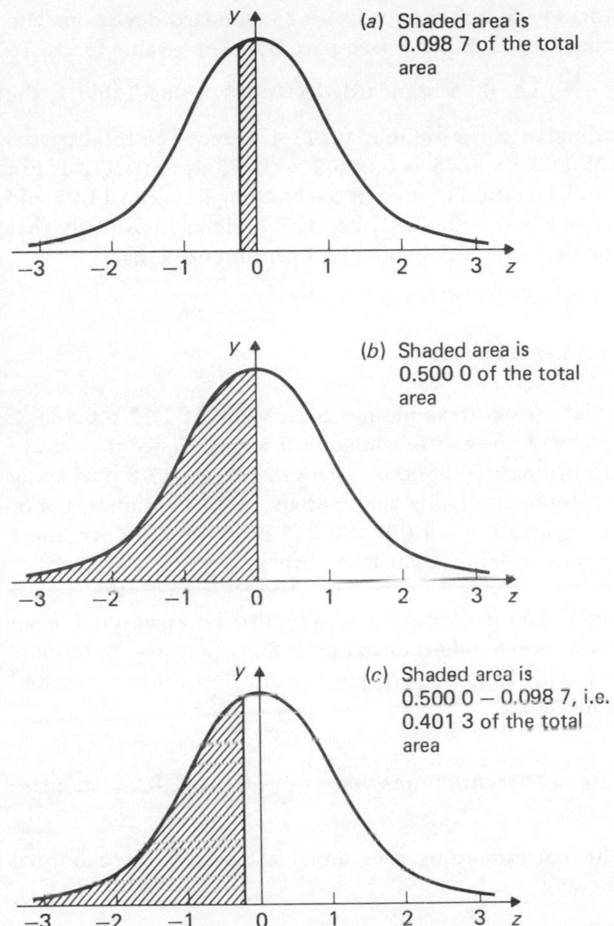

Figure 6 Partial areas under the standardised normal curve (Problem 1)

Since the standardised normal curve is symmetrical about the ordinate drawn through $z = 0$, then the total area to the left of this line is 0.500 0, shown shaded in Fig. 6 (b). The required area giving the probability of a component being less than 14.95 centimetres long is the area to the left of the $z = -0.25$ ordinate and is shown as the shaded area in Fig. 6 (c). This area is the shaded area in Fig. 6 (b) minus the shaded area in Fig. 6 (a), that is, 0.500 0 – 0.098 7 or 0.401 3. But the area under the standardised normal curve is proportional to probability, so the probability that a component will be less than 14.95 centimetres long is 0.401 3. For 1 000 components, the expectation is 1 000 × 0.401 3, i.e. 401.3. Thus, there are likely to be **402 components** having a length of 14.95 centimetres or less, results usually being 'rounded-up' in statistics.

(b) The z-value for 14.95 centimetres is −0.25 standard deviations and the corresponding partial area is 0.098 7 (see part (a)). The z-value for 15.15 centimetres is $\frac{15.15 - 15}{0.2}$, i.e. 0.75 standard deviations. Using Table 1, the partial area corresponding to this z-value is 0.273 4. Hence the total partial area between $z = -0.25$ and $z = 0.75$ is 0.098 7 + 0.273 4, i.e. 0.372 1. For 1 000 components, the expectation of a component being between 14.95 and 15.15 centimetres long is 1 000 × 0.372 1, i.e. 372.1. Hence it is likely that **373 components** will be between 14.95 and 15.15 centimetres long.

(c) The z-value of 15.43 centimetres is given by

$$z = \frac{15.43 - 15}{0.2} = 2.15 \text{ standard deviations.}$$

From Table 1, the partial area corresponding to a z-value of 2.15 is 0.484 2. The total area to the right of the $z = 0$ ordinate is 0.500 0. Hence the area to the right of the $z = 2.15$ ordinate is 0.500 0 − 0.484 2, i.e. 0.015 8, this being the probability of a component being larger than 15.43 centimetres. For 1 000 components, the expectation is 1 000 × 0.015 8, i.e. 15.8. Hence, there are likely to be **16 components** larger than 15.43 centimetres.

Problem 2. The heights of 500 people are normally distributed about a mean value of 172 centimetres. The standard deviation is 8 centimetres. Determine how many people: (a) are likely to be between 174 and 189 centimetres; and (b) smaller than 149 centimetres.

(a) The z-value for 174 centimetres is $\frac{174 - 172}{8}$, i.e. 0.25 standard deviations.

From Table 1, the corresponding area under the standardised normal curve is 0.098 7.

The z-value for 189 centimetres is $\frac{189 - 172}{8}$, i.e. 2.125 or 2.13 standard deviations correct to 2 decimal places. From Table 1, the corresponding area under the standardised normal curve for a z-value of 2.13 is 0.483 4. The area under the standardised normal curve between z-values of 2.13 and 0.25 is 0.483 4 − 0.098 7, i.e. 0.384 7. For 500 people, the expectation of being between 174 and 189 centimetres tall is 500 × 0.384 7, i.e. 192.35. Hence, there are likely to be **193 people** having heights between 174 and 189 centimetres.

(b) The z-value for 149 centimetres is $\frac{149 - 172}{8}$, i.e. $\frac{-23}{8}$ or −2.875, i.e. −2.88 correct to 2 decimal places. From Table 1, the partial area corresponding to a z-value of 2.88 is 0.498 0. The area under the standardised normal curve for the z-values of less than −2.88 is 0.500 0 − 0.498 0, i.e. 0.002. For 500 people, the expectation of being less than 149 centimetres tall is 500 × 0.002, i.e. 1 person. That is, **one person** is likely to be less than 149 centimetres tall.

Further problems on the standardised normal curve may be found in Section 4 (Problems 7–22).

3. Testing the normality of a distribution

In the practical application of statistics, it should never be assumed that because a set of data is continuous, it is also normally distributed, or even approximates to a normal distribution. Various tests can be carried out to determine how accurately a set of data approximates to a normal distribution and the simplest test is to plot the data on specially ruled paper called **normal probability paper** or just **probability paper.**

This has a linear scale usually taken to be the horizontal axis, and probability is usually taken to be the vertical axis. The divisions of the probability scale are so spaced that when the percentage cumulative frequency is plotted against the upper class-boundaries of the variable, a straight line results when the data is normally distributed. The use of normal probability paper for testing the normality of a distribution is shown in the following worked problems.

Worked problems on testing the normality of a distribution

Problem 1. Use normal probability paper to test whether the data given below is normally distributed.

Variable (class mid-point)	40	50	60	70	80	90
Frequency	40	80	120	170	180	160

Variable (class mid-point)	100	110	120	130
Frequency	110	80	40	20

Upper class-boundary	Cumulative frequency	Percentage cumulative frequency
Less than 45	40	4
Less than 55	120	12
Less than 65	240	24
Less than 75	410	41
Less than 85	590	59
Less than 95	750	75
Less than 105	860	86
Less than 115	940	94
Less than 125	980	98
Less than 135	1 000	100

The upper class-boundaries are 45, 55, 65, . . . and the corresponding cumulative frequencies ('less than' basis) are 40, 40 + 80 = 120, 120 + 120 = 240, 240 + 170 = 410, and so on. The sum of all the frequencies is 1 000 and the percentage cumulative frequencies are obtained by dividing the cumulative frequencies by 1 000 (to give the relative frequencies) and multiplying by 100 to bring these values to a percentage. This gives 4, 12, 24, 45, and so on. The table on the previous page shows these results:

Using normal probability paper, shown in Fig. 7, a suitable scale is selected for the upper class-boundary values on the horizontal linear axis and the vertical probability axis represents the percentage cumulative-frequency scale. The points are plotted and eight of the nine points are found to lie in a reasonably straight line. (The 100% cumulative frequency cannot be plotted on normal probability paper.) Hence **the data given does approximate to a normal distribution.** (The divisions on a probability scale are in direct proportion to the ordinate heights of the standardised normal distribution curve. If a straight line occurs when percentage cumulative frequency is plotted against upper class-boundary on normal probability paper, this indicates that the points do lie on the standardised normal curve.)

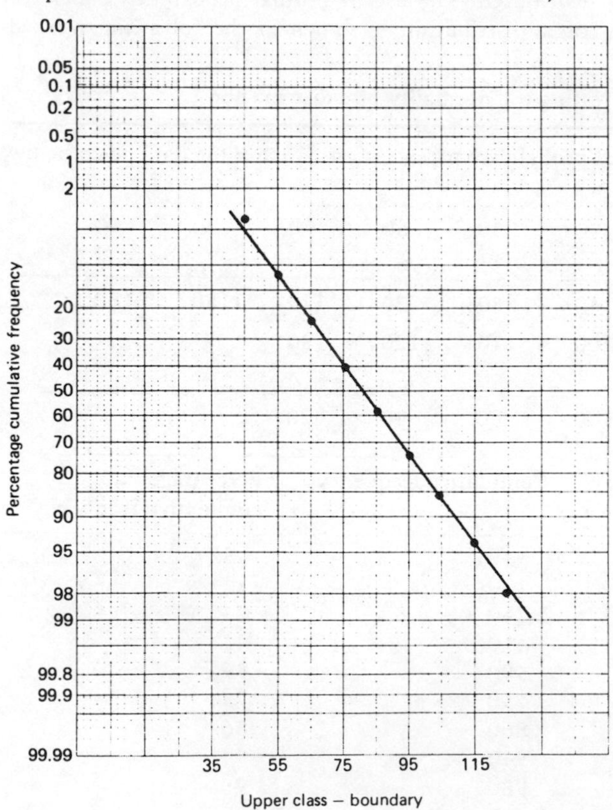

Figure 7

Problem 2. Each of a sample of 150 components produced by a machine is weighed and the results are as shown below, the masses being corrected to the nearest gram. Determine if the sample is normally distributed.

Mass (grams)	326	327	328	329	330	331	332	333	334	335
Frequency	5	9	18	22	42	29	9	7	6	3

The table below shows the values of upper class-boundaries and the corresponding percentage cumulative frequencies, rounded up to the nearest whole number.

Upper class-boundary	Cumulative frequency	Percentage cumulative frequency $\left(\text{cumulative frequency} \times \frac{100}{150}\right)$
Less than 326.5	5	3
Less than 327.5	14	9
Less than 328.5	32	21
Less than 329.5	54	36
Less than 330.5	96	64
Less than 331.5	125	83
Less than 332.5	134	89
Less than 333.5	141	94
Less than 334.5	147	98
Less than 335.5	150	100

Using normal probability paper, shown in Fig. 8, values of percentage cumulative frequency are plotted against upper class-boundary values.

A straight line can be drawn through six of the nine points, but a group of three points for masses of 330.5 to 332.5 grams does not lie on the line. This indicates that **the data given is not normally distributed.**

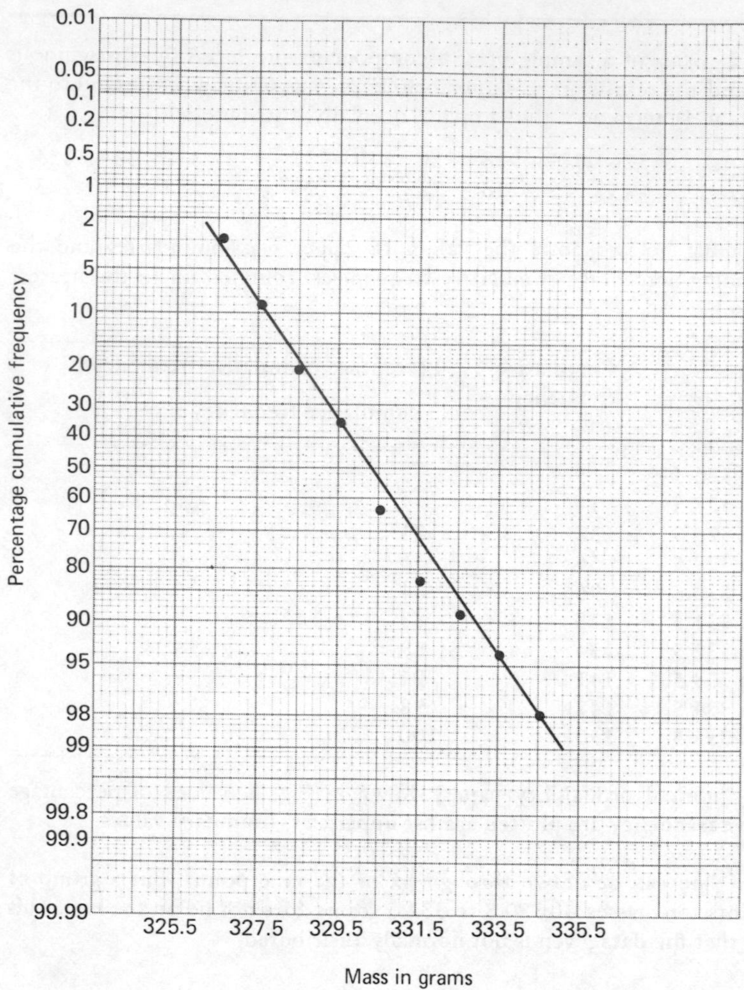

Figure 8

Further problems on testing the normality of a distribution may be found in the following section (4) (Problems 23–28).

4. Further problems

The normal probability curve

1. The average mark obtained in an examination was 55 and the standard deviation was $7\frac{1}{2}$. Determine how many candidates failed the examination if 400 took it and a pass mark is 40. Assume the marks are normally distributed. [10 candidates]

2. A component is defective if it is oversized. A sample of 500 components produced by a machine have a mean size of 7.31 cm and a standard deviation of 0.1 cm. If the maximum size acceptable is 7.41 cm, determine how many components are defective, assuming a normal distribution. [84]
3. The mean value of a large sample is 130 g and the standard deviation is 7 g. Determine the percentage of the sample which will be between 123 and 151 g, assuming the sample is normally distributed. [$83\frac{5}{24}$% or 83.21%]
4. The mean diameter of 400 ball-bearings is 0.322 cm and the standard deviation of the sample is 0.001 5 cm. Assuming a normal distribution, determine:
   (a) the number of ball-bearings having a diameter of between 0.319 0 and 0.323 5 cm;
   (b) the number of ball-bearings having a diameter greater than 0.325 0 cm; and
   (c) the number of ball-bearings having a diameter of less than 0.317 5 cm.
   (a) [324] (b) [10] (c) [1]
5. The heights of 1 000 people are normally distributed with a mean of 1.72 m and a standard deviation of 8 cm. Determine how many people will:
   (a) have heights between 1.48 and 1.96 m;
   (b) have heights of less than 1.72 m;
   (c) have heights of more than 1.80 m; and
   (d) have heights between 1.56 and 1.64 m.
   (a) [998] (b) [500] (c) [167] (d) [142]
6. A set of 100 measurements is normally distributed with a mean of 157.8 m and a standard deviation of 1.4 m. Determine how many of the measurements will be:
   (a) between 155 and 162 m;
   (b) between 159.2 and 160.6 m;
   (c) more than 157.8 m; and
   (d) less than 156.4 m.
   (a) [98] (b) [15] (c) [50] (d) [17]

The standardised normal probability curve

In Problems 7–13, use Table 1 to determine the partial areas under the standardised normal curve for the z-values given.

7. Between $z = 0$ and $z = 1.35$. [0.411 5]
8. Between $z = -2.34$ and $z = 0$. [0.490 4]
9. Between $z = -0.46$ and $z = 2.21$. [0.663 6]
10. Between $z = 0.81$ and $z = 1.94$. [0.182 8]
11. Larger than $z = -1.28$. [0.899 7]
12. Less than $z = -0.6$. [0.274 3]
13. Less than $z = -1.44$ and larger than $z = 2.05$. [0.095 1]
14. The mean mass of a set of components is 151 kg and the standard deviation is 15 kg. Assuming the masses are normally distributed about

the mean, determine for a sample of 500 components, how many are likely to have masses of:

(a) between 120 and 155 kg;
(b) more than 185 kg; and
(c) less than 128 kg.
(a) [294] (b) [6] (c) [32]

15. The mean diameter of a batch of bolts is 0.502 cm and the standard deviation of the batch is 0.05 mm. Bolts outside of the range of diameters 0.496 to 0.508 cm are rejected. Determine the percentage of bolts rejected assuming the diameters are normally distributed. [23.02%]
16. Containers having a fluid for innoculation are filled with an average of 68 cm^3 of a fluid and the standard volume deviation is 3 cm^3. Assuming the contents of the containers to be normally distributed, determine in a sample of 300 containers, how many are likely to contain:
(a) more than 72 cm^3;
(b) less than 64 cm^3; and
(c) between 65 and 71 cm^3.
(a) [28] (b) [28] (c) [205]
17. Tablets contain on average 0.614 g of a drug, the standard deviation being 2.5 mg. Assuming the masses of the drug are normally distributed, determine the percentage of the tablets having:
(a) more than 0.617 g of the drug;
(b) less than 0.608 g of the drug; and
(c) between 0.610 and 0.618 g of the drug.
(a) [11.51%] (b) [0.82%] (c) [89.04%]
18. A set of measurements is normally distributed. Determine the percentage which differ from the mean by:
(a) more than half of the standard deviation, and
(b) by less than three-quarters of the standard deviation.
(a) [62.70%] (b) [54.68%]
19. The contents of milk bottles are measured and the amount of milk the bottles contain is found to be normally distributed. Determine the percentage of bottles containing milk:
(a) within the range: mean ±2 standard deviations;
(b) outside of the range ±1.2 standard deviations; and
(c) greater than the mean −1.5 standard deviations.
(a) [95.44%] (b) [23.02%] (c) [93.32%]
20. Tins are packed with an average of 1.0 kg of a compound and the masses are normally distributed about the average value. The standard deviation of a sample of the contents of the tins is 12 g. Determine the percentage of tins containing:
(a) less than 985 g;
(b) more than 1 030 g; and
(c) between 985 and 1 030 g.
(a) [10.56%] (b) [0.62%] (c) [88.82%]
21. The weights of tablets in a bottle are normally distributed with a mean of 20 g and standard deviation of 2 g. Using the area under the standard

normal curve, calculate the probability of a bottle containing tablets weighing between: (a) 16 and 18 g; (b) 18 and 23 g.

Note: extract from normal table:

Units of s. dev. from mean	0	0.5	1.0	1.5	2.0	2.5
Area under normal curve	0	0.191 5	0.341 3	0.433 2	0.477 2	0.493 8

(a) [0.135 9] (b) [0.774 5]

22. A chemical manufacturer produces aspirin tablets having a mean mass of 4 g and a standard deviation of 0.2 g. Assuming that the masses are normally distributed and that a tablet is chosen at random what is the probability that it: (a) has mass between 3.55 and 3.85 g; (b) differs from the mean by less than 0.35 g?
(c) If the tablets are packed in cartons of 400, how many in each carton may be expected to have a mass of less than 3.7 g?
(a) [0.214 4] (b) [0.919 8] (c) [27]

Testing the normality of a distribution

In problems 23–27 use normal probability paper to determine whether the distributions given are normally distributed.

23. The diameter of a sample of rivets produced by an automatic process.

Diameter mm	4.011	4.015	4.019	4.023	4.027	4.031	4.035
Frequency	13	40	79	114	116	82	40

[Yes]

24. The resistance of a sample of carbon resistors.

Resistance MΩ	1.28	1.29	1.30	1.31	1.32	1.33	1.34	1.35	1.36
Frequency	10	18	36	44	64	40	24	16	12

[No]

25. The mass of a sample of components produced by a casting process.

Mass kg	4.00	4.05	4.10	4.15	4.20	4.25
Frequency	22	78	150	260	280	440

Mass kg	4.30	4.35	4.40	4.45	4.50	4.55
Frequency	360	180	140	56	18	4

[Yes]

26. The volume output of a bottling machine.

Volume l	1.007	1.010	1.013	1.016	1.019	1.022	1.025	1.028
Frequency	2	5	12	17	14	6	3	1

[Yes]

27. The capacitance of a sample of rolled metal foil capacitors of nominal capacitance 3.75 μF.

Capacitance μF	3.72	3.73	3.74	3.75	3.76	3.77
Frequency	2	6	8	15	32	48

Capacitance μF	3.78	3.79	3.80	3.81	3.82	3.83
Frequency	39	25	18	12	4	1

[No]

28. The amount of lead in a given foodstuff was determined with the following results:

40–44 ppm 1 analysis
45–49 ppm 4
50–54 ppm 14
55–59 ppm 22
60–64 ppm 31
65–69 ppm 20
70–74 ppm 6
75–79 ppm 2

Prove graphically that the distribution is normal and find the mean and standard deviation.
[straight line with probability paper, mean = 60.6, standard deviation = 6.79]

Table 1 Partial areas under the standardised normal curve

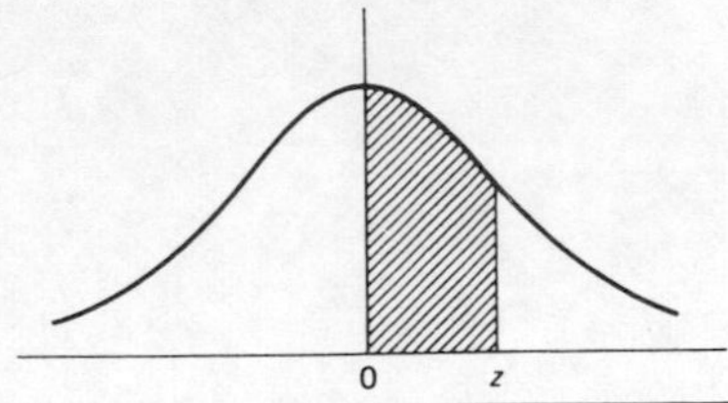

$z = \dfrac{x - \bar{x}}{\sigma}$	0	1	2	3	4	5	6	7	8	9
0.0	0.0000	0.0040	0.0080	0.0120	0.0159	0.0199	0.0239	0.0279	0.0319	0.0359
0.1	0.0398	0.0438	0.0478	0.0517	0.0557	0.0596	0.0636	0.0678	0.0714	0.0753
0.2	0.0793	0.0832	0.0871	0.0910	0.0948	0.0987	0.1026	0.1064	0.1103	0.1141
0.3	0.1179	0.1217	0.1255	0.1293	0.1331	0.1388	0.1406	0.1443	0.1480	0.1517
0.4	0.1554	0.1891	0.1628	0.1664	0.1700	0.1736	0.1772	0.1808	0.1844	0.1879
0.5	0.1915	0.1950	0.1985	0.2019	0.2054	0.2086	0.2123	0.2157	0.2190	0.2224
0.6	0.2257	0.2291	0.2324	0.2357	0.2389	0.2422	0.2454	0.2486	0.2517	0.2549
0.7	0.2580	0.2611	0.2642	0.2673	0.2704	0.2734	0.2760	0.2794	0.2823	0.2852
0.8	0.2881	0.2910	0.2939	0.2967	0.2995	0.3023	0.3051	0.3078	0.3106	0.3133
0.9	0.3159	0.3186	0.3212	0.3238	0.3264	0.3289	0.3215	0.3340	0.3365	0.3389
1.0	0.3413	0.3438	0.3451	0.3485	0.3508	0.3531	0.3554	0.3577	0.3599	0.3621
1.1	0.3643	0.3665	0.3686	0.3708	0.3729	0.3749	0.3770	0.3790	0.3810	0.3830
1.2	0.3849	0.3869	0.3888	0.3907	0.3925	0.3944	0.3962	0.3980	0.3997	0.4015
1.3	0.4032	0.4049	0.4066	0.4082	0.4099	0.4115	0.4131	0.4147	0.4162	0.4177
1.4	0.4192	0.4207	0.4222	0.4236	0.4251	0.4265	0.4279	0.4292	0.4306	0.4319
1.5	0.4332	0.4345	0.4357	0.4370	0.4382	0.4394	0.4406	0.4418	0.4430	0.4441
1.6	0.4452	0.4463	0.4474	0.4484	0.4495	0.4505	0.4515	0.4525	0.4535	0.4545
1.7	0.4554	0.4564	0.4573	0.4582	0.4591	0.4599	0.4608	0.4616	0.4625	0.4633
1.8	0.4641	0.4649	0.4656	0.4664	0.4671	0.4678	0.4686	0.4693	0.4699	0.4706
1.9	0.4713	0.4719	0.4726	0.4732	0.4738	0.4744	0.4750	0.4756	0.4762	0.4767
2.0	0.4772	0.4778	0.4783	0.4785	0.4793	0.4798	0.4803	0.4808	0.4812	0.4817
2.1	0.4821	0.4826	0.4830	0.4834	0.4838	0.4842	0.4846	0.4850	0.4854	0.4857
2.2	0.4861	0.4864	0.4868	0.4871	0.4875	0.4878	0.4881	0.4884	0.4882	0.4890
2.3	0.4893	0.4896	0.4898	0.4901	0.4904	0.4906	0.4909	0.4911	0.4913	0.4916
2.4	0.4918	0.4920	0.4922	0.4925	0.4927	0.4929	0.4931	0.4932	0.4934	0.4936
2.5	0.4938	0.4940	0.4941	0.4943	0.4945	0.4946	0.4948	0.4949	0.4951	0.4952
2.6	0.4953	0.4955	0.4956	0.4957	0.4959	0.4960	0.4961	0.4962	0.4963	0.4964
2.7	0.4965	0.4966	0.4967	0.4968	0.4969	0.4970	0.4971	0.4972	0.4973	0.4974
2.8	0.4974	0.4975	0.4976	0.4977	0.4977	0.4978	0.4979	0.4980	0.4980	0.4981
2.9	0.4981	0.4982	0.4982	0.4983	0.4984	0.4984	0.4985	0.4985	0.4986	0.4986
3.0	0.4987	0.4987	0.4987	0.4988	0.4988	0.4989	0.4989	0.4989	0.4990	0.4990
3.1	0.4990	0.4991	0.4991	0.4991	0.4992	0.4992	0.4992	0.4992	0.4993	0.4993
3.2	0.4993	0.4993	0.4994	0.4994	0.4994	0.4994	0.4994	0.4995	0.4995	0.4995
3.3	0.4995	0.4995	0.4995	0.4996	0.4996	0.4996	0.4996	0.4996	0.4996	0.4997
3.4	0.4997	0.4997	0.4997	0.4997	0.4997	0.4997	0.4997	0.4997	0.4997	0.4998
3.5	0.4998	0.4998	0.4998	0.4998	0.4998	0.4998	0.4998	0.4998	0.4998	0.4998
3.6	0.4998	0.4998	0.4999	0.4999	0.4999	0.4999	0.4999	0.4999	0.4999	0.4999
3.7	0.4999	0.4999	0.4999	0.4999	0.4999	0.4999	0.4999	0.4999	0.4999	0.4999
3.8	0.4999	0.4999	0.4999	0.4999	0.4999	0.4999	0.4999	0.4999	0.4999	0.4999
3.9	0.5000	0.5000	0.5000	0.5000	0.5000	0.5000	0.5000	0.5000	0.5000	0.5000

Index